Excel

SCIENCE STUDY GUIDE

Year 9

Get the Results You Want!

Jim Stamell & Geoffrey Thickett

PASCAL PRESS

Reprinted 2016, 2017, 2021, 2023

ISBN 978 1 74125 393 1

Pascal Press
PO Box 250
Glebe NSW 2037
(02) 9198 1748
www.pascalpress.com.au

Publisher: Vivienne Joannou
Project editor: Mark Dixon
Edited by Leanne Poll
Indexed by Mary Coe
Cover and page design by DiZign Pty Ltd
Typset by Precision Typesetting (Barbara Nilsson)
Printed by Vivar Printing/Green Giant Press
Photos by Dreamstime and iStockphoto. Other photo credits: Figure 1.23, p. 18, Cindee Madison and Susan Landau, UC Berkeley; Figure 2.40, p. 76, © Dr. Donald Fawcett/Visuals Unlimited/Corbis; Figure 3.5, p. 92, © NASA; Figure 4.30, p. 142, image courtesy of Bionic Vision Australia, © Beth Croce, CMI.

Contents

How to use this book

The Australian Curriculum

This study guide covers the complete course of the Year 9 Science Australian Curriculum including:

- all core knowledge
- all required problem-solving skills
- all aspects of the scientific method.

The book is divided into sections based on the Australian Curriculum's three learning strands and substrands. The three learning strands in the Australian Curriculum are:

- Science Understandings (Biological Sciences; Chemical Sciences, Earth and Space Sciences; Physical Sciences)
- Science as a Human Endeavour
- Science Inquiry Skills.

The Science as a Human Endeavour strand is covered in Chapters 1 to 4, plus each substrand of the Science Understandings strand is treated as follows:

- Chapter 1—Biological Sciences
- Chapter 2—Chemical Sciences
- Chapter 3—Earth and Space Sciences
- Chapter 4—Physical Sciences.

Chapter 5 covers the Science Inquiry Skills strand.

Tips for tests and examinations

Preparation

In order to prepare for your school exams, you will need to examine the Contents section at the front of the book and then identify the relevant sections of this book before you begin to revise and practise exam-style questions.

Each topic in your school programs for Year 9 may have integrated content from more than one of these different areas of study. It is unlikely that you will have studied the content in this order.

You need to allow sufficient time for this thorough revision—at least one week for a class test and at least three weeks for a major examination. Revise the content of the chapter or section and attempt all questions. Use the supplied answers to determine which areas you need further work in.

Test/exam questions

Multiple-choice questions

In order to answer these sorts of questions, make sure that you do the following.

- Read the stem of the question thoroughly.
- Look carefully at any diagrams, flow charts or tables, and interpret them thoroughly.
- Choose the letter of the best response and not just a correct independent statement; if you don't know the answer, make the most logical choice you can.

Free-response questions

In order to answer this type of question effectively, follow the guidelines below.

- Highlight the verbs and key words in the question and respond accordingly.
- Don't waste time by restating the question; keep your answers concise.
- Label any diagrams that you draw with a pencil and ruler.
- Line graphs should occupy more than 80% of the available grid space.
- All graphs must have a title and the axes should have linear scales and be appropriately labelled with titles and units.
- Any experimental methods must be written as a series of numbered sentences in the present or past tense.
- Repeating the experiment at least five times or more can improve the reliability of experimental results.
- For questions involving the scientific method, ensure that you state the dependent and independent variables, the variables you controlled (kept the same) and the experiment that you used as a control.
- When drawing a table of data, ensure that the table is fully bounded by lines to create columns and rows. The column headings and units should occupy the first row.

Features of this book

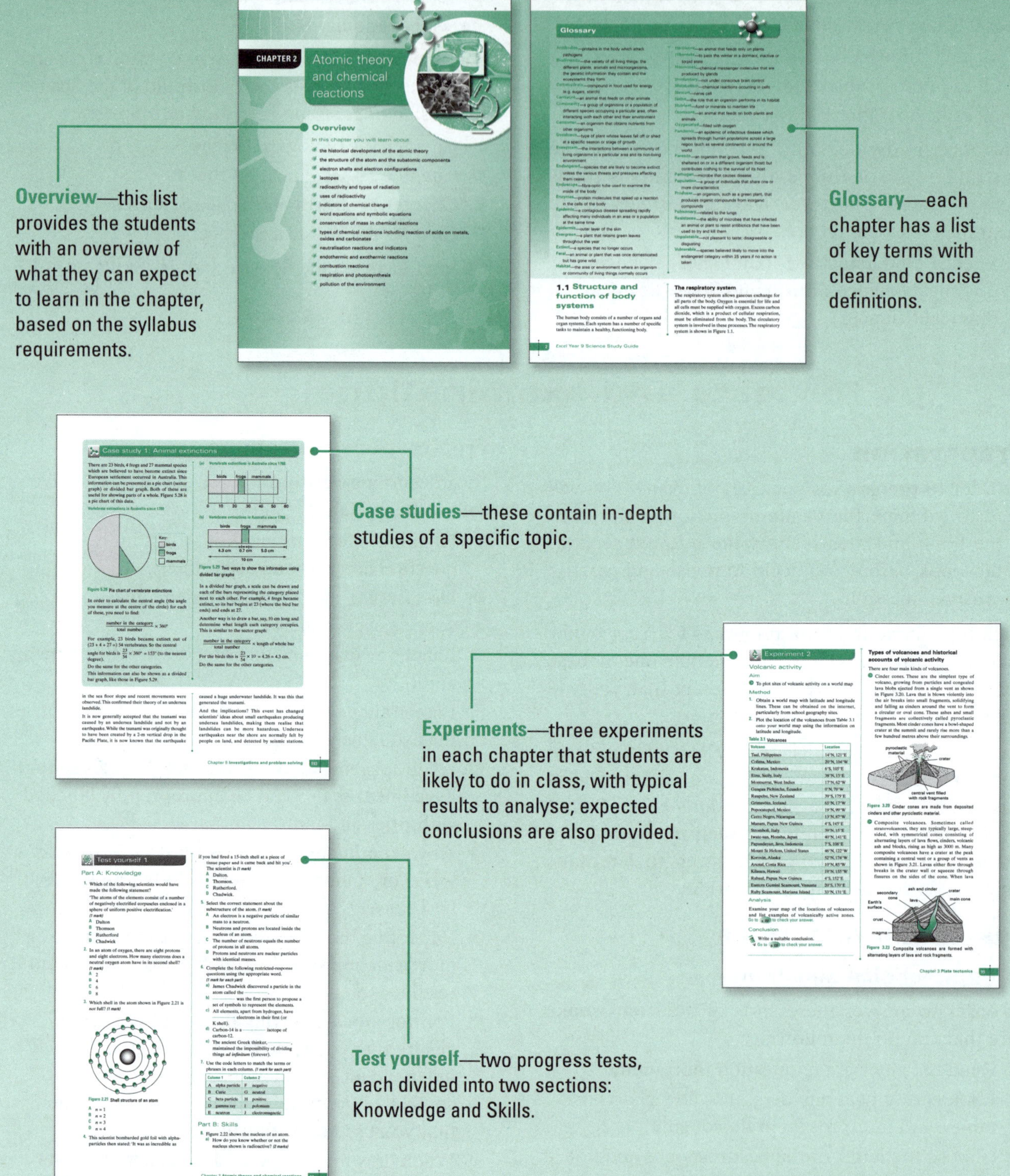

You will be thoroughly prepared for examinations and tests when you use this study guide. This guide is an effective revision and study program for exams and class tests in Year 9.

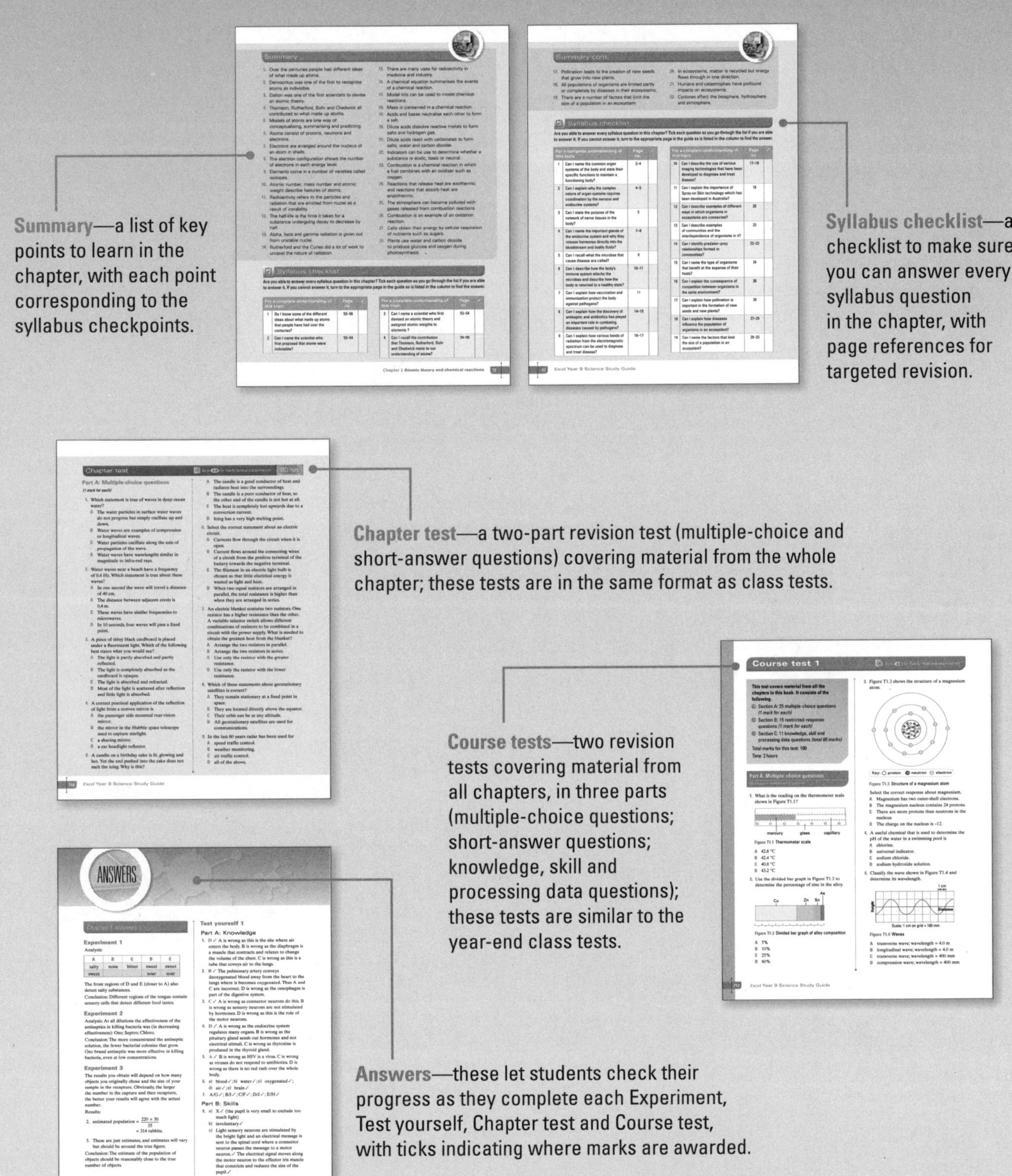

Summary—a list of key points to learn in the chapter, with each point corresponding to the syllabus checkpoints.

Syllabus checklist—a checklist to make sure you can answer every syllabus question in the chapter, with page references for targeted revision.

Chapter test—a two-part revision test (multiple-choice and short-answer questions) covering material from the whole chapter; these tests are in the same format as class tests.

Course tests—two revision tests covering material from all chapters, in three parts (multiple-choice questions; short-answer questions; knowledge, skill and processing data questions); these tests are similar to the year-end class tests.

Answers—these let students check their progress as they complete each Experiment, Test yourself, Chapter test and Course test, with ticks indicating where marks are awarded.

CHAPTER 1 Body systems and ecosystems

Overview

In this chapter you will learn about:

- the requirements of life and how various body systems provide these requirements
- the coordination of body systems
- the nervous and endocrine systems
- the effects of microorganisms on the human body
- the effect of electromagnetic radiation on the human body
- interactions between organisms
- factors that influence population size
- energy flow in an ecosystem
- how ecosystems change as the result of natural catastrophes.

Glossary

Antibodies—proteins in the body which attack pathogens

Biodiversity—the variety of all living things: the different plants, animals and microorganisms, the genetic information they contain and the ecosystems they form

Carbohydrate—compound in food used for energy (e.g. sugars, starch)

Carnivore—an animal that feeds on other animals

Community—a group of organisms or a population of different species occupying a particular area, often interacting with each other and their environment

Consumer—an organism that obtains nutrients from other organisms

Deciduous—type of plant whose leaves fall off or shed at a specific season or stage of growth

Ecosystem—the interactions between a community of living organisms in a particular area and its non-living environment

Endangered—species that are likely to become extinct unless the various threats and pressures affecting them cease

Endoscope—fibre-optic tube used to examine the inside of the body

Enzymes—protein molecules that speed up a reaction in the cells of the body

Epidemic—a contagious disease spreading rapidly affecting many individuals in an area or a population at the same time

Evergreen—a plant that retains green leaves throughout the year

Extinct—a species that no longer occurs

Feral—an animal or plant that was once domesticated but has gone wild

Habitat—the area or environment where an organism or community of living things normally occurs

Herbivore—an animal that feeds only on plants

Hibernate—to pass the winter in a dormant, inactive or torpid state

Hormones—chemical messenger molecules that are produced by glands

Involuntary—not under conscious brain control

Metabolism—chemical reactions occurring in cells

Neuron—nerve cell

Niche—the role that an organism performs in its habitat

Nutrient—food or minerals to maintain life

Omnivore—an animal that feeds on both plants and animals

Oxygenated—filled with oxygen

Pandemic—an epidemic of infectious disease which spreads through human populations across a large region (such as several continents) or around the world

Parasite—an organism that grows, feeds and is sheltered on or in a different organism (host) but contributes nothing to the survival of its host

Pathogen—microbe that causes disease

Population—a group of individuals that share one or more characteristics

Producer—an organism, such as a green plant, that produces organic compounds from inorganic compounds

Pulmonary—related to the lungs

Resistance—the ability of microbes that have infected an animal or plant to resist antibiotics that have been used to try and kill them

Unpalatable—not pleasant to taste; disagreeable or disgusting

Vulnerable—species believed likely to move into the endangered category within 25 years if no action is taken

1.1 Structure and function of body systems

The human body consists of a number of organs and organ systems. Each system has a number of specific tasks to maintain a healthy, functioning body.

The respiratory system

The respiratory system allows gaseous exchange for all parts of the body. Oxygen is essential for life and all cells must be supplied with oxygen. Excess carbon dioxide, which is a product of cellular respiration, must be eliminated from the body. The circulatory system is involved in these processes. The respiratory system is shown in Figure 1.1.

The circulatory system

The circulatory system transports water, nutrients, gases and other chemicals to the cells of the body and transports wastes away from cells. This system is shown in Figure 1.2.

The digestive system

The digestive system breaks down food into nutrients that can be absorbed into the bloodstream, where they are then transported to body cells. Indigestible material is removed from the body. The digestive system is shown in Figure 1.3.

The excretory system

The excretory system removes liquid and gaseous wastes produced by metabolic processes. This system is shown in Figure 1.4.

Systems working together

The four body systems just outlined collaborate and work efficiently together to maintain a properly functioning organism. Consider the following points.

- Each cell of your body needs a constant supply of nutrients. These nutrients come from the food we eat.
- The food we eat contains many different types of chemical compounds including carbohydrates, fats, proteins, minerals, vitamins and water.
- Most of the food we eat needs to be broken down into small particles so it can be absorbed. This is the job of your digestive system. This process starts in the mouth and is then completed in the small intestine.
- The broken-down nutrients must be transported to all cells of the body. This is the job of the circulatory system.
- Any indigestible food is stored as solid waste before it is removed from the digestive system as faeces.
- Our body cells need oxygen. The absorption of oxygen is the job of the respiratory system.
- Once the oxygen is absorbed, it is transported to all the body cells via the blood flowing through the circulatory system.

Structures

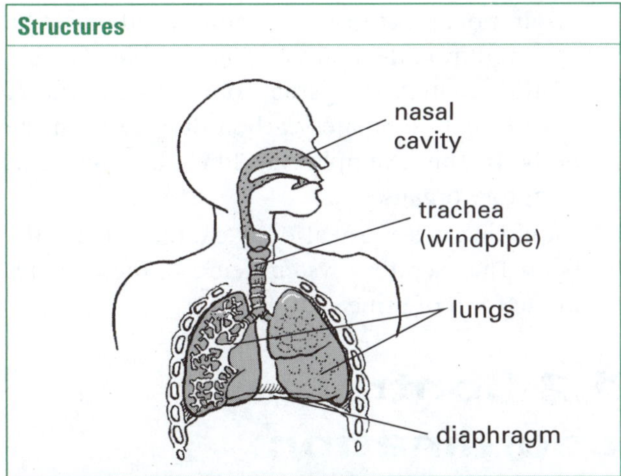

Function

- nasal cavity: air is inhaled and exhaled through the nose
- trachea: tube through which the air moves between the nasal cavity and the lungs
- lungs: site of gaseous exchange with the blood stream; oxygen is absorbed by the blood and carbon dioxide is released from the blood
- diaphragm: muscle that contracts/relaxes to change volume of chest cavity, allowing the lungs to expand and contract

Figure 1.1 Respiratory system

Structures

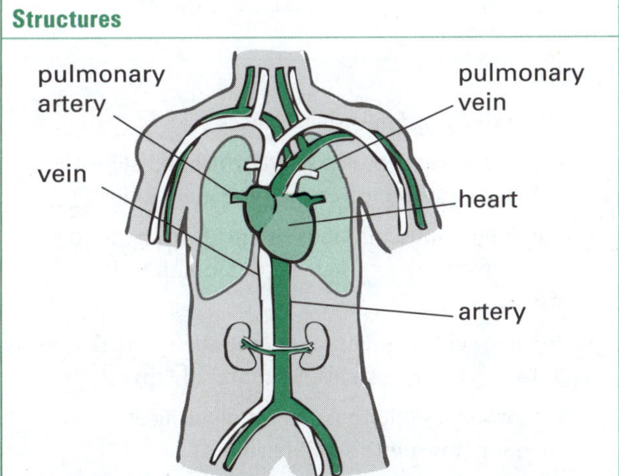

Functions

- heart: pumps blood around the body
- arteries: transport oxygenated blood (filled with oxygen) away from the heart to the organs of the body
- pulmonary artery: transports deoxygenated blood to the lungs
- veins: transport deoxygenated blood from the body back to the heart
- pulmonary vein: transports oxygenated blood back to the heart from the lungs
- blood: carries gases (oxygen and carbon dioxide), electrolytes, nutrients and waste products

Figure 1.2 Circulatory system

Structures

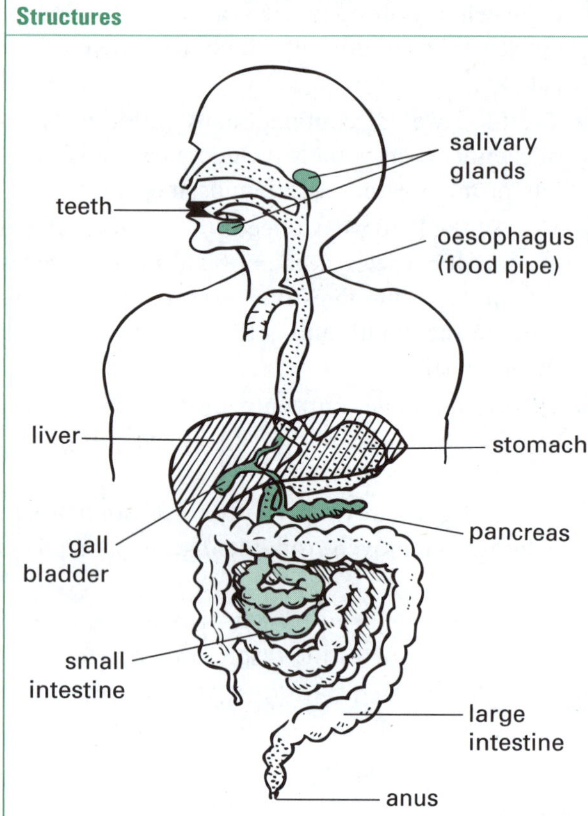

Function

- teeth: grind food into small pieces
- salivary glands: produce saliva which moistens the food and starts the digestion of starchy food
- oesophagus: muscular movements help the food move through this tube from the mouth to the stomach
- stomach: acids and enzymes continue to break down the food to form a semisolid material (chyme)
- liver: produces bile to aid in fat emulsification (breaking down into small particles)
- gall bladder: stores bile until it is released into the small intestine
- pancreas: produces digestive enzymes and juices that neutralise the acidic material leaving the stomach
- small intestine: digestion is completed and the nutrients pass through the intestinal wall into the bloodstream
- large intestine: water is removed from the remaining solid waste.
- anus: opening for solid waste to be removed from the body

Figure 1.3 Digestive system

Structures

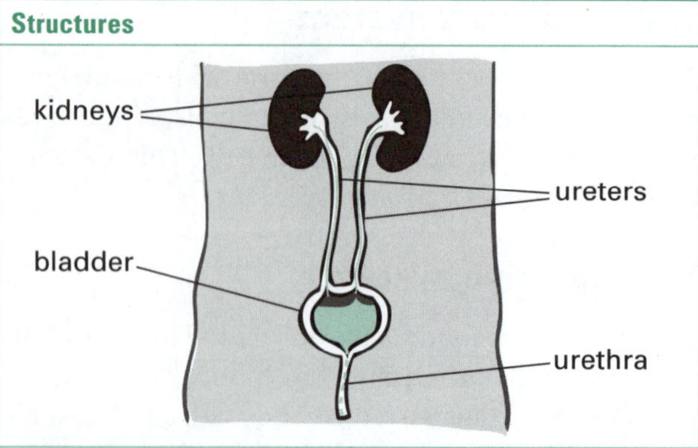

Function

- lungs: remove carbon dioxide during exhalation (see respiratory system)
- kidneys: filter the blood to remove nitrogenous wastes, salts and excess water
- ureters: tubes carrying urine from the kidneys to the bladder
- bladder: stores urine until it is eliminated
- urethra: tube through which urine is eliminated from the body

Figure 1.4 Excretory system

- Carbon dioxide is a waste product of cellular respiration. Most of this carbon dioxide must be removed from the body. It is the combined action of the circulatory system and the respiratory system that eliminates carbon dioxide from the body. In this example, the lungs are acting as excretory organs.
- Liquid wastes must also be excreted from the body. The excretory system removes these wastes in the form of urine.

1.2 Control and coordination: the nervous and endocrine systems

The human body is composed of tissues, organs and organ systems. These organ systems collectively function to satisfy the needs of all cells. Such a complex organisation requires coordination; otherwise the human body cannot function efficiently. Communication and control systems are needed to ensure that all body systems are coordinated. The following are the two important control and coordination systems.

- Nervous system. The brain is connected to the rest of the body by nerve fibres that transmit and receive information as electrical signals
- Endocrine system. Various glands produce chemical messenger molecules called hormones that control various bodily processes.

The nervous system

The nervous system consists of a network of nerve tissues which transmit electrical information from one site to another. Nervous tissues are composed of nerve cells called neurons (also spelt neurones). The nervous system is composed of two parts. These are the:

- central nervous system (CNS), which consists of the brain and spinal nerve cord
- peripheral nervous system, which consists of nerves that connect the CNS to the rest of the body.

Structure and types of neurons

Each neuron consists of a cell body with radiating fibres called dendrites. Many types of neurons contain one longer and thicker fibre called the axon. The axon is covered in a fatty insulating layer called the myelin sheath. It prevents nervous impulses from crossing over to neighbouring neurons.

- Axons conduct electrical impulses *away* from the cell body.
- Dendrites conduct electrical impulses received from another neuron *towards* the cell body.

Neurons can be classified into three main types. (See also Figure 1.5.)

- Sensory neurons carry electrical impulses towards the central nervous system. Our sense organs (e.g. our tongue, eyes, ears and skin) contain sensory receptors that contain many sensory neurons. They transmit information into the spinal cord.
- Connector neurons (or interneurons) are part of the CNS and are located in the spinal cord. They receive information from the sensory neurons. Some of this information is relayed up the spinal cord to the brain. Other information may be relayed immediately to muscles (or glands) along motor neurons.
- Motor neurons carry electrical impulses away from the CNS towards muscles or glands which then respond. Muscles contract and glands release their hormones.

Structure of the central nervous system

The central nervous system consists of the brain and spinal nerve cord. Layers of fluid, membranes and bone tissue protect them.

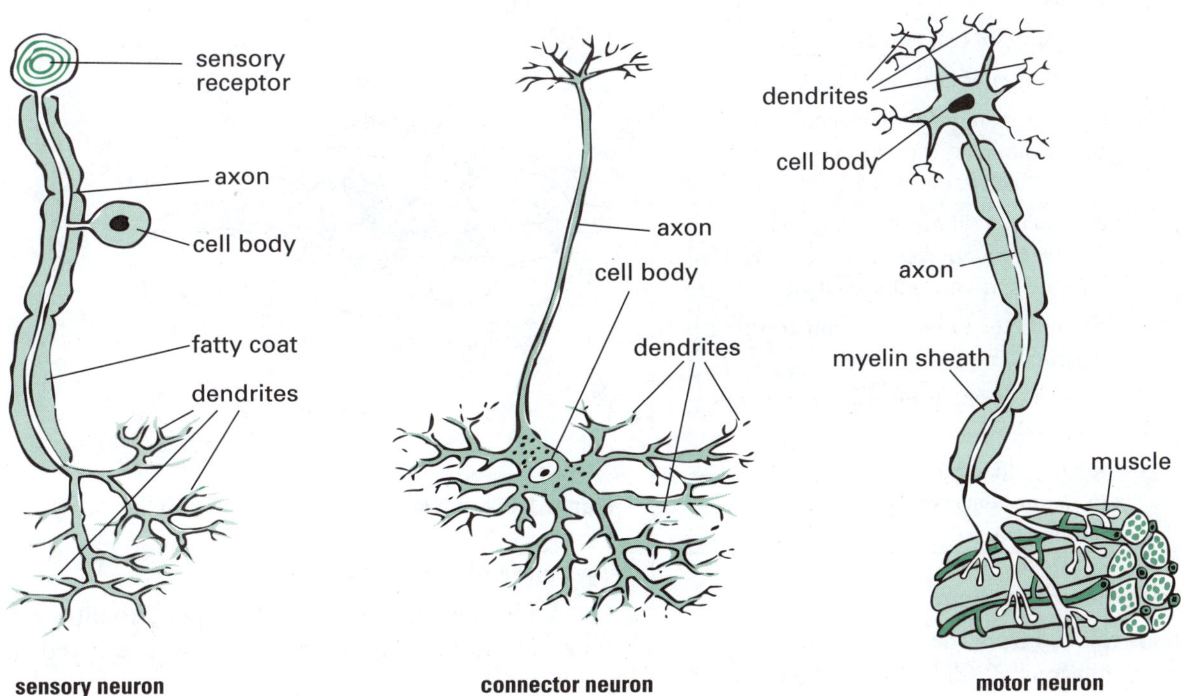

Figure 1.5 Types of neurons

- The skull protects the brain.
- The vertebral bones protect the spinal nerve cord—nerves from the body enter the spinal cord through small gaps between the vertebrae.

The brain is a control centre. Information received is processed and messages are sent by the spinal cord to various muscles and glands. Figure 1.6 shows some of the important parts of the brain.

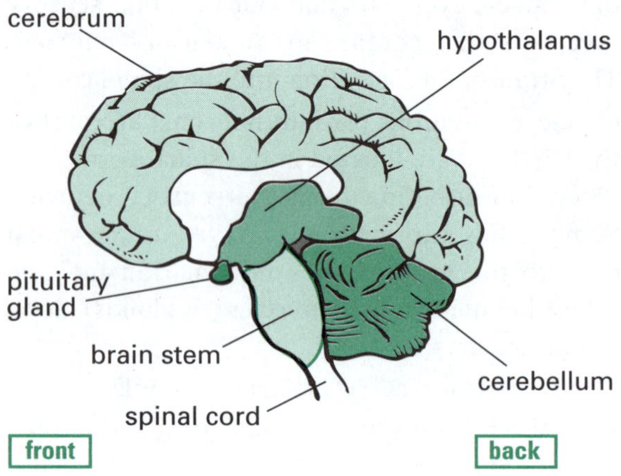

Figure 1.6 Plan diagram of the brain

Functions of the parts of the brain

Table 1.1 lists the major functions of the parts of the brain.

Table 1.1 Brain functions

Part	Function
cerebrum	• composed of two halves • controls voluntary movements • controls memory, intelligence, behaviour, emotions, speech, vision, smell, touch and hearing
cerebellum	• controls muscles involved in involuntary movements such as balance and fine motor control
brain stem	• connects the brain to the spinal cord; information sorting centre • control centre for breathing, heart rate and swallowing • the hypothalamus (at the top of the brainstem) controls thirst and temperature

The peripheral nervous system

Information is relayed from the CNS to the body via the peripheral nervous system. Part of this system involves voluntary movements and the rest involves involuntary (automatic) movements.

- The 43 pairs of voluntary nerves from the brain and spine connect to the muscles and sense organs in the head and body. These nerves control voluntary movements of the arms, legs and head.
- The system of involuntary nerves regulates many functions including:
  - heart muscle control
  - iris muscle control
  - bladder and bowel muscle control (can be controlled with training)
  - responses to danger.
- The body has various automatic (involuntary) movements called reflex arcs. If a sensory neuron is stimulated, messages are sent via the connector neurons in the spinal cord directly to the motor neuron and an effector muscle or gland. This leads to a rapid response that is important in many situations.

For example, if you tread on a sharp thorn with bare feet your body immediately responds by raising your foot away from the danger. The brain does not control this response, although it does register the event and the pain associated with it.

Figure 1.7 shows the movement of nerve impulses in a reflex arc. A classic example of the reflex arc is the patellar reflex, shown in Figure 1.8. When the patellar tendon is tapped just below the knee, the quadriceps muscle contracts and the lower leg kicks forward.

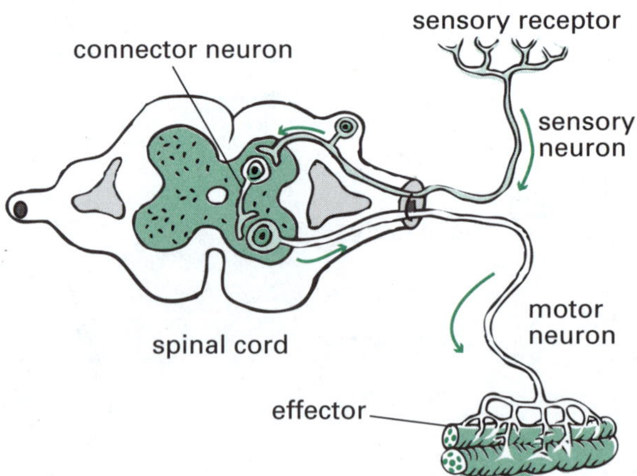

Figure 1.7 Nerve impulse movement in a reflex arc

The sense organs and receptors

Our body has sense organs that respond to different types of stimuli. These sense organs are listed in Table 1.2.

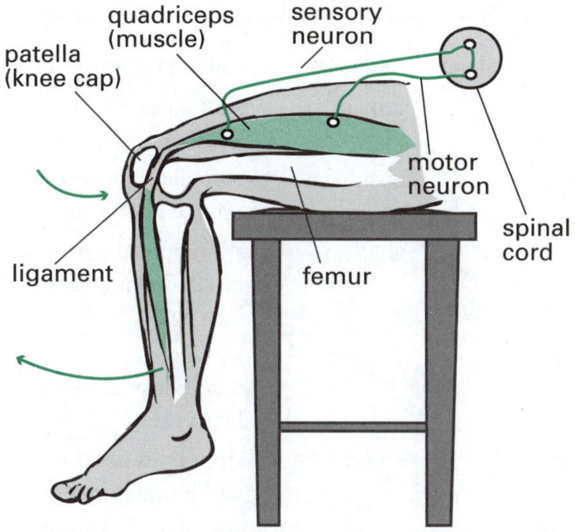

Figure 1.8 Patellar reflex

Table 1.2 Sense organs and their location

Sense organ	Location of sensory receptors
taste	tongue surface
smell	top of nasal cavity
vision	retina in the eyes
hearing	ear drum and auditory nerve
touch/pressure	skin
pain	skin; throughout body
temperature	skin; throughout body

Experiment 1

Stimulus and response

Aim

- To examine stimulus and response in taste testing

Method

1. Choose four different solutions (salty, sweet, sour and bitter).
2. Using fresh toothpicks or cotton buds each time, place drops of these solutions on different locations of your tongue. Use a mirror to observe your tongue. Figure 1.9 shows the positions on the tongue to test.
3. Record whether or not each zone will detect each of the four solutions.

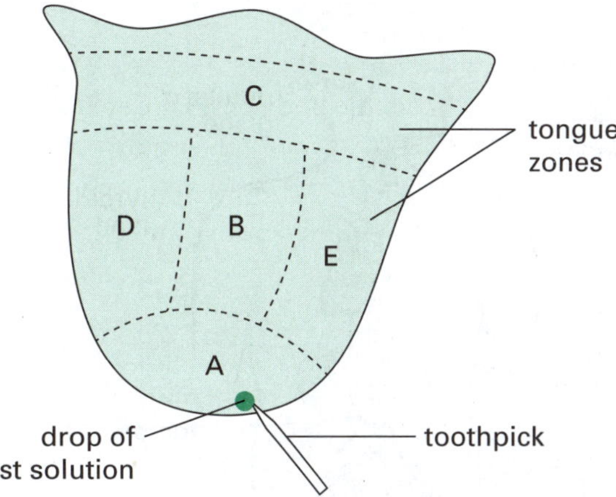

Figure 1.9 Taste-testing zones on the tongue

Analysis

Once you have completed this experiment and recorded your results, determine which zones of the tongue are more sensitive to each chemical. Go to p. 224 to check your answer.

Conclusion

Write a suitable conclusion.

Go to p. 224 to check your answer.

The endocrine system

The human body has a second system that is involved in control and coordination. This is the endocrine system that consists of various glands that release hormones (special chemical messengers) directly into the bloodstream and bodily fluids (vascular system). The vascular system carries these hormones to various organs or cells around the body, which are then stimulated to respond. The endocrine system is important in controlling growth, metabolism and reproduction. Metabolism refers to all the biochemical processes that keep an organism alive. Figure 1.10 shows the location of some important endocrine glands. The pituitary gland at the base of the brain is an important gland in that it controls and stimulates many other glands. It is often referred to as the master gland. The pituitary releases many hormones including ones that regulate skin pigmentation, re-absorption of water in the kidneys as well as excretion of milk in nursing mothers.

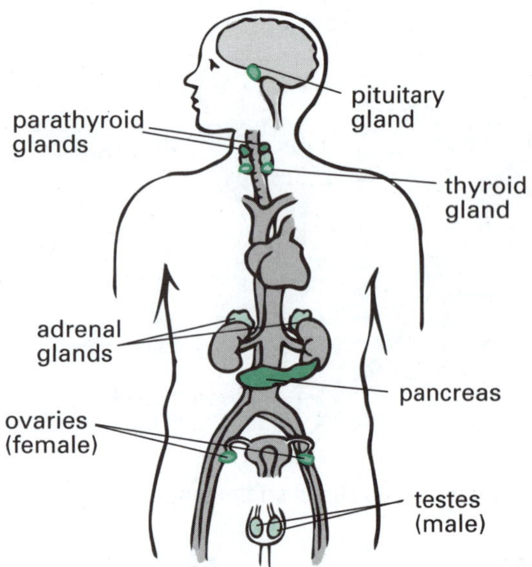

Figure 1.10 Locations of some endocrine glands

Table 1.3 lists some important endocrine glands and some of the functions controlled or affected by the hormones they produce.

Endocrine glands secrete their hormone products directly into the blood rather than through a duct. Some organs not so well known for their endocrine activity are the stomach and the hypothalamus. Exocrine glands secrete their products (including digestive enzymes) into ducts leading directly to the external environment. These include sweat glands, salivary glands, mammary glands, liver and pancreas.

Table 1.3 Some important functions of endocrine glands

Endocrine gland	Hormone	Function of hormone
pituitary	growth hormone	stimulates growth and DNA synthesis
pancreas	insulin	stimulates glucose uptake in all cells
	glucagon	stimulates the liver to break down glycogen into glucose
thyroid	thyroxine	stimulates the metabolism and the heart rate
adrenal	adrenaline	stimulates heart rate and blood pressure
parathyroid	parathyroid hormone	stimulates calcium ion release in bones

Example 1: Controlling body temperature

Heat is produced in the body by metabolic processes that occur in the cells. Hormones produced in the thyroid gland control the production of heat (see Figure 1.11).

The control of body temperature is called thermoregulation. This type of regulation is considered a negative feedback as the regulation involves reducing the activity or output of an organ or gland. Thermoregulation is controlled by the

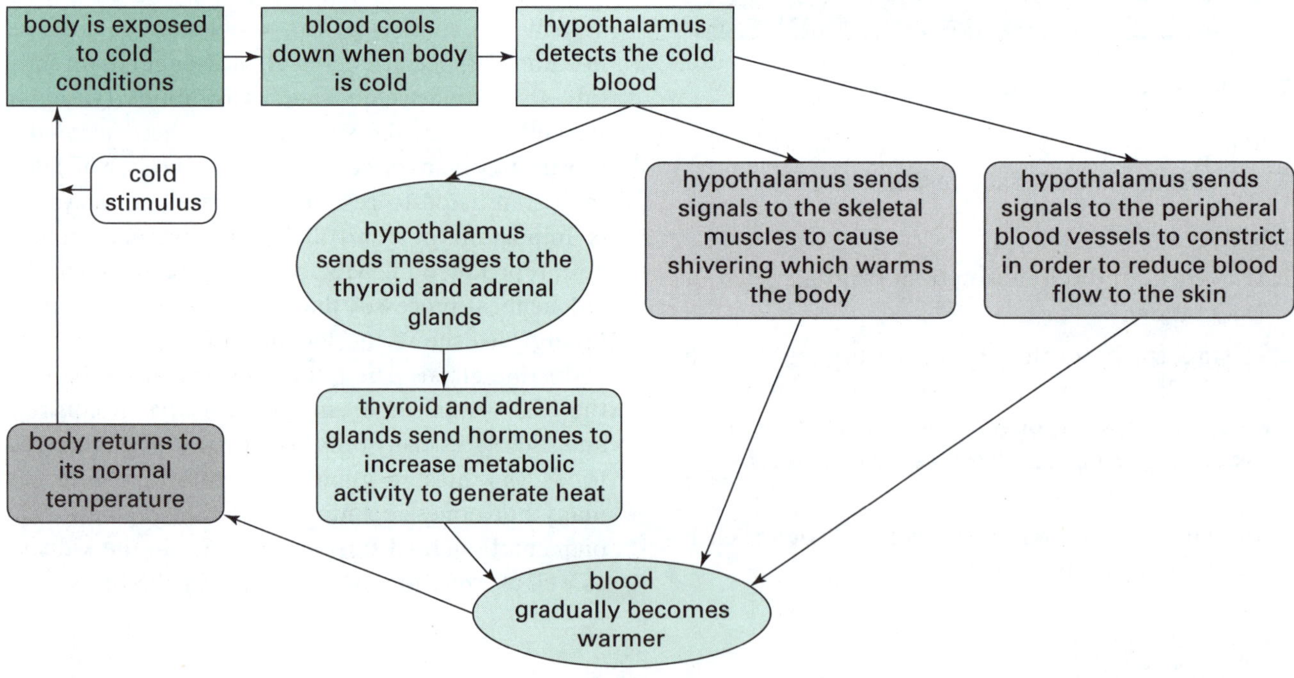

Figure 1.11 What happens if you get cold?

hypothalamus in the brain. There are a number of ways in which the temperature of the body is controlled.

- Sweating is a process that helps to regulate the temperature of the body. When it is hot the body sweats and is cooled by the evaporation of sweat from the skin. On cold days, sweating stops to prevent heat loss.
- Fine hairs cover our body. On hot days the hairs lie flat against the skin and allow air flow to remove heat from the body by convection. On cold days, muscles under the skin contract and raise the hairs and so create a still layer of air next to the skin which acts as an insulator against heat loss.
- Increasing blood flow to the surface blood capillaries also assists heat to be lost from the body on hot days. This is achieved by relaxing the smooth muscles in arteriole walls, leading to greater blood flow.
- Shivering is a response to cold conditions. The brain coordinates the shivering process of the muscles. This process produces heat to warm you up. It is illustrated in Figure 1.11.

Example 2: Controlling your water balance

Your kidneys control the water balance of your body. This water balance is connected to the role of the kidneys in removing urea (waste) from the blood. The steps in this process are as follows.

1. Blood enters the kidneys via blood capillaries.
2. The blood is filtered to remove the urea. This happens in the nephrons inside the kidney. At the same time other substances including glucose, ions (salts) and water are also removed from the blood.
3. Because your body needs a balance of glucose for cellular respiration, the glucose is selectively reabsorbed back into the blood.
4. Some ions and water are also reabsorbed to maintain a balance.
5. The urea and some ions and water pass out of the kidney in the form of urine. The urine is stored in the bladder.
6. If your body has too much water, your urine will be dilute as it contains more water.
7. If your body has too little water, your urine is concentrated as it contains less water.

Example 3: Controlling glucose levels in the blood

The pancreas produces two important hormones called insulin and glucagon. Glucagon is produced by alpha cells, and insulin by beta cells in the pancreas.

- Normal glucose levels in the blood fall in the range 0.7 to 1.0 g/L.
- If the glucose levels rise higher or lower, the body responds to return the glucose levels to the normal range or set point.
- Insulin reduces glucose levels in the blood by stimulating body cells to take up glucose.
- Glucagon acts in the reverse way. When blood glucose levels are too low, glucagon stimulates the liver to break down its stored glycogen into glucose which is then released into the blood and body fluids.
- Glucagon and insulin are never released at the same time.
- Together these two hormones keep glucose levels regulated. The disease called Type 2 diabetes is caused by a failure of cells in the pancreas to produce the correct levels of these glucose-regulating hormones. (Note: Type 1 diabetes involves the destruction of insulin-producing cells by a malfunction of the immune system.)

Figure 1.12 shows the cycle required to keep your blood glucose in balance.

1.3 Microorganisms and disease

A large number of microscopic organisms are alive in your body. Microbes are found all over you, including on your skin, in your mouth and in your intestines. Microbes that cause disease are called pathogens.

Not all microbes are pathogens. The bacterium *Escherichia coli*, for example, is found in your intestines and provides you with vitamin K and some B-group vitamins. In fact there are over 700 species of bacteria that perform a variety of useful functions in your large intestine. Another bacterium, *Propionibacterium acne*, lives on your skin and produces a chemical called propanoic acid which stops the growth of unwanted microbes.

Disease-causing microbes belong to several kingdoms of living things.

1. Kingdom monera (e.g. tuberculosis bacteria)

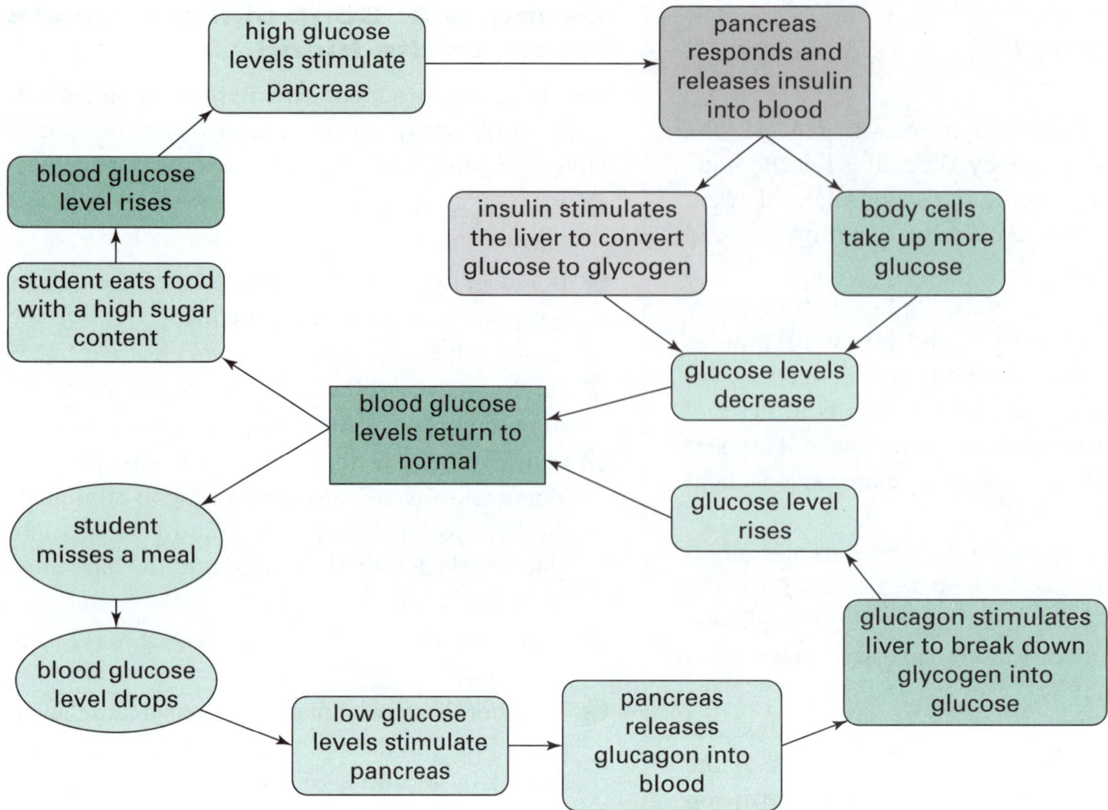

Figure 1.12 Keeping your blood glucose in balance

2. Kingdom protista (e.g. malaria protozoa)
3. Kingdom fungi (e.g. tinea)

Figure 1.13 shows the cellular structure of two pathogenic bacteria.

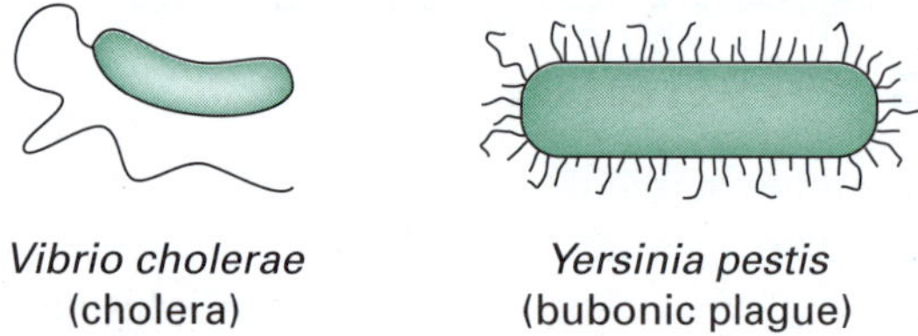

Figure 1.13 Pathogenic bacteria

Viruses are not cellular life forms, but are microbes and they do cause disease. Viruses are bodies that are much smaller than cells, and are composed of some genetic material (codes required for reproduction and control) surrounded by a protein coat. Viruses cannot reproduce unless they first invade a host cell. Influenza, mumps, poliomyelitis, HIV and even the common cold are caused by viruses (see Figure 1.14).

When HIV (human immunodeficiency virus) invades our body, it attacks the immune system and produces a disease called AIDS (acquired immune deficiency syndrome). This process is shown in Figure 1.15. Although treatments for AIDS and HIV can slow the course of the disease, there is as yet no known cure or vaccine. Viral drugs have been developed to reduce the damaging effects of the virus on the body.

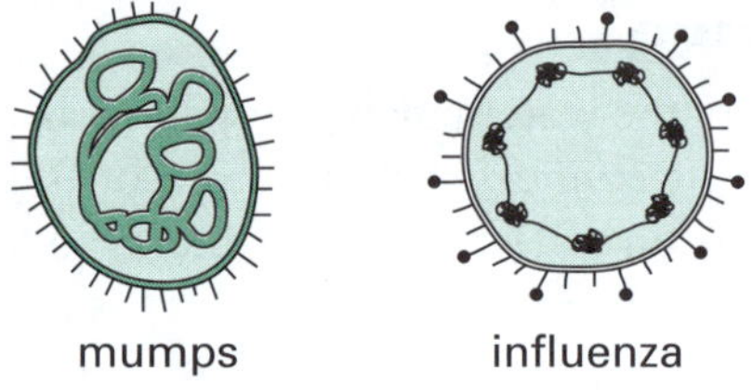

Figure 1.14 Examples of viruses

Infection and immunity

Microbes invade our bodies to find a site to reproduce. When a pathogen enters your body, there is an immediate response.

The symptoms of an infection vary from one microbial infection to another. Some typical responses to microbe infection are as follows.

- Bacterial infections: localised redness of the skin; swelling and/or pain in certain areas of the body; production of pus from a cut.

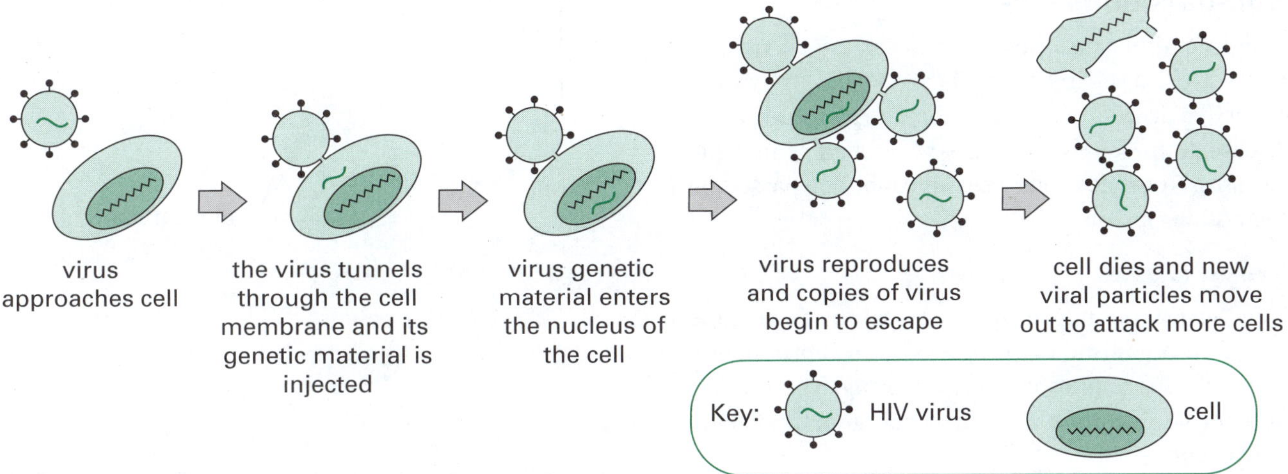

Figure 1.15 HIV infection

- Viral infections: rise in temperature; coughing; muscular aches; sinus congestion and nasal discharge.
- Fungal infections: red, cracked, itchy skin.

Upon infection, your body's defences attempt to destroy the invader. Once the danger is over, your body has acquired immunity to that particular disease.

If that microbe attacks again, your body has special protein molecules called antibodies that are ready to deal with it. You may only experience slight symptoms instead of the full effects of the disease. Immunisation is an active process that seeks to protect you before you are attacked by certain microbes.

Vaccination and immunisation

Scientists have developed vaccines from dead or ineffective microbes. If these vaccines are injected into your blood, you develop immunity to that disease, that is, you are protected from it. Active immunisation has been developed for bacterial diseases such as diphtheria and tuberculosis as well as viral diseases such as polio and measles.

Some vaccines, such as the measles vaccine, give life-long immunity; others, such as the polio vaccine, are given to children in a series of vaccinations over 15 years. Some people have a higher risk of catching certain diseases. A farmer can, for example, have an injection of tetanus antibodies to provide immediate but short-term protection against this organism found in the soil. This type of protection is called *passive* immunity. However, while it provides immediate protection the body does not develop memory of it, and so the patient is at risk of being infected by the same pathogen later.

Immunisation: good or bad?

Since the introduction of mass vaccination programs around the world, millions of lives have been saved. Despite this fact, there are a variety of opinions in society about its benefits. The anti-immunisation opponents are very active on the internet and they often back their arguments with misleading and selective use of data.

Here are some points made by people on each side of the argument.

For

1. Vaccination helps to eliminate disease from the population and protects individuals from getting the disease.
2. We need compulsory immunisation programs so as to protect babies until they are old enough to be vaccinated.

Against

1. Some people (but very few) get adverse reactions from vaccinations and some have even died.
2. Compulsory programs take away our right to choose.

After vaccination campaigns throughout the 19th and 20th centuries, the World Health Organisation officially declared the eradication of smallpox in 1979. The other infectious disease to have been eliminated from this planet is rinderpest, a viral disease of cattle. This became official in 2011.

Non-infectious disease

Non-infectious diseases have a variety of causes. Some are caused by faulty DNA leading to faulty genes. Toxic chemicals can cause organ and tissue disease. Some diseases are also classified as lifestyle diseases. Non-infectious diseases cannot be passed on to another person.

Genetic diseases

Genetic diseases may have several causes. During the life of an organism, the process of cell reproduction and cell differentiation may go wrong. There are a number of different ways in which genetic disease may occur:

- gene mutation caused by insertion, substitution or deletion of nucleobases in the DNA code
- deletion of one or more genes from a chromosome
- whole chromosome(s) are missing or extra chromosomes are present.

Causes of genetic disease

Gene mutations

A gene is a segment of the DNA molecule which codes for the production of a protein. Genes are composed of four chemicals called nucleobases arranged in a specific order. This order is called the genetic code. The nucleobases are represented by the first letters of their names (A = adenine; C = cytosine; G = guanine; T = thymine). Genes may mutate by the insertion, substitution or deletion of one or more nucleobases.

Sickle-cell anaemia is a blood disorder caused by a substitution of an A by a T nucleobase on chromosome 11, leading to a defective haemoglobin molecule (see Figure 1.16).

Cystic fibrosis is a hereditary disease caused by the deletion of three nucleobases on chromosome 7. The faulty protein that results leads to difficulties in regulation of digestive juices and mucus formation. Serious lung problems often lead to an early death.

Gene deletion

This type of genetic diseases results from the deletion of whole genes from a chromosome. This may happen during replication (copying) of the chromosome.

Cri du chat (also known as Le Jeune's syndrome) is a genetic disease caused by the partial deletion of one of the short arms of chromosome 5. Affected children make cat-like cries early in life. They suffer speech and behavioural problems as well as intellectual disabilities.

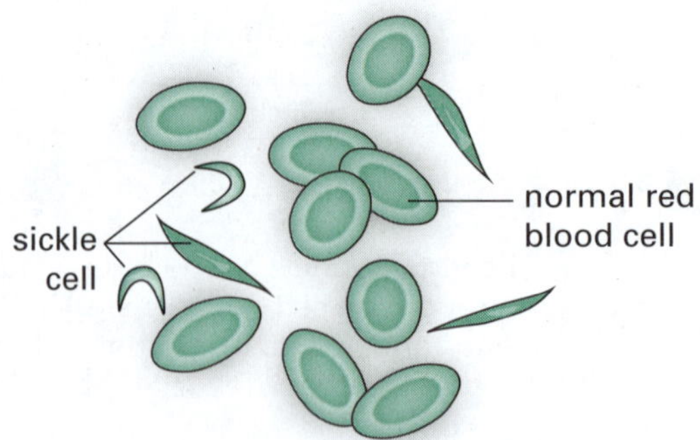

Figure 1.16 Sickle-cell anaemia is an inherited disease where the normal red blood cells become sickle shaped, rigid and move through blood vessels with difficulty. Sickle blood cells die off early and leave a shortage of red blood cells to carry oxygen.

Missing chromosomes or extra chromosomes

Sometimes chromosomes do not divide correctly and some cells inherit additional copies while others receive less than the normal number. The normal number of chromosomes in human cells is 46.

Down syndrome is caused by cells receiving an extra copy of chromosome 21 attached to the end of chromosome 15. Thus people with Down syndrome have 47 chromosomes. Affected people have mild to moderate developmental difficulties. Down syndrome is characterised by a rounded face, very small chin and an oversized tongue.

Turner syndrome is caused by a missing sex chromosome. Normal females have two copies of the sex chromosomes (i.e. XX). Females with Turner syndrome only have one X chromosome. Affected women are infertile due to the absence of a menstrual cycle. Affected women also tend to suffer from diabetes and vision problems.

Diseases caused by environmental agents or lifestyle

Environmental agents (e.g. toxic chemicals, high-energy radiation, and viruses) may damage the chromosomes, leading to changes or mutations in genes.

Lifestyle diseases are non-infectious. They are caused by organs or other parts of the body failing to work properly. Heart disease and obesity can be caused by lifestyle choices. The chemicals in tobacco smoke, for example, can cause emphysema, coronary heart disease, strokes and lung cancer.

Public health: smoking and cancer

Smoking is a major public health issue. Diseases that result from smoking have very high social costs including the costs of medicines and hospitalisation. Data from the 2004–05 report by the Australian Department of Health showed that smoking was responsible for about 80% of all lung cancer deaths and 20% of all cancer deaths. Anti-smoking campaigns have led to significant decreases in the rate of smoking and consequently a huge saving in health costs. In the last 60 years the smoking rate for men has decreased from 72% to about 18%. For women the rate has dropped from 26% to about 15%.

Why do you think symbols like that shown below in Figure 1.17 are common inside workplaces, restaurants and public buildings?

Figure 1.17 'No smoking' sign

Second-hand smoke, or passive smoking, is inhaling smoke from tobacco products used by others. This tobacco smoke lingers in any environment, and is breathed in by other people within that environment. Being exposed to second-hand tobacco smoke can cause disease, disability and even death.

Health, diet and exercise

Many scientific studies have shown the medical benefits of a healthy lifestyle. Lifestyle diseases such as obesity, diabetes and heart disease can be reduced in frequency by changes in what we eat, how we exercise and by eliminating high-risk behaviours such as drug taking and binge drinking.

The recommendations for a healthy diet include:

- eat more vegetables, wholegrains and fruit as these contain more complex carbohydrates, vitamins, minerals and fibre
- reduce saturated fats and total fat intake and replace with smaller amounts of unsaturated fats to control blood cholesterol
- reduce simple sugar intake and fast foods containing them.

Regular daily exercise (e.g. walking, bike riding, swimming and running) helps to control weight and strengthen muscles and the cardiovascular system.

1.4 History of disease

The cause of disease was not known for most of human history.

Medieval times

In medieval times (between the 5th and 15th centuries) most people viewed disease as a punishment from God. In medical terms a disease was thought to be caused by an imbalance of four forces or humours (phlegm, yellow bile, black bile and blood). People believed that bringing these back into balance could then restore health. Blood letting was often used to try and cure the sick. Sometimes leeches were used to draw blood from a patient.

Doctors used many plants and herbs in these medieval times to cure ailments. Burns, for example, were treated with lavender leaves. Potions made from various herbs such as henbane and hemlock were used to treat jaundice and vomiting.

Great disease plagues caused death on a massive scale in the medieval period. The Black Death (1348–1350) was a pandemic caused by an outbreak of bubonic plague (caused by the *Yersinia pestis* bacterium). A pandemic is an epidemic that spreads quickly over very large areas including continents. It has been estimated that between 30 and 60% of the population of Europe died of this disease. Now we know that bacteria and the lack of hygienic practices caused the disease at that time and led to its rapid spread. Fleas, living on black rats, carried the disease and their bites transferred the bacteria to humans. Figure 1.18 shows the areas over which the Black Death spread.

The first written records of an outbreak of syphilis in Europe occurred in the late 15th century in Naples, Italy, during a French invasion. And throughout the 17th century, syphilis was more widespread and devastating than AIDS is now. Syphilis is a sexually transmitted infection caused by the spirochete bacteria *Treponema pallidum*. But, of course, no one knew that at the time. A variety of useless 'cures' were tried. Gum from the wood of the guaiacum tree was once used to treat syphilis. Another treatment was mercury and isolation, but this was often worse than the disease. It gave rise to the famous line at the time: 'A night with Venus means a lifetime with Mercury'. Nowadays syphilis can be effectively treated with antibiotics, specifically penicillin G, but there is no effective preventative vaccine.

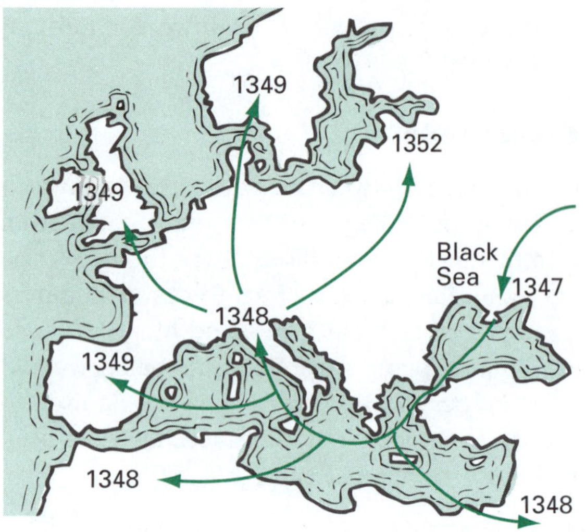

Figure 1.18 The Black Death seems to have started somewhere in Asia and was brought to Europe by traders, where it quickly spread.

The 16th to 18th centuries

Following the Black Death, bubonic and pneumonic plagues regularly returned to Europe for the next 300 years. These plagues mainly affected people living in crowded conditions in cities. In 1665–1666 a Great Plague hit London and over 30 000 people died. Many people, including Isaac Newton, fled to the countryside to avoid the disease. Because the disease affected so many different types of people, many educated people believed that the plague had origins other than the wrath of God. The idea of 'putrid air and vapours' was used to explain the origins of the disease. As a consequence sanitation was gradually improved in an attempt to clean up the foul smells from sewers in an overcrowded city. This in turn reduced the habitat of the black rat which carried the fleas that harboured the disease. By 1720 the conditions for plague largely disappeared.

Vaccination

The Chinese used an early form of vaccination (called variolation) in the 10th century. They exposed healthy individuals to pus obtained from smallpox survivors. In 1716 Mary Montagu observed that Turkish people gained protection from smallpox by deliberately introducing a small amount of pus from an infected person into a small wound made in the skin of a healthy person. Those inoculated had a mild case of the disease, recovered and then became resistant to any further attack by smallpox. However, inoculation with live smallpox virus was dangerous as too large a dose would kill the patient. In 1796 Edward Jenner inoculated a young boy with the pus from a person suffering from cowpox (a much milder, non-fatal disease). This inoculation was shown to protect the child from the deadly smallpox. These experiments were the beginnings of the technique called vaccination. Jenner first used the word *vaccination*, from the Latin *vacca* meaning 'cow'. It was so named because the first vaccine was derived from a virus affecting cows.

Microscopy and microbes

In 1683 Anton van Leeuwenhoek observed microorganisms (microbes) in samples of water using a microscope. It was suggested that these microbes were formed spontaneously when wastes putrefied. By the 19th century, however, it had been shown that these microbes were present naturally and reproduced in large amounts under certain conditions.

By the 1860s it became apparent that disease could be spread by human contact, by water and via contaminated food. In 1857 Louis Pasteur showed that microbes caused milk to become sour by fermentation. Pasteur then went on to prove that if microbes were excluded or killed then the fermentation was prevented. This work led to the pasteurisation of milk. Pasteur also showed that a microbe was responsible for a disease of silkworms. Pasteur was the first person to develop a vaccine for rabies (a viral disease) which contained a weakened form of the virus.

In 1876 Robert Koch proved that bacteria caused anthrax (a disease of cattle). In 1882 Koch isolated tuberculosis bacteria and showed they could cause the disease in uninfected animals. He therefore proved that tuberculosis was contagious, as it was spread by sneezing and coughing.

Antiseptics and antibiotics

Surgery in hospitals often led to the death of patients due to infection of wounds by bacteria. In the first 50 years of the 19th century, 50% of surgery patients died from infection. In 1867 Joseph Lister introduced the use of antiseptics in surgery and the survival rate from operations increased greatly. Carbolic acid (containing phenol) was the first antiseptic used in hospitals.

Sulfonamides were the first antibiotic drugs. In 1932 Prontosil was developed and found to be effective against *Streptococci* bacteria that caused infections of the blood. In 1928 Alexander Fleming accidentally discovered that an active compound produced by the *Penicillium* mould killed cultured bacterial colonies (see Figure 1.19). He failed to isolate the active compound called penicillin. This was achieved by a team lead by the Australian scientist Howard Florey. By 1945 penicillin antibiotics were in mass production. Since then many different types of antibiotics have been developed for use against many different bacterial diseases.

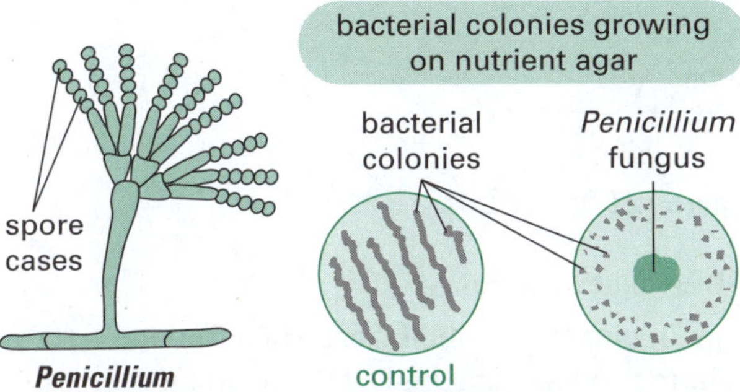

Figure 1.19 *Penicillium* mould produces the antibiotic penicillin which kills bacteria.

Experiment 2

Testing antiseptics

Aim

- To use second-hand data to determine which brand of antiseptics is best at killing bacteria

Method

The method for this experiment is set out below. Read the method that was used.

1. Select three commercial brands of antiseptics.
2. Prepare various dilutions (1%, 2% and 4%) of each antiseptic in water.
3. Prepare 10 nutrient agar culture plates of one strain of bacteria.
4. Add 1 mL of each dilution of the three antiseptics to the nine bacterial plates. Leave one plate as the control for each antiseptic tested.
5. Leave the plates at 25 °C for two days and then count the number of bacterial colonies on each plate.

Results

The results of this experiment are shown in Table 1.4.

Table 1.4 Average count of bacterial colonies per plate

Antispetic	Dilution			
	No antispetic	1% solution	2% solution	4% solution
Chloro	50	45	40	30
Ozo	50	5	3	1
Septro	50	38	26	2

Analysis

Analyse these second-hand results and compare your analysis to the answer provided in the back of the book. Go to p. 224 to check your answer.

Conclusion

Write a suitable conclusion.

Go to p. 224 to check your answer.

1.5 Technology and the human body

Our knowledge of the way organs and organ systems work in our bodies has been extended by the development of various technologies.

Properties of electromagnetic radiation

Visible light is just one example of a larger group of waves called electromagnetic radiation. Figure 1.20 is called the electromagnetic spectrum and it shows all the forms of light radiation. Light is a wave motion that consists of moving electric and magnetic fields. This is why it is called an electromagnetic wave.

The visible part of the electromagnetic spectrum contains light of various colours. The colours (from longest wavelength to shortest wavelength) are red, orange, yellow, green, blue, indigo and violet. This order is also from the lowest energy wave (red) to the highest energy wave (violet). The highest energy waves are gamma rays. The lowest energy waves are radio waves.

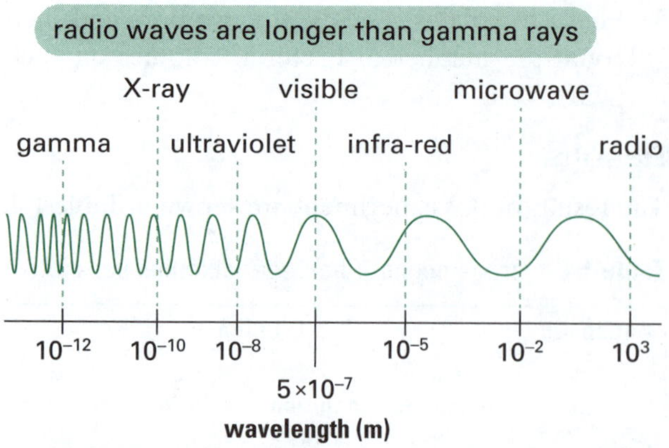

radio waves are longer than gamma rays

X-ray visible microwave

gamma ultraviolet infra-red radio

10^{-12} 10^{-10} 10^{-8} 10^{-5} 10^{-2} 10^{3}

5×10^{-7}

wavelength (m)

Figure 1.20 Electromagnetic spectrum

All electromagnetic waves have the following properties.

- They do not require a medium for their propagation.
- They travel at 300 000 km/s through empty space.
- They reflect off surfaces.
- They refract or bend when waves move from one medium to another.

Uses of electromagnetic radiation in medicine

Various bands of the electromagnetic spectrum are used in detecting and diagnosing disease in the body as well as in their treatment.

Radio wave radiation

Radio waves can be used to transmit medical images and information from one location to another.

Microwave radiation

Microwaves can be used in the detection of cancerous tumours in the breast.

Infrared radiation

When infrared radiation is absorbed by the body, it is converted to heat energy. This property is used in infrared (or heat) lamps for treating sore muscles and stiff joints as well as promoting blood circulation. Infrared cameras can be used to detect cool and hot regions of the skin. Hot regions may indicate underlying infection or disease.

Visible radiation

Exposure of the body to visible light rays is called phototherapy. Blue-violet light has been found to be useful in treating acne as it helps to kill bacteria in the skin pores. Eczema and some fungal infections of the skin can also be treated with phototherapy.

Endoscopes are tools used to examine the inside of the body using visible light. They consist of flexible fibre optic tubes through which light rays can travel by reflection off the inner wall (see Figure 1.21). These endoscopes can be inserted into the body so that doctors can examine internal organs and also remove samples for testing.

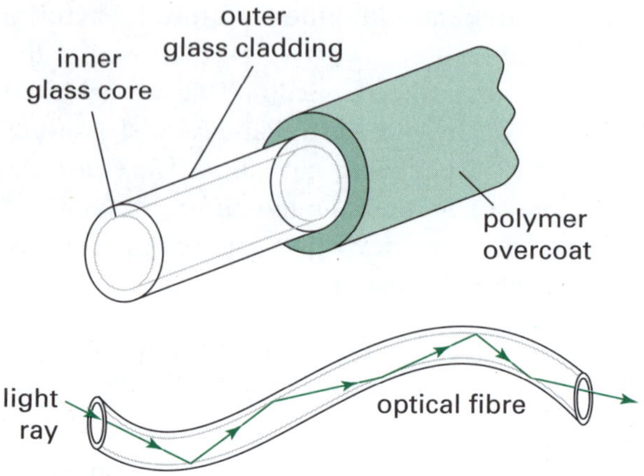

Figure 1.21 Optic fibres

Ultraviolet radiation

Our bodies need small doses of ultraviolet (UV) light to produce vitamin D. UV radiation is useful in sterilisation of instruments in hospitals. Bacteria and viruses are killed by UV exposure. UV lamps can be used to diagnose fungal skin infections and foreign objects in the eye. UV phototherapy is used to treat psoriasis of the skin.

X-radiation

X-rays are used to detect and diagnose injuries or disease in the body. X-rays can penetrate the body and images of bones and soft tissue can be made (see Figure 1.22). For example, lung cancer and the presence of gall stones can be demonstrated using X-rays. High-dose X-rays can also be used in therapy as cancerous cells can be killed by X-ray treatment. Radiology is an important career in the field of medical science.

Gamma radiation

Gamma rays can be used in sterilisation of medical instruments as well as gamma irradiation of some foods to kill microbes that cause decay. Gamma-emitting radioactive isotopes can be used to detect

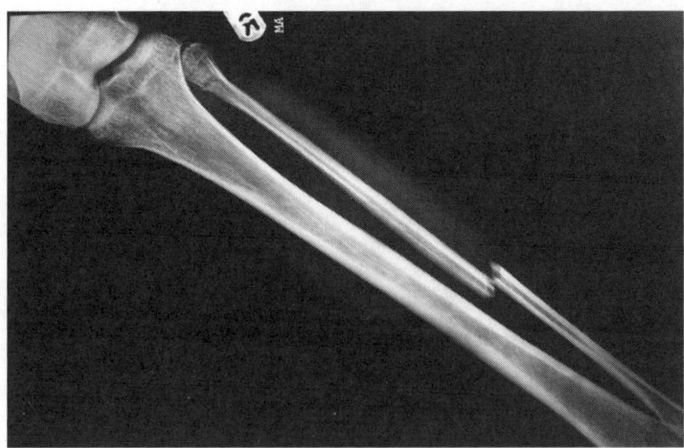

Figure 1.22 X-ray of a broken arm

abnormalities in body organs. These radioisotopes produce low doses of gamma radiation which cause minimal damage to the body. The radioisotopes are injected into the body and gamma cameras scan the body to produce an image that can be studied to detect abnormalities.

Dangers of electromagnetic radiation

The following are some of the dangers associated with different types of electromagnetic radiation.

- Microwaves. Microwaves are converted to heat when various molecules such as water absorb them. This is how a microwave oven cooks food. Microwave ovens are sealed but if they are damaged then microwaves can leak out and cause burns to people nearby. Microwave radiation is non-ionising and consequently does not have the cancer risks of X-rays and UV rays.
- Ultraviolet light. Excess exposure of the skin produces sunburn and skin aging. Skin cancers are also the result of excess exposure to UV.
- X-rays. High doses of X-rays cause DNA mutations, cancer, anaemia and impairments of the immune system.
- Gamma rays. In high doses, exposure can produce DNA mutations and cancer. Higher dosage produces organ failure, bleeding and death.

Imaging technologies

Imaging technologies are used to detect bone fractures, cancers and infections. Many people now train for careers as radiographers in these imaging procedures. Radiologists are specialist medical doctors who have had specific training in performing and interpreting diagnostic imaging tests and intervention procedures or treatments that involve the use of X-ray, ultrasound and magnetic resonance imaging equipment.

X-rays

Accelerated electron beams that collide with a tungsten target produce X-rays. About 1% of the kinetic energy of the electrons is converted to X-rays, and these have a range of wavelengths. Hard X-rays have wavelengths of about 0.01 nm and are strongly penetrating. Soft X-rays have longer wavelengths of about 1 nm and have less penetration.

A radiograph produced by passing X-rays through a part of the patient's body is an image produced on photographic film. Bone tissue is dense and so the radiograph is clear as less X-rays reach the film. Soft tissue is less dense and this produces darker (opaque) images on the film.

CAT scans

The term CAT is an acronym for computerised axial tomography. It is also called computerised tomography (CT). This imaging technology uses X-rays and generates clearer images of soft tissues. In this technique X-rays are fired from a rotating gantry at the organ beings scanned. The X-rays that pass through to detectors are analysed by a computer to produce cross-sectional slices of the organ or body on a computer screen.

The advantage of CAT scans over conventional X-rays is that the image of the body can be manipulated by removing bone images to allow soft tissues to be more readily seen. It is also much more sensitive and so allows the early diagnosis of cancer.

PET scans

The term PET is an acronym for positron emission tomography. This imaging technology uses positrons (positively charged electrons) to diagnose metabolic activity in various organs of the body. The patient is injected with a radioisotope attached to a modified glucose molecule. This combination is called a radiopharmaceutical. The radiopharmaceutical enters tissues or organs, where it is then metabolised. The speed at which it is metabolised is indicated by the rate of positron production in that tissue. Positrons that are emitted combine with electrons to generate gamma rays that escape from the body and are detected by gamma detectors arranged in a ring around the patient. Computers analyse the results to produce images of the tissue or organ. They are displayed as sections or slices on a computer monitor (see Figure 1.23).

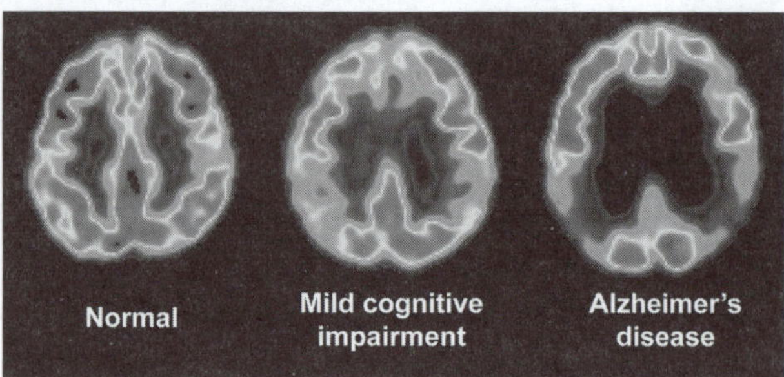

Figure 1.23 PET scans of a normal brain and brains with developing Alzheimer's disease

EOS orthopaedic imaging system

EOS was developed from a Nobel prize-winning technology by a team of scientists and doctors as a complete orthopaedic imaging solution. This scanner dramatically cuts the amount of radiation patients are exposed to, emitting up to 10 times less radiation than X-rays and up to 1000 times less than a CT scan. This makes it safer for patients requiring several scans, such as those with spinal deformities like scoliosis, each year. Its two-dimensional and three-dimensional scans can capture a full body image while a person is standing or sitting in less than 20 seconds. Computer technology automatically provides over 100 clinical parameters to the orthopaedic surgeon for pre- and post-operative surgical planning. Doctors can now examine one continuous digital image of a patient instead of having to piece together various X-rays. EOS also allows for three-dimensional reconstruction of individual bone positions, rotation and orientation. Because these images provide a much clearer picture of the spine, doctors can see and document the results of treatment. The first EOS scanner in the southern hemisphere began operation in Sydney in June 2011.

What radiation level is safe?

No medical studies have definitively shown what radiation dose is safest over a person's lifetime. Ultrasound and magnetic resonance imaging (MRI), do not deliver radiation doses to the patient, but both have their drawbacks as well. Ultrasound is used to typically diagnose problems in the liver, spleen, gall bladder, kidneys, ovaries and testes. However, it won't produce an image of anything with air, such as lungs, because it can't 'see' through it. MRI also doesn't use radiation and can produce high-quality images. It provides good contrast between the different soft tissues of the body and this is especially useful in imaging the brain, muscles, heart and cancers compared with other medical imaging techniques. However, patients must stay still for long periods. A typical MRI takes 25 to 30 minutes, in 3- to 5-minute segments.

Development of artificial skin

Maria Stoner is an Australian scientist and Fiona Wood is an Australian plastic surgeon. Together they have developed a product called CellSpray or Spray-on Skin. This technique is used to replace skin on patients with severe burns. They won the 2005 Clunies Ross Award for their contributions to Medical Science in Australia.

Spray-on Skin reduces the time to culture skin cells from 21 days to 5 days. This short time has been shown to significantly reduce the extent of scarring in burns victims. Traditional methods involved harvesting skin cells and culturing them so they formed thin sheets of skin tissue. Dr Wood discovered that if these sheets had holes in them, then the degree of scarring was reduced. Using this observation, epidermal cells were harvested from some other part of the patient and grown to produce a cell suspension which could be sprayed on the wound. Once these cells make contact with the wound, they grow more rapidly than the cells in the culture medium.

In addition to these treatments, the same method can be used to harvest pigment cells that can then be cultured and applied to patients who have patches of skin that lack pigmentation.

Nanotechnology and pharmaceutical delivery in medicine

Nanomedicine is an emerging technology used in medicine for diagnosis, the delivery of pharmaceuticals and enhanced imaging of the body. Nanoparticles have diameters usually between 10 nm and 50 nm. They are so small that they can penetrate through cell membranes. These particles contain drugs that may normally be insoluble or toxic if injected or swallowed. Once inside the target tissue or organ, the coating of the nanoparticles dissolves and the drug is released (see Figure 1.24). The polymeric coating of the nanoparticle protects the drug from the action of enzymes and acids in the digestive tract.

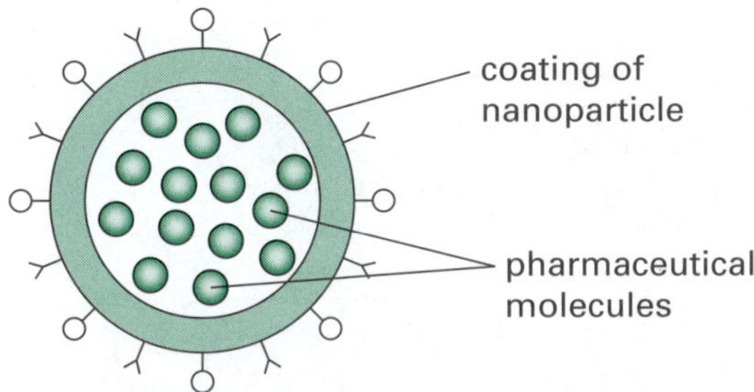

Figure 1.24 Nanoparticles can carry drugs into the body.

Test yourself 1

Part A: Knowledge

1. The site of gaseous exchange with the bloodstream is the *(1 mark)*
   - **A** nasal cavity.
   - **B** diaphragm.
   - **C** trachea.
   - **D** lungs.

2. The vessel that transports deoxygenated blood to the lungs is the *(1 mark)*
   - **A** heart.
   - **B** pulmonary artery.
   - **C** pulmonary vein.
   - **D** oesophagus.

3. Sensory neurons *(1 mark)*
   - **A** transfer information inside the spinal cord.
   - **B** are stimulated by hormones and transmit information through the blood stream.
   - **C** carry electrical messages towards the central nervous system from our sense organs.
   - **D** carry electrical impulses away from the central nervous system towards the peripheral nervous system.

4. Select the statement that is true about the endocrine system. *(1 mark)*
   - **A** The endocrine system is only involved in reproduction and the control of sugar metabolism.
   - **B** The pituitary gland controls all other glands by sending electrical stimuli via the central nervous system.
   - **C** Insulin is a hormone produced by the thyroid gland.
   - **D** Diabetes is caused by a failure of pancreatic cells to produce the correct levels of glucose-regulating hormones.

5. Select the correct statement concerning a common infectious disease. *(1 mark)*
   - **A** Tuberculosis is a bacterial disease that causes the formation of lesions in the lungs.
   - **B** Malaria and HIV are both infectious diseases caused by protozoans.
   - **C** Measles is a viral disease that responds readily to antibiotic treatment.
   - **D** Tinea is a fungal disease characterised by a red rash that covers most of the body.

6. Complete the following restricted-response questions using the appropriate word. *(1 mark for each part)*
   - **a)** Kidneys filter the .... to remove nitrogenous wastes, salts and excess water.
   - **b)** In the large intestine, ... is removed from the remaining solid waste.
   - **c)** Arteries transport .... blood away from the heart to the organs of the body.
   - **d)** The trachea is a tube through which .... moves between the nasal cavity and the lungs.
   - **e)** The central nervous system (CNS) consists of the .... and the spinal nerve cord.

7. Use the code letters to match the terms or phrases in each column. *(1 mark for each part)*

Column 1	Column 2
A motor neurons	F pituitary gland
B axons	G muscles and glands
C growth hormone	H influenza
D blood pressure	I dendrites
E virus	J adrenaline

Part B: Skills

8. Figure 1.25 shows two diagrams of eyes (X and Y) under different lighting conditions.

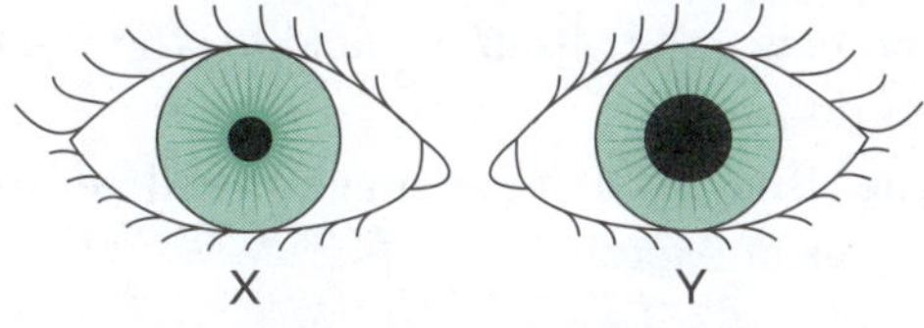

Figure 1.25 Eyes under different lighting conditions

   - **a)** Which eye corresponds to conditions of bright sunlight? *(1 mark)*

b) Is this change in the pupil of the eye an example of a voluntary or involuntary movement? *(1 mark)*

c) Explain in terms of nerves the sequence of events that leads to a change in pupil size on walking from a darkened room into bright sunlight. *(2 marks)*

9. Hormones are produced by various glands. Various hormonal controls ensure that their production is regulated. Figure 1.26 shows the effects of two of these hormones on the control of the menstrual cycle. The organs and glands involved are the uterus, ovary and pituitary gland. *(3 marks)*

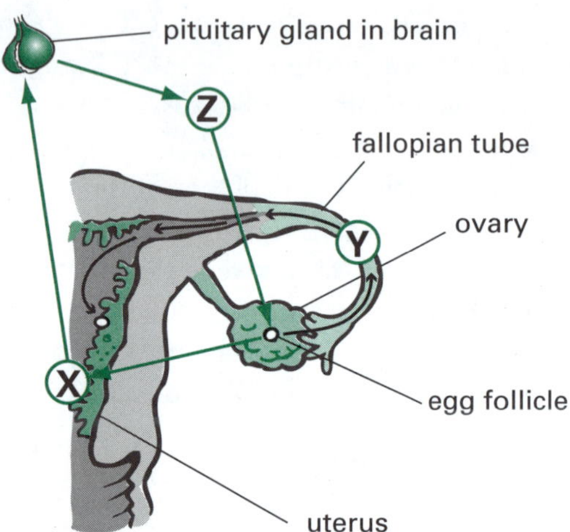

Figure 1.26 Hormones

Match the code labels in the figure to one of the statements below.

Statements

A. Follicles in the ovary containing unripe eggs are stimulated to grow by the follicle-stimulating hormone (FSH) released from the pituitary gland.

B. Oestrogen is released into the bloodstream during the growth of the egg follicle in the ovary. The rising level of oestrogen stimulates the pituitary gland to stop FSH production.

C. The rising level of oestrogen stimulates the uterus lining to grow.

10. Viruses are a non-cellular life form. They use the host's DNA for their reproduction. Figure 1.27 shows the jumbled steps in the reproductive cycle of a typical virus. Use the code letters to list the jumbled diagrams in their correct sequence. *(2 marks)*

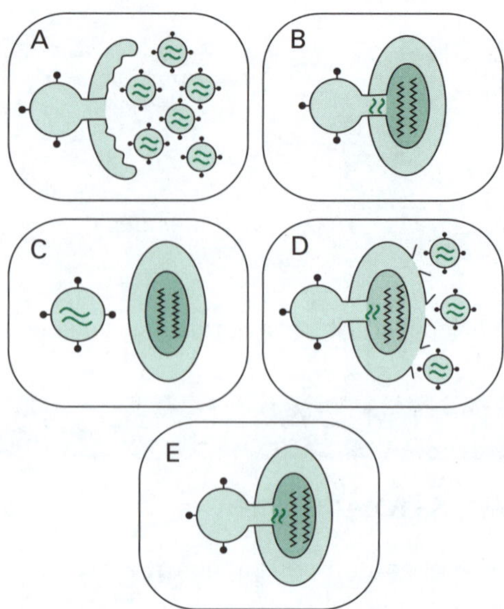

Figure 1.27 Virus

11. Figure 1.28 shows various microbes. Use the *visible features* of these microbes to develop a written dichotomous key to classify and identify these selected microbes. *(4 marks)*

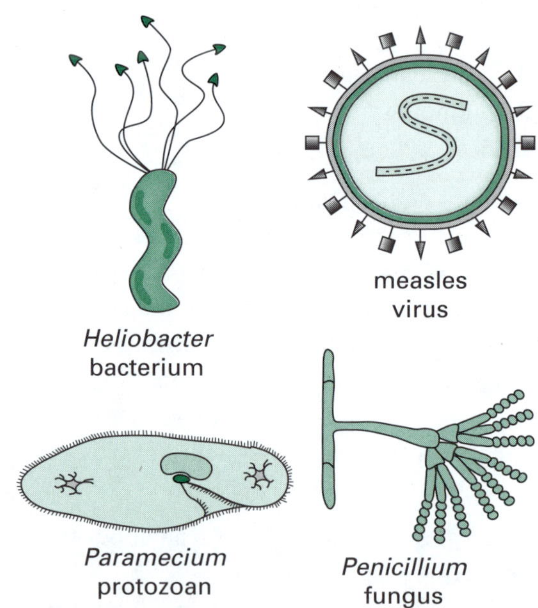

Figure 1.28 Microbes

12. Louis Pasteur proved experimentally that if microbes were excluded or killed then fermentation of a food or nutrient broth was prevented. A student repeated the Pasteur broth fermentation experiments using the apparatus shown in Figure 1.29. The clear nutrient broth was placed in the flask and the broth boiled till steam emerged from the plastic tube. The apparatus was cooled, placed in a warm water bath at 37 °C and then left for a week. Observations were made daily.

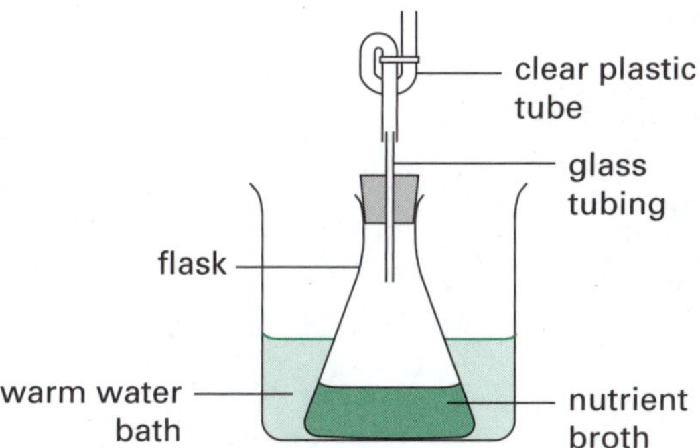

Figure 1.29 Pasteur's experiment

a) Explain the purpose of boiling the clear broth before the experiment started. *(1 mark)*

b) Why was the broth held at 37 °C during the experiment? *(1 mark)*

c) The student also performed a control experiment. Explain what apparatus is used in the control. *(1 mark)*

d) The student observed that the control broth went cloudy and the test broth stayed clear. Explain these observations. *(1 mark)*

13. An experiment was performed to assess the effectiveness of three antibiotics on the growth of bacteria in nutrient agar culture plates. The antibiotics (A, B and C) were impregnated into three paper discs that were placed on the surface of the nutrient agar. The plate was inoculated with three different bacteria (X, Y and Z). This is shown in Figure 1.30. After several days in an incubator, the plate was examined and the pattern of bacterial colonies noted.

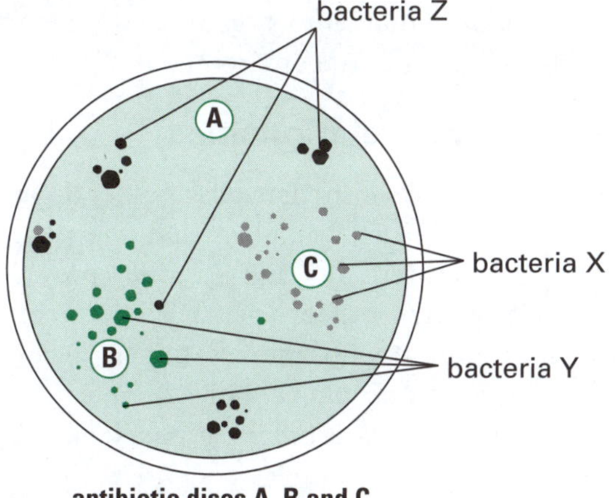

Figure 1.30 Culture plate

Discuss the effectiveness of the three different antibiotics on the growth of each of the three different types of bacteria. *(3 marks)*

14. Match the stimulus in column 1 to the response in column 2. *(5 marks)*

Stimulus		Response	
A	smell or sight of food	F	sneezing
B	a sudden explosion	G	production of saliva
C	peppering your food	H	jerking of the body
D	touching a hot surface	I	shivering
E	sudden drop in temperature to zero degrees	J	hand pulls away suddenly

Go to pp. 224–225 to check your answers.

1.6 Interactions between organisms

Everything in the natural world is connected in some way. A group of living and non-living things that interact with each other is called an ecosystem. Ecosystems vary in size, being as large as a desert, ocean or lake, or as small as a garden, tree or terrarium.

Like a properly functioning organism, the components of an ecosystem all work together to make a balanced system. For example, if there isn't enough light or water or if the soil doesn't have the right nutrients, or if it's too hot or too cold, plants will die. This means that any animals that depend on them for food or shelter will also die or be forced to migrate. This has a cascading effect on any other animals that depend on these animals for their survival.

In a healthy ecosystem such as that shown in Figure 1.31, there are many species and this diversity is less likely to be seriously damaged by human interaction, natural disasters or climate changes.

Communities of interdependent organisms

There are a number of terms that relate to ecosystems.

- A population is a group of living organisms of the same species living in the same place and at the same time.

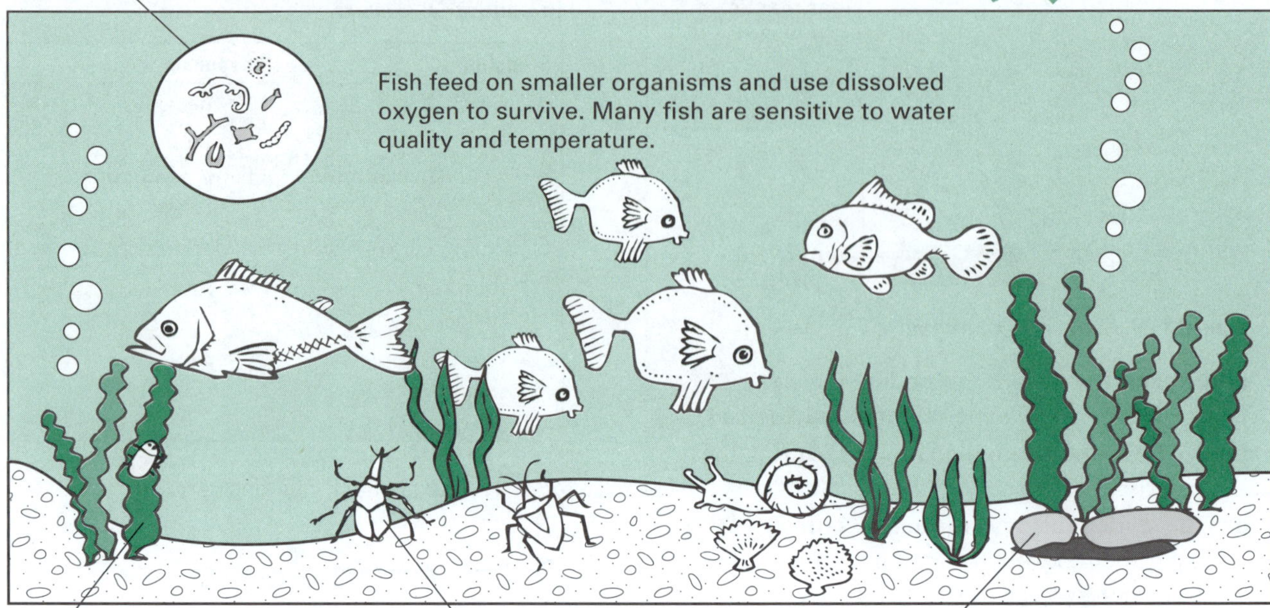

Figure 1.31 A healthy pond or lake ecosystem

- A habitat is the place within an ecosystem where a population lives. Habitats are specific to a population, and each population has its own habitat. Habitats may vary in size, but must supply the needs of organisms, such as food, water, oxygen, temperature and minerals. Several populations may share a habitat.
- All of the populations interact to form a community.
- The community of living things interacts with the non-living world around it to form the ecosystem.
- A niche is the function or role that a species performs within nature. Two different populations cannot occupy the same niche at the same time.

There are many different living organisms that interact with each other in an ecosystem. These living organisms can be categorised as follows.

- Producers: the green plants, algae or phytoplankton; they make their own food.
- Consumers: animals that get their energy from producers or from organisms that eat producers. These include herbivores (animals that eat plants); carnivores (animals that eat herbivores or other carnivores); and omnivores (animals that eat plants and other animals).
- Decomposers: organisms that break down dead plants and animals into organic materials to be recycled back into the environment.

Predator–prey relationships

A predator is an organism that eats another organism, while the prey is the organism that is eaten. For example, a fox (predator) and a rabbit (prey), or a lion (predator) and a deer (prey). These terms are almost always used with animals, but they can also apply to animals and plants. For example, a rabbit eating a carrot, or a caterpillar eating a leaf.

Plant–animal relationships

While plants cannot outrun the animals that will eat them, they can make themselves more unpalatable. Here are some examples.

- Many herbivores avoid the plant morning glory, as it contains chemicals that makes it not very tasty, if not toxic. However, some species are used as food plants by the caterpillars of certain butterflies and moths.
- *Melaleuca* and *Eucalyptus* species possess oil glands within their leaves that produce a pungent, volatile oil that makes them unpalatable to most herbivores. But koalas live almost exclusively on these leaves, being able to neutralise their toxic and indigestible compounds (see Figure 1.32).

Figure 1.32 Koalas spend three of their five active hours each day eating an average of 500 g of eucalypt leaves, usually at night. Their livers deactivate the toxic components in the leaves.

- Some plants, such as cactuses, raspberries and roses, have thorns, spines or prickles to prevent (or at least make it difficult) for animals to eat them.
- The leaves of holly plants are very smooth and slippery, making feeding difficult. Other plants produce sap that traps insects.
- Some plants, such as coconut palms, protect their fruit by surrounding it with multiple layers of armour.
- Many plants have microbial organisms (bacteria or fungi) that live inside them. While some cause disease, others protect plants from herbivores and diseases. They help the plant by creating toxins harmful to other organisms that would attack the plant.
- Some plants naturally produce chemicals that deter or are toxic to insect pests. For example, the marigolds shown in Figure 1.33 produce a smell that deters aphids but is attractive to hoverflies (a predator of aphids). This makes them a useful plant to grow near crops in order to deter pests.

Figure 1.33 Marigolds are often grown near crops suffering from aphids.

- Some passion flowers develop small coloured growths that look like butterfly eggs. This seems to fool butterflies into thinking too many eggs have already been deposited on a plant, and this deters them from laying more eggs. Many *Passiflora* species also produce sweet, nutrient-rich liquid from glands on their leaf stems. These fluids attract ants that will kill and eat many pests which may be feeding on the passion flowers.

While browsers learn to avoid unpalatable plants, they do not always benefit. For example, in an effort to eat other plants nearby, deer can create unfavourable conditions for unpalatable plants through effects such as trampling or indirectly influencing soil fertility.

Animal–animal relationships

In any population, individuals exhibit a range of characteristics. For instance, the fastest tigers are better able to catch their food and eat. So they are more likely to survive and reproduce. And over a long period of time faster tigers will make up more and more of the population. Of course, the fastest zebras, gazelles or deer are able to escape the tigers, and so they survive to reproduce. Gradually, faster zebras, gazelles or deer make up more and more of their populations. As both organisms become faster to adapt to their environments, their relationship remains the same; they are both getting faster. Neither gets faster in relation to the other. This is true in all predator–prey relationships.

Figure 1.34 shows a frequency graph of a variety of characteristics that affect population numbers.

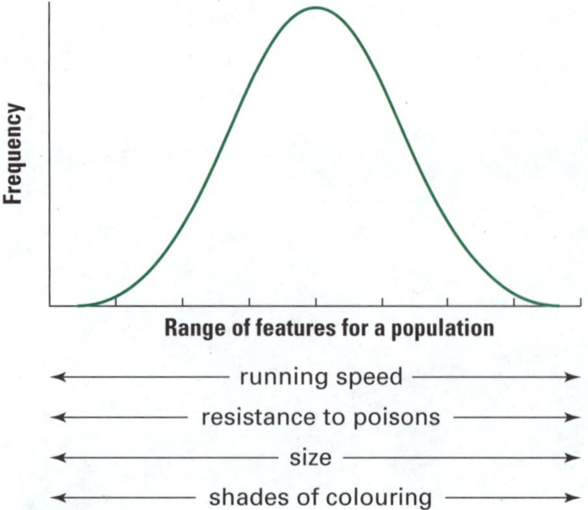

Figure 1.34 In any population, individuals exhibit slightly different characteristics which may or may not be to their advantage.

This works for other characteristics, too. A lion that can stealthily creep up to a deer has a better chance of catching it. Similarly, a deer with more acute hearing will have a head start and avoid being eaten.

Other examples include the following.

- Camouflage or protective colouration allows the animal to blend in with its environment to avoid being detected. In snowy environments a white polar bear avoids being noticed as it approaches a seal, and the seal pup is white to avoid being noticed by the bear.
- Trickery and false features, such as enormous eyes or appendages, can serve to ward off potential predators.
- Mimicking an animal that is dangerous to a predator is another effective means of avoiding being eaten.
- Some animals have physical features that make them undesirable. The extremely sharp quills on spiny anteaters and porcupines make it very difficult for predators. Similarly, predators would have a tough time trying to get to a turtle or tortoise through its protective shell.
- Chemical features are effective. Skunks release a chemical that is a not-so-pleasant aroma for an attacker. The Queensland cane toad uses chemicals squirted from glands to deter attackers. Any animals that eat these toads, or swallow the poison, are likely to get very sick or die (see Figure 1.35).

Figure 1.35 The Queensland cane toad (*Bufo marinus*) is an introduced pest that has invaded north-eastern Australia. On its head behind each 'ear' are the parotid glands, which can squirt a poisonous milky toxin up to 1 m away.

Without predators, certain species of prey would increase in numbers and drive other species to become extinct through competition. Extinction occurs when the population of a species drops to zero. Without prey, there would be no predators. Hence, this relationship is necessary to the existence of life on this planet.

Parasites

Parasitism is a type of relationship between organisms of different species where one organism, the parasite, benefits at the expense of the other, the host. Parasites, unlike predators, are generally much smaller than their host. They are highly specialised for their mode of life, and reproduce at a faster rate than their hosts.

There are various ways to classify parasites. A classification scheme based on how the parasites interact with their hosts is shown in Figure 1.36. Figure 1.37 shows a parasite called a tick. They suck blood from a variety of animals.

Head lice are tiny insects that live on the scalp and lay eggs (called nits) around the roots and on a person's hair. The lice live, breed and feed on the scalp of infected humans. They are highly contagious, moving freely from person to person if those two people have close head-to-head contact. They are particularly common among groups of children who play closely with each other. Treatment is effective

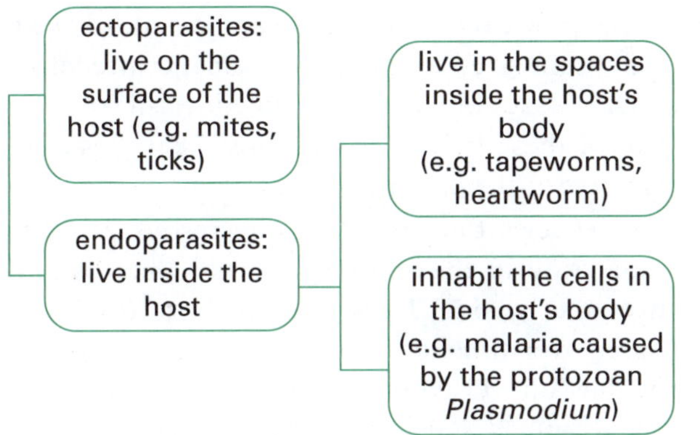

Figure 1.36 Parasites may be classified on how they interact with their hosts.

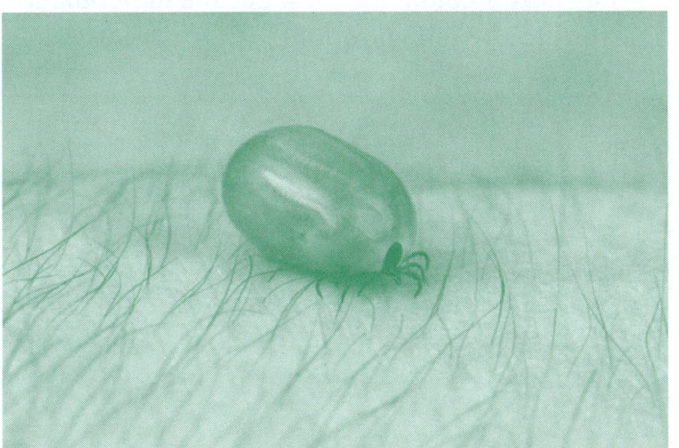

Figure 1.37 Ticks are blood-sucking parasites living and feeding on mammals, birds and reptiles. They are known carries of diseases such as Lyme disease and tularaemia.

with chemical insecticides specifically prepared for head lice.

Bacteria, protozoans and viruses are common parasites. The parasite has to be in its host to live, grow and multiply, and it rarely kills its host. Hookworms, which are nematodes (roundworms), are a common parasite of humans and their pets. They attach themselves to the lining of the small intestine, causing diseases and malnutrition as they eat the nutrients and keep them from going to the host.

The guinea worm (the nematode *Dracunculus medinensis*) is a parasite that enters humans when a person drinks stagnant water contaminated with the larvae of the guinea worm. Fewer than 2000 cases were reported in 2010, mainly in Sudan. The guinea worm isn't vulnerable to vaccines or medication. Eradication efforts are largely through education. This roundworm can grow up to a metre long in humans. It moves downwards, usually to the lower leg, through tissues below the skin, leading to intense localised pain in its path of travel. Public health officials hope to eradicate this disease completely in the next few years. Figure 1.38 shows the lifecycle of a tapeworm. Part of this lifecycle involves humans.

In 1998 Sydney's water supply was contaminated by microscopic cryptosporidium and giardia (both

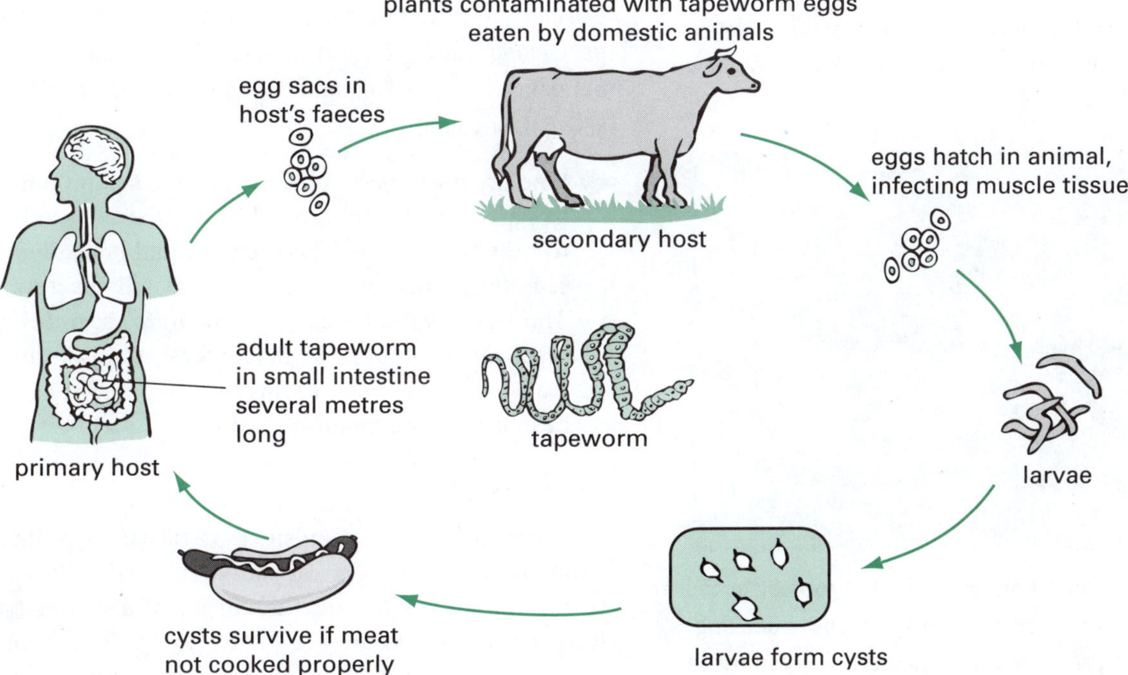

Figure 1.38 The lifecycle of the tapeworm. Each of the 1000 to 2000 body segments contains from 80 000 to 100 000 eggs.

protozoan pathogens). Low-quality raw water entering Warragamba dam caused this, and a 'boil water' alert (where residents were instructed to boil their tap water before using) was declared for many districts. Affected individuals might experience diarrhoea and abdominal pain lasting for about a week. Soon afterwards Sydney Water Corporation announced that water was safe to drink again.

Competitors

Competition occurs naturally between living organisms that coexist in the same environment. It occurs between organisms or species where the fitness of one is lowered by the presence of another. For competition to occur, both organisms must rely on a limited supply of a resource such as food, water and territory. The following are some examples of competition.

- Competition can occur when members of the same species need the same resources in an ecosystem. For example, two trees growing close together will compete for light above ground, and water and nutrients from the ground. Being close together, they will usually perform less well than if they grew some distance apart. An established tree often prevents a seedling from growing near it, as the seedling is often deprived of light and nutrients.
- Competition between individual animals using aggression is often seen when the individuals interfere with foraging, survival, reproduction of others (such as finding a mate), and territory. (See Figure 1.39.)

Figure 1.39 There can only be one adult rooster in the hen house. Introducing another rooster leads to pecking-order fights that can be violent and more likely to result in injury or death.

- Large predators, such as dogs and dingoes, have a significant size advantage over smaller predators such as foxes or weasels. A violent clash between two of these species would most likely result in victory for the larger predator.
- Two species that use the same resource in the same way in the same space and time cannot live side by side. They must diverge from each other over time, or one of the species migrates or dies out. Usually one species often exhibits an advantage in using the resource. This superior species will out-compete the other by using the limited free source more efficiently. As a result, the inferior competitor will decline in population over time. (See Figure 1.40.)

Figure 1.40 The Indian myna (*Acridotheres tristis*), native to Asia, has strong territorial instincts and has adapted extremely well to urban environments. It is an invasive pest along the east coast of Australia and has edged out native birds, kicking other birds from their nests and killing their young.

- One organism may release a chemical compound as part of its normal metabolism that is harmful to other organisms. The bread mould *penicillium* secretes penicillin, a chemical that kills bacteria. The black walnut tree (*Juglans nigra*) secretes a respiratory inhibitor from its roots, nut husks and leaves; this chemical harms or kills some species of neighbouring plants.

Pollinators

A pollinator is an organism that transfers pollen from the male anthers on the stamen of a flower to the female stigma on the carpel of a flower so fertilisation can occur (see Figure 1.41). Pollination is very important as it leads to the creation of new seeds that grow into new plants.

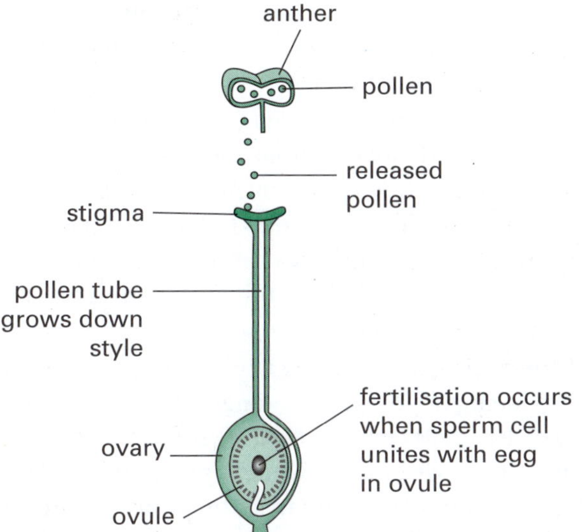

Figure 1.41 Pollination occurs when pollen is transferred to the receptive stigma. Fertilisation occurs somewhat later when male sperm fuses with the eggs in the ovule within the ovary.

Bees are the most obvious creature adapted to pollination. Bees are fuzzy and carry an electrostatic charge, and these features help pollen grains adhere to their bodies. They also have specialised pollen-carrying structures (pollen baskets) on their hind legs as shown in Figure 1.42. Most bees gather nectar, (a concentrated energy source) and pollen (a high-protein food) to nurture their young. They inadvertently transfer some pollen among the flowers as they move from flower to flower of the same species.

Figure 1.42 Honey bee with pollen visible in its pollen basket

Apiarists (bee-keepers) often contract out their hives as pollinators to plant farmers. This helps farmers to have their crops pollinated effectively, and allows the apiarist to derive honey in the process. Honey bees are by far the most important commercial pollinating agents.

Insects are not the only pollinators. Here are some other examples.

- Bats pollinate some tropical flowers.
- Flowers pollinated by birds produce large amounts of sweet nectar as a lure (and much less pollen than those pollinated by wind). As birds have little sense of smell, such plants are rarely fragrant. Also, their flowers are often red or orange since birds see these colours very clearly.
- Wind can also pollinate plants that have inconspicuous flowers, to the bane of hayfever sufferers. As wind pollination is very inefficient, large quantities of pollen are produced which are then combed from the air by female flowers.
- Butterflies and moths also pollinate to some extent, but are not major pollinators of food crops. However, various moths are important for some wildflowers or for other crops such as tobacco.
- Other insects, such as beetles, midges and even thrips or ants, can sometimes pollinate flowers.

Many gardeners hand-pollinate garden vegetables and flowers as a way to keep a strain genetically pure. This can involve using a small brush or cotton swab to move pollen.

Pollination is one of the most important mechanisms in maintaining and promoting biodiversity and life on Earth. Biodiversity means that the great variety of gene variants that exist in a natural population is maintained. Many ecosystems, including many agricultural ventures, depend on the diversity of pollinators to maintain overall biological health.

Disease

A disease is an abnormal condition of some part, organ or system of an organism resulting from various causes, such as infection, genetic defect or environmental stress. It can be indicated by an identifiable group of symptoms or signs.

In some cultures a condition may be considered to be a disease but not in others. For example, while obesity is a scourge of affluent western society, it can represent abundance and wealth and be considered a status symbol in famine-prone regions. Similarly, epilepsy is regarded as a sign of spiritual awareness among the Hmong people, an Asian ethnic group from the mountainous areas of China and Vietnam.

All populations of organisms are limited partly or completely by diseases in their ecosystems. The prevalence of disease in populations is influenced by a number of factors:

- infectious organisms (such as fungi, bacteria and viruses)
- pollutants such as chemical and biological wastes
- shortages of food and nutrients
- poor hygiene.

As more individuals (people, animals or plants) live in close proximity to each other, there is an increase in the spread of infectious organisms among them. When the spread is rapid, we then have an epidemic of that disease. For example, human plagues or epidemics such as the Black Death, tuberculosis (TB), cholera and HIV are essentially problems of urban density. The following are some other examples.

- The mosquito *Aedes aegypti* spreads dengue fever. The mosquito breeds in tin cans, old tyres and puddles. Dengue fever is expanding rapidly in crowded tropical cities, especially in those without adequate sanitation, with over 60 million infections now occurring each year
- Many people in developing nations lack clean, safe water. Most household and industrial wastes are dumped directly into rivers and lakes without treatment, which contributes to the rapid increase in waterborne diseases in humans. Villages downstream depend on this water for growing crops, drinking and washing.
- Monoculture is an agricultural practice of producing or growing one single crop over a wide area. Examples include orange groves, wheat fields and grape vineyards. Farmers prefer their crops to be genetically similar as they have uniform growing requirements and patterns, resulting in greater yields on less land. However, if a crop is struck by a disease to which it has no resistance, entire crop populations can be destroyed since all plants in a monoculture are genetically similar.

Case study 1: The Irish potato famine

The potato was originally introduced to Ireland as a garden crop, but by the late 17th century it had become widespread as a supplementary food. Over time, it became a staple food throughout the year for farmers and the poor.

By the 1840s, about one-third of the Irish population was entirely dependent on the potato for food. Then in 1845 (lasting until 1852) the potato blight disease (*Phytophthora infestans,* a fungus-like microorganism) ravaged potato crops. This created the Great Famine (or the Great Hunger) where around a million people died and another million emigrated. This reduced the island's population to around 75 to 80% of what it had been before. Figure 1.43 shows that potato production in Ireland barely recovered even after 1852.

Phytophthora infestans is spread by the wind, and diseased plants wither very quickly. When potatoes were dug up for harvesting, they were found to be rotting. Potatoes turn black and have a soggy consistency with a bad smell. These farmers also found that the potatoes they could normally store for up to 6 months were not edible.

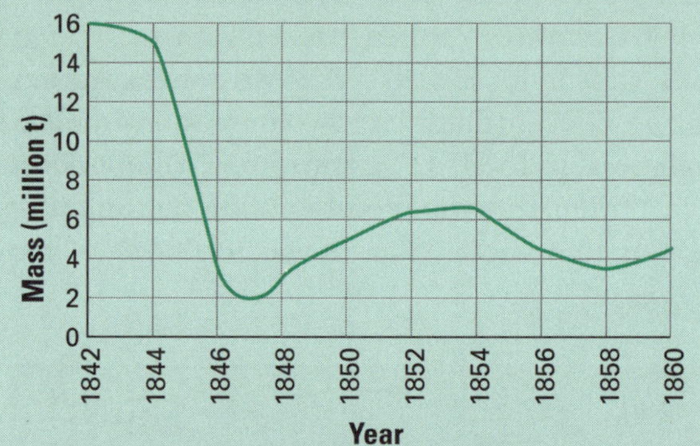

Figure 1.43 Estimated potato production (Ireland) during the Great Famine

It was not until 1882 that scientists discovered a cure for *Phytophthora infestans*, but at the time of the famine there was nothing that farmers could do to save their crops. Today potato farmers routinely spray plants to prevent blight, but in the 1840s the blight was not well understood and many unfounded and incorrect rumours spread as to the cause of the disease.

The famine permanently changed Ireland's population, and its political and cultural environment. This is what can happen if you rely too heavily on one particular crop.

Case study 2: Malaria

Malaria is one of the world's leading infectious diseases caused by a parasite (the protozoan *Plasmodium*) transmitted by mosquitoes (more specifically, female mosquitoes of the *Anopheles* genus). The potentially deadly disease results in recurrent attacks of chills and fever. While it has been virtually eradicated in countries with temperate climates, it is still found in the poorer tropical and subtropical countries of Asia, the Middle East, Africa, and South and Central America.

Malaria is transmitted by the bites of infected mosquitoes. In the human body, the parasites multiply in the liver and then infect red blood cells. An infected pregnant woman can transmit malaria to her unborn child. This process is shown in Figure 1.44.

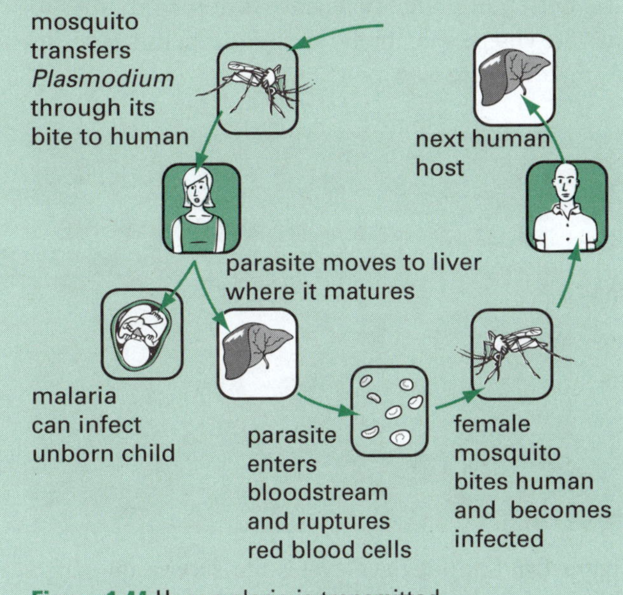

Figure 1.44 How malaria is transmitted

- Minamata disease was first discovered in Minamata city, Japan, in 1956. A chemical factory upstream kept releasing methyl mercury in its industrial wastewater. This continued from 1932 to 1968. The highly toxic chemical built up in shellfish and fish in the bay and surrounding sea. When eaten by the locals, it resulted in mercury poisoning. Over 2000 victims have been recognised, of which some 80% died.

1.7 Factors affecting ecosystems and population size

There are a number of factors that affect the population size in an ecosystem. For example, if the birth rate is greater than the death rate, the population increases. Similarly, if immigration (organisms entering into an area) is greater than the emigration (organisms migrating out of an area), the population increases.

A simple formula to show this is:

population growth = (birth + immigration) – (death + emigration)

Limiting factors are any environmental factors that cause a population to stop increasing. These can include availability of food, space to grow, weather conditions, disease, sunlight and so on. While a factor might be limiting to one population, it may not necessarily be limiting to another. Similarly, a factor that is not presently limiting may become limiting over time.

Biotic and abiotic factors in ecosystems

Abiotic factors are the non-living chemical and physical factors in the environment. These include temperature, rainfall, wind, light intensity and humidity. They can affect the density of a population.

Biotic factors are produced by, or caused by, living organisms. These include organisms of the same or other species living in the same area. They affect the population since they are involved in different types of food relationships.

If, for instance, a population of organisms increases in numbers, it prompts an increase in its predator numbers (there is more prey for them to eat) or a decrease in the amount of available food (more mouths to feed but limited food supply).

In nature, factors such as predators, food scarcity and diseases prevent a population from growing indefinitely. These factors regulate the population size. Different populations have different abilities to tolerate changes in abiotic and biotic factors (see, for example, Figure 1.45).

Figure 1.45 Cockles have evolved to survive the harsh seashore environment.

Seasonal change

Populations are not stable and always exhibit changes in response to changes in environmental factors. The types of animals that inhabit particular areas change from season to season. Birds, for example, seem to disappear in the cold winter months only to reappear in the spring. Winter can be very difficult for both plants and animals: freezing temperatures, lack of water and fewer hours of daylight. Some animals overcome this problem by hibernation. Hibernation involves animals spending the winter in a dormant or inactive state (often in a burrow) until spring. During this time they live off body fat.

Like animal populations, plant populations vary greatly in the course of a year. While animals can migrate (or hibernate in an inactive state in their burrows), plants cannot move. In winter, as there is often not enough light or water for photosynthesis, plants live off stored food made during summer. As the green chlorophyll disappears from the leaves, yellow and orange colours appear before leaves are shed. Deciduous plants lose their leaves to conserve water or to better survive harsh winter weather conditions. Most Australian native plants, however, are evergreen as they do not shed their leaves in the winter.

Spring is a time for new growth. Plants will grow new foliage and flower, and animals take the opportunity to reproduce. Many animals and birds have their young at this time. There is plenty of food available and, with the days getting longer, they have more time to find food for their offspring.

Honey bees swarm as part of their natural reproductive life. This often occurs between September and December. The old queen and about half of the bees from the existing colony leave the hive and cluster on a nearby object such as a tree or fence (see Figure 1.46). The swarm soon moves to a new site and establishes a new colony.

Figure 1.46 While a swarm looks terrifying, bees are generally not inclined to sting provided they are left alone.

Habitat destruction

A drastic change in the environment can destabilise or even exterminate a population. Natural disasters such as earthquakes, fires, floods, climate change or volcanic eruptions can lead to the destruction of resources necessary for survival. Nature's role is well documented in the fossil record, but natural events are not the only things driving this change.

Habitat destruction occurs when a natural habitat becomes unable to support the species present. The organisms that previously used the site are displaced or destroyed and this reduces biodiversity. Humans are the worst culprits in habitat destruction. This is mainly to harvest natural resources for industry (including mining and logging) and urbanisation. Clearing habitats for agriculture is a leading factor in habitat destruction. It is currently ranked as the primary cause of worldwide species extinction.

Habitats have not only been destroyed on land. Deep-sea trawling is devastating corals and pristine marine habitats. These trawlers use giant, heavy-duty nets that are dragged across the sea floor over a kilometre below sea level. The nets are fitted with rubber rollers called 'rock hoppers', which destroy the corals that provide habitats for fish and other marine organisms. Among the most threatened sites are cold water coral reefs in temperate regions.

European settlement in Australia has significantly altered its natural landscape and biodiversity. Along the east coast some 90% of native vegetation has been cleared for agriculture, industry, transport, and towns and cities. Around 50% of Australia's rainforests have been removed with considerable effect.

Habitat destruction results in living things becoming vulnerable and endangered. When population numbers start to decline, scientists classify them as vulnerable. If their population numbers continue to decline even further as there are increasingly more deaths than births, then they become endangered. If the process continues, the species may become extinct (when there are none left). Figure 1.47 shows the breakdown of these in different classifications in Australia.

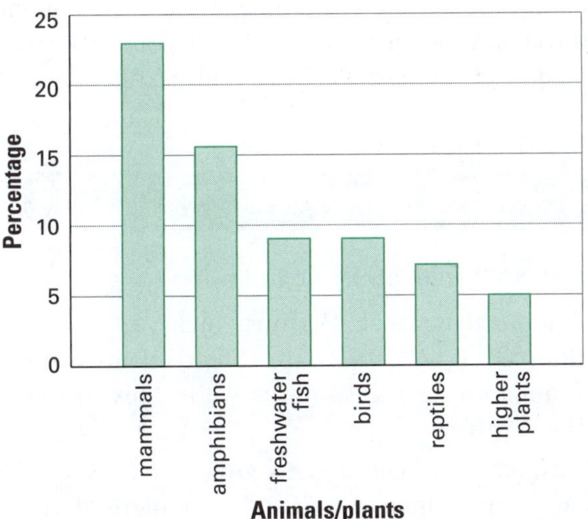

Figure 1.47 Percentage of animals and plants listed as extinct, endangered or vulnerable in Australia

The impacts of habitat destruction include:

- death and extinction of native wildlife and destruction of biodiversity
- farmlands, building foundations, roads and water supplies become salt-ridden as groundwater rises caused by land clearing
- land clearing contributing 10 to 15% of Australia's total greenhouse gas emissions.

Australia has lost more plants and mammals to extinction than any other country and has more threatened animals than almost every other country.

Regions of unsustainable agriculture and/or unstable governments typically experience high rates of habitat destruction. This includes parts of Sub-Saharan Africa, Central America and the Amazon tropical rainforest. Of the some 16 million km^2 of tropical rainforest habitat that originally existed worldwide, less than 50% remains today.

Introduced species

An introduced or non-native species is an organism living outside its native distribution range. It has arrived there by human activity, either deliberately or accidentally. While some introduced species can damage the ecosystem they are introduced into, others may have no negative effect.

Many introduced animals and plants are successful because they are better at breeding, eating and competing compared to native animals and plants. This puts pressure on native plants and wildlife, reducing their numbers and even making them extinct in that locality. Further, many of these introduced species have few natural predators. Table 1.5 shows some of the plants and animals that have been introduced into Australia.

Table 1.5 Some introduced species in Australia

Plants	Animals
lantana	rabbits
prickly pear	foxes
Paterson's curse	cats
bitou bush	mice
	cane toads
	camels

These invaders cause massive damage to the Australian environment, not to mention the cost to the economy. Plant species not native to Australia now account for about 15% of our total flora.

The following are some plant examples in Australia.

- Para grass (*Brachiaria mutica*, semi-aquatic) was originally planted for pond pasture but is now spreading into other areas, destroying waterbird breeding habitats, choking tropical streams and displacing native vegetation.
- Rubber vine (*Cryptostegia grandiflora*, vine/shrub) smothers trees and shrubs, shades out the

ground layer preventing growth, destroys riparian vegetation including forests and threatens the animals that live off them, and forms impenetrable thickets in Queensland's gulf river systems.

- Water hyacinth (*Eichhornia crassipes*, aquatic) and salvinia (*Salvinia molesta*, aquatic) invade open water aggressively and have the potential for very rapid growth, altering aquatic ecosystems. They are still spreading in Australia despite extensive control measures.
- Japanese kelp (*Undaria pinnatifida*, marine kelp) is spreading along the southern coastline at a rate of 10 km/year, displacing native marine plants.

Many examples involve feral animals. A feral animal is one that is domesticated and which then escapes into the wild environment and breeds there. The following are some animal examples in Australia.

- Feral goats (*Capra hircus*) have established populations in a variety of habitats. They compete with native fauna and cause land degradation, threatening plant and animal species and ecological communities.
- Feral cats (*Felis catus*) are found in most habitats. They have caused the extinction of some species on islands and are thought to have contributed to the disappearance of many ground-dwelling mammals and birds on the mainland.
- Feral camels (*Camelus dromedarius*) are well adapted to arid and semi-arid conditions and are capable of inflicting enormous damage to desert ecosystems. (See Figure 1.48.)

Figure 1.48 Feral camels (represented by the green area) are just one example of an introduced animal that is causing damage to Australian ecosystems.

- Feral pigs (*Sus scrofa*) damage the environment through wallowing, rooting for food and selective feeding. They destroy crops and pasture, as well

Case study 3: The European rabbit

The European rabbit (*Oryctolagus cuniculus*) is one of the most widely distributed and abundant mammals in Australia. It causes severe damage to the natural environment and to agriculture, competing with native wildlife, damaging vegetation and degrading the land. The rabbits ringbark trees and shrubs, and prevent regeneration by eating seeds and seedlings, roots and all. They have contributed to the decline in numbers of many native plants and animals, if not the extinction of several.

Rabbits are night-time grazers, preferring green grass and herbs. They also dig below grasses to reach roots and seeds. They are now spread over most of the country except the extreme north. This is in spite of the Western Australian Government building a 1700-km rabbit-proof fence between 1901 and 1907. Before adequate controls, rabbit numbers in Australia are believed to have reached the 200 million mark.

Rabbit control involves the following.

- Chemical control. Poisons, such as sodium fluoroacetate (1080), are effective; pressure fumigation can kill rabbits while they are in their warrens.
- Physical control. Destroying warrens can be a cost-effective and efficient method for decreasing rabbit numbers and preventing re-invasion as it deprives rabbits of a safe breeding; less widely used are fencing, shooting, trapping and explosives to destroy warrens.
- Biological control. The myxoma virus was introduced in 1950 causing myxomatosis in rabbits, which is often fatal. Some rabbits have now developed a resistance to this pathogen. Rabbit calicivirus disease (released 1995) is more effective in the wetter regions, but rabbits are now starting to develop a resistance to it, too.

as habitats for native plants and animals. This animal can spread environmental weeds and could spread exotic diseases.

- European red foxes (*Vulpes vulpes*) have spread across most of the country, playing a major role in the decline of a number of species of native animals. They also prey on newborn lambs.

There are different ways of controlling the numbers of introduced species. Poisons are sometimes ineffective and can damage the environment. Physical control (through shooting or removing individuals) requires a significant amount of workers and needs to be ongoing. Scientists have employed biological control using other living organisms to control the spread of the feral species.

1.8 Energy flow and food webs

Ecosystems basically contain two kinds of materials.

- Matter. Nutrients cycle through the biotic and abiotic parts of the ecosystem and are available for repeated use by the organisms in the community.
- Energy. This enters an ecosystem by converting low-energy carbon dioxide into high-energy carbohydrate. It then passes through one or more of the organisms of the community and finally is lost to the ecosystem. Eventually, all of the energy that enters the ecosystem is lost in the form of heat.

Food webs

In previous years you learnt about food chains. These follow just one path of energy as animals find food. A food web is more complex and shows how plants and animals are connected in many ways to help them all survive. Essentially it is a map that indicates who eats whom (the flow of energy) in an ecological community.

In a food web (as in a food chain) the arrow → is a link and means 'is eaten by', which also means 'energy flows to'. As only a small amount of energy passes from one level to the next, there cannot be too many links from producer organisms leading to the top consumer organism. Figure 1.49 shows part of a typical Australian food web.

Food webs are complicated by the fact that many species feed at various levels. Some living things can be a first-order consumer in one part of the food web and be second-order consumers in another part. Studying food webs is also important in understanding the route by which pollutants get concentrated up the food chain. Also knowing how species interrelate is necessary to understanding how natural and man-made environmental pressures affect ecosystems.

Energy pyramids

As energy moves up the food chain from producer to animal to animal, there is less and less of it available for animals to use. This means towards the top of the food chain, the total mass of animals becomes less. This might mean fewer animals, if the animals get pretty big. On the other hand, if the organism is fairly small (think of a dog supporting thousands of fleas) there might be more. However, the total mass of those animals is much less than the total mass of the producers. Figure 1.50 shows an example of an energy pyramid.

Food webs have trophic levels and positions. In the simplest scheme, plants occupy the first level and

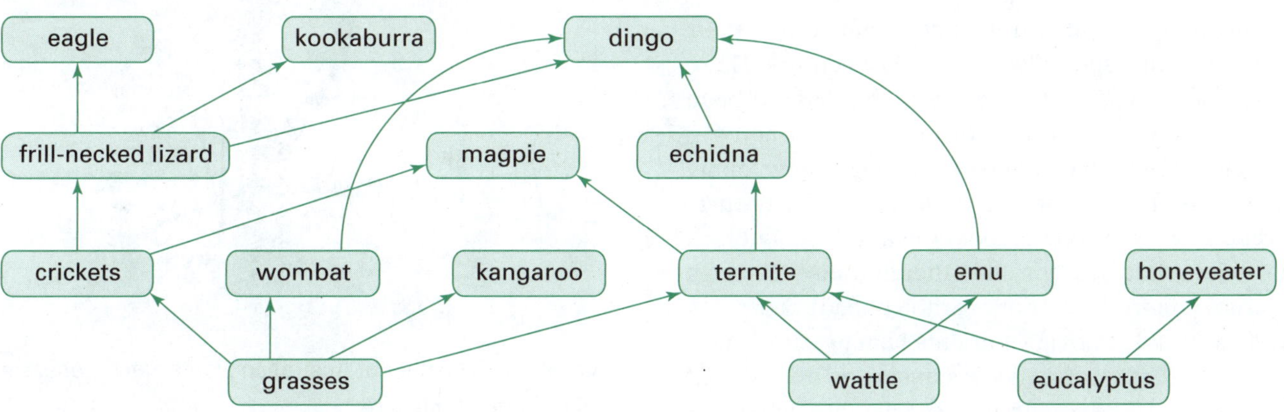

Figure 1.49 Part of a typical food web in an Australian grassland

herbivores occupy the second level, followed by carnivores (level 3 and beyond).

Each organism requires energy to grow, move and maintain all its bodily functions. This means that even though a lot of energy may be taken in at any level, the energy that ends up being stored by the organism is far less. It is this stored food that becomes available to the next level. An average of 90% of the energy entering each step of the food chain is 'lost' this way. But the total amount in the system remains unchanged. The energy that enters a community is ultimately lost to the living world as heat.

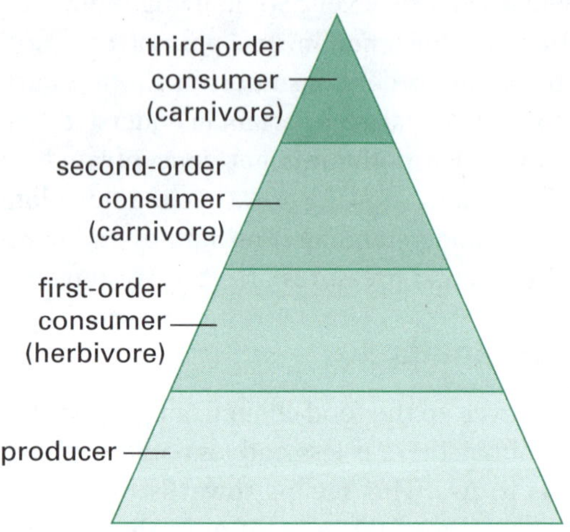

Figure 1.50 An energy pyramid is a way of showing that there is less and less energy available as it moves up the food chain.

1.9 The effect of humans and catastrophes on ecosystems

Humans have had a tremendous impact on their environment, especially over the last century. There is virtually no ecosystem on Earth that has not been significantly transformed through human activity, with the fastest changes now occurring in developing countries. Ecosystems rely on important natural cycles such as the continuous circulation of materials including water, carbon and other nutrients. Through increases in freshwater use, carbon dioxide emissions, fertilisers and removing nutrients, human actions have modified these cycles. For instance, it is no accident that a quarter of mammal species are currently threatened with extinction. And, increasingly, some species (think dogs, cats, cattle, chickens, horses) are found virtually all over the planet, having been physically taken there by humans. This results in the overall biodiversity decreasing, because some rare species are lost while common ones are spread to new areas.

Effect of bushfires

Fire has been a part of the environment in Australian ecosystems for millions of years. Many species rely on it to regenerate and flourish. Frequent low-heat bushfires, generally started by lightning, destroyed and changed many ecosystems. Those plants and animals that could not adapt to them gradually died out from the area. Other native plants and animals adapted by developing special features that allow them to survive during and after such fires. However, ecosystems such as rainforests, alpine moss beds and peat bogs are sensitive to fire. (Of course, in a very intense bushfire all organisms would be lost.)

A fire destroys some adult trees, burning branches, trunks and leaves, and also burns the undergrowth and accumulated leaf litter. So now that more sunlight can reach the soil, seeds that were lying dormant can germinate after a fire releases them from their thick protective coats. Seedlings can thrive because of an increase in space and light.

Common eucalypts can cope with fire as they have a thick insulating bark layer around the lower trunk. If severely scorched by fire they can also re-sprout from the base and body of the trunk and canopy branches (see Figure 1.51). However, other species such as turkey bushes and cypress pines are relatively fire-sensitive and may be killed by a single fire if the canopy is completely scorched.

Figure 1.51 A number of Australian plants can regenerate quickly after a bushfire.

While fires alter the structure of the soil, stored nutrients from burnt vegetation now act as fertiliser. This makes it easier for seeds to germinate but, as the soil is finer after a fire, it can be washed or blown away. Sediment and ash can foul waterways, damage food supplies and destroy habitats.

A bushfire reduces the number of animals in a region. Those animals that could not escape or burrow deep are killed. Even surviving animals will not be able to remain as their food sources are gone and it may take years for the land to recover.

Effect of drought

Australia occupies the driest inhabited continent, so consequently droughts have a greater effect than either fires or floods. They can devastate large areas of land, destroying both agricultural and natural ecosystems, and degrading the land, affecting its future use.

As there is less vegetation, there is not much anchoring the soil so it is more vulnerable to erosion and dust storms. Less moisture leads to an increase in bushfires. Pressures put on farmers by drought can have long-term implications for agricultural practices (see Figure 1.52).

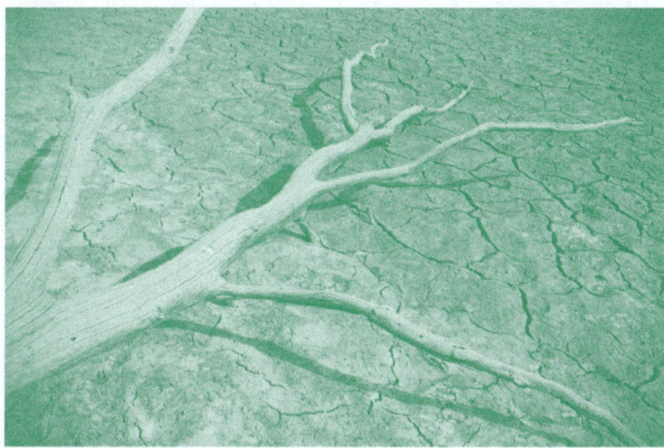

Figure 1.52 Droughts cause plants to wither, animals to die and the ground to dry out and crack.

When creeks and rivers dry up, the organisms relying on the water either perish or migrate if they can. This leads to food shortages for animals higher up the food chain.

Effect of floods

Floods cause loss of life and damage to property and habitats. Many animals that can't escape, drown. Since plant life and other food sources are usually destroyed, surviving animals are at risk of dying from starvation.

However, floods do have some positive ecological benefits.

- Floodwater soaks into the soil, replenishing groundwater supplies. This can then provide a long-term supply of water for native plants and agricultural crops.
- Floods clean soils by flushing out excess salt and chemicals.
- Soils can be replenished with nutrients enabling fish, birds and other organisms to breed and multiply. The flooding of the Nile River in Egypt each year brings in silt-laden waters. When the waters recede, the silt remaining behind fertilises the land allowing crops to grow.
- Desert plants rely on floods to regenerate and bloom. Their life cycles are tuned to high activity in a short timeframe. There is an increase in animal activity as a consequence; this population decreases when the waters recede and plants die off.
- Some ecosystems have adapted to regular flooding. River red gums growing along the Murray River use flooding to disperse seeds and deposit fertile soil. (However, humans have altered this periodic flooding by building dams and weirs, and reducing river flow through irrigation.)
- Lake Eyre is the largest of Australia's dry salt pans covering some 9300 km^2 (see Figure 1.53) and is covered with water only on average twice a century. Plants that grow here have adaptations enabling them to withstand long droughts. When water comes, plants spring to life supporting a variety of animal life.

Figure 1.53 Lake Eyre

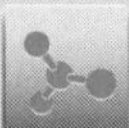

Case study 4: Modelling the effect of environmental changes on animal populations

Kangaroo bodies lay scattered by the road and through what once was bush while wombats that survived the ravaging wildfire's onslaught emerged from their underground burrows finding only blackened earth and nothing to eat.

Stories like this are usual in newspapers after reporters are allowed in to view the devastation. Animal populations are affected by the results of fire and floods.

But what exactly are the sizes of these populations? It is impossible to catch and count every kangaroo or rabbit, and animals don't fill in census forms.

Models are devised to estimate populations. Rabbit numbers are determined by systematically counting them at dawn or dusk or at night (when they are most active) and using this to estimate the population in a given area. But no matter how hard you try, it is very unlikely you have counted every rabbit. So a model is developed by comparing rabbit counts with the actual number of rabbits present, from a known population of animals.

Kangaroo numbers may be estimated by flying over the countryside, counting some of the animals visible from the air, and using a computer model to 'estimate' kangaroo numbers Australia-wide.

For example, in developing a model to determine rabbit or kangaroo numbers, a number of factors need to be considered.

- How many rabbits or kangaroos are present in the area?
- How much food and water is available for these animals?
- How many offspring does each pair of animals produce each year?
- What predators, if any, are present and how do they impact on animal numbers?
- How long do these animals live for?
- Are there animals coming into or leaving the area?
- How are these animal numbers affected during environmental changes, such as fires and floods?

Figure 1.54 shows the changes in fox and rabbit populations over time. When the rabbit (prey) population increases, there is more food for the foxes (predator) so their population also increases. Eventually there will be too many foxes and not enough rabbits, so their numbers will decrease through starvation. This allows the rabbit numbers to increase again and so the cycle continues.

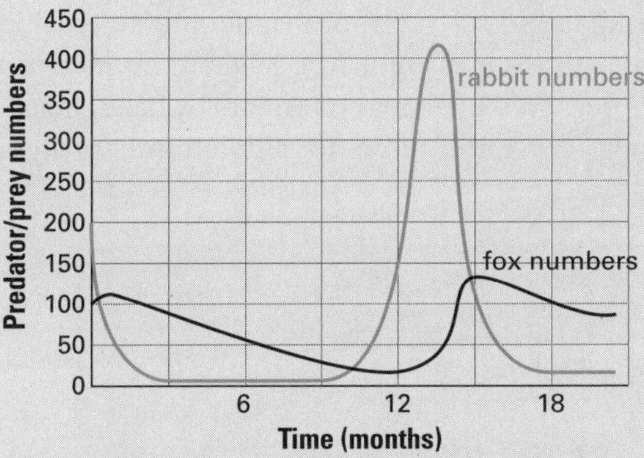

Figure 1.54 Fox and rabbit populations change over time.

A simple mathematical equation used as a model is easy to solve but not very realistic. The predator–prey model provides a truer picture but is harder to solve. Ecologists use advanced models and computers to study natural habitats and learn more about the animal kingdom and how environmental conditions affect their numbers.

Capture–recapture

The capture–recapture technique is used to estimate a population of animals in an area. A sample of animals is caught, tagged and counted. They are then released back into their habitat. Sometime later another sample is captured and the proportion of tagged animals in this sample is counted. This technique works best for a closed population, that is, animals that don't leave or enter an area such as the population of fish in a lake. This technique is shown in Figure 1.55.

The population can now be estimated:

$$\text{total population} = \frac{\text{number tagged} \times \text{total recaptured}}{\text{number tagged in recapture}}$$

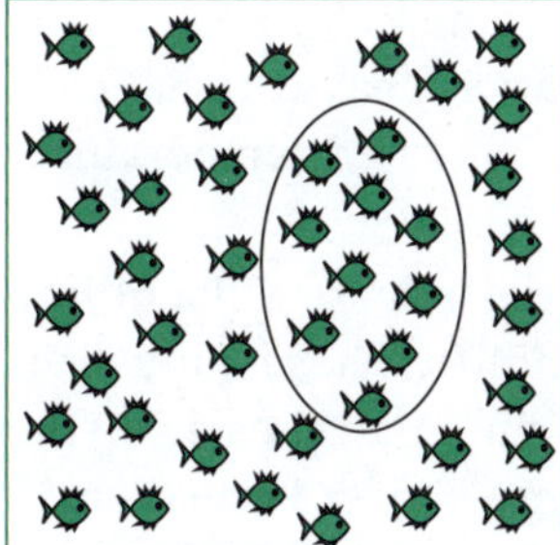

1. A random sample of the population is captured and tagged.

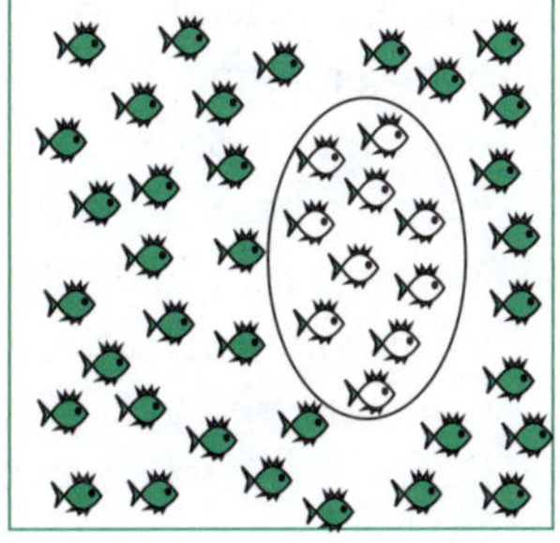

2. This tagged population is released back into its environment.

3. Time is allowed for a thorough mixing of tagged and untagged individuals.

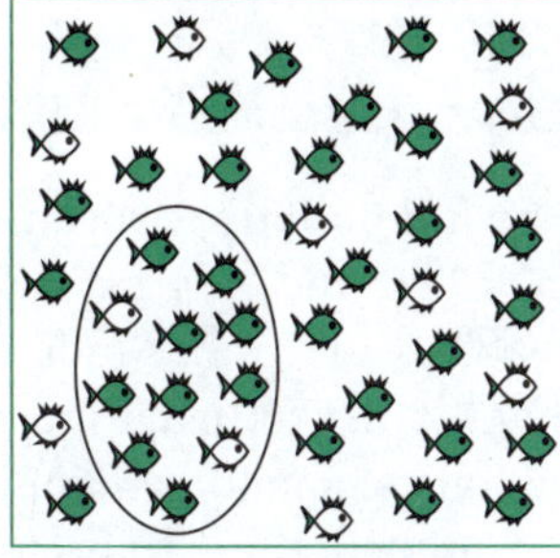

4. Another random sample is captured (not necessarily the same size) and the numbers tagged and untagged recorded.

Figure 1.55 The capture–recapture technique

Example: Capture–recapture

Suppose 35 beetles were originally tagged. They were released and a few days later 30 beetles were recaptured of which 16 were observed to be tagged. The estimate for the total population is:

$$\text{population} = \frac{35 \times 30}{16} = 65.6 = 66 \text{ beetles}$$

Of course, this is only an estimate. The actual population could vary by a few individuals either way, but it gives an indication of the size of the population and allows for changes, such as those caused by changing environments, to be monitored over time.

Experiment 3

Using the capture–recapture technique

Aim

- To determine the approximate size of a population using the capture–recapture technique

Method

1. Use a jar to place a few hundred matchsticks (or beans, counters or rice grains). The actual number doesn't matter—these objects will simulate animals in the wild.
2. Draw out a small handful and mark them with a marker.
3. Return these to the jar and shake it up.
4. From the jar, recapture another handful of objects.
5. Record how many of these you drew out, and how many are marked.
6. Repeat steps 4 and 5 in this experiment several more times.
7. Now count the total number of objects.

Results

1. Record your results in the table, leaving the last column of the table blank.

Trial number	Total objects marked	Total number recaptured	Number tagged in recapture	Population estimate
1				
2				
3				
4				
5				

2. Practice first with this example.
   Fifty rabbits are caught, tagged and released. Later, 220 rabbits are caught of which 35 are tagged. What would be the total population estimate?
3. Complete the last column in the table by calculating the population estimate.
4. Obtain an average of these five estimates.
5. Why is each of these population calculations not necessarily the same?

Analysis

How does your population estimate compare with the actual number of objects you used?
Go to p. 224 to check your answer.

Conclusion

Write a suitable conclusion.
Go to p. 224 to check your answer.

Human impacts on ecosystems and conservation

Humans have impacted on their environment, altering both biotic and abiotic factors, in a number of ways. Several have already been mentioned in these notes. Some other examples are as follows.

- Clearing bush and wooded areas for pastures and crops has destroyed the habitat of many organisms.
- Cities (basically concrete jungles with very little plant or animal life and variety) have dramatically reduced biodiversity but increased temperatures around them.
- Mining and other industries release effluent into waterways, affecting wildlife. These industries can also release gases and particles into the atmosphere, which can spread over large areas.
- Farming practices have increased the number of domesticated animals and plants, along with pest

Case study 5: The Exxon Valdez oil spill

In 1989 the oil tanker Exxon Valdez spilt some 260 000 barrels of oil in Prince William Sound, Alaska. This catastrophe is considered to be one of the most devastating human-caused environmental disasters. The remote region is a habitat for salmon, sea otters, seals and seabirds. In the clean-up, populations of plankton were displaced. Many of these organisms form the basis of the coastal marine food web.

The immediate consequences of the oil spill included the deaths of approximately 250 000 seabirds and thousands of sea otters, as well as many other animals, including the destruction of billions of salmon and herring eggs. The effects of the spill continue to be felt and various estimates suggest that it will take decades to fully recover.

Case study 6: Introducing *Melaleuca* to Americans

The paperbark tree *Melaleuca quinquenervia* (see Figure 1.56) is an invasive tree that was introduced into south Florida, United States, in the late 19th century. It was used as an ornamental plant and to stabilise the soil in swampy areas near lakes and canals. No one at the time realised that this Australian native would displace native plants and animals, draw up water from wetlands and create a fire hazard. It causes up to $200 million in environmental losses every year and takes over around 6 hectares each day.

In order to combat this, Americans have tried spraying the trees with herbicide and cutting them down. A biological control agent, the melaleuca leaf weevil (*Oxyops vitiosa*) was imported from Australia and released in 1997. Another predator, the aphid-like psyllid (*Boreioglycaspis melaleucae*), and finally the melaleuca bud gall fly (*Fergusonina turneri*), also from Australia, have since been introduced and are making inroads into this costly mistake.

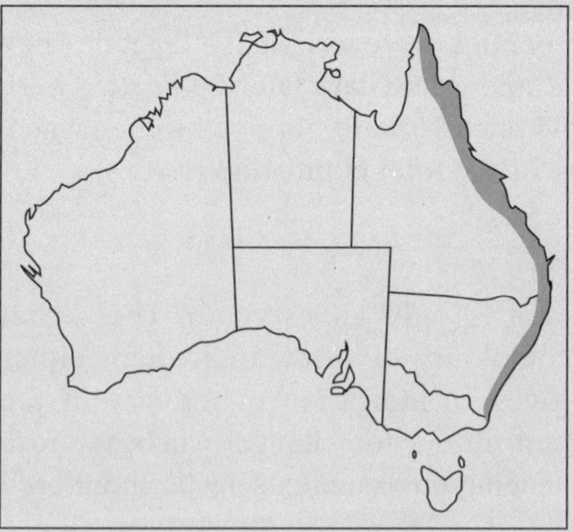

Figure 1.56 *Melaleuca quinquenervia* is a very common native small to medium-sized tree along coastal streams and swamps of eastern Australia and is widely cultivated.

Like the introduction of the cane toad or the rabbit into Australia, there can be catastrophic consequences when organisms are introduced into foreign ecosystems.

species, and excess fertiliser and pesticide use can end up in natural water courses.

Two examples are detailed in the Case studies on the previous page. The first involves an unintended consequence of human activity. The second is a misguided attempt to improve the environment.

The effect of cyclones on the biosphere, hydrosphere and atmosphere

Tropical cyclones are systems of low pressure that form over warm tropical waters. They can produce gale-force winds (winds of over 63 km/h with gusts exceeding 90 km/h, sometimes reaching 280 km/h) near their centres. These winds travel clockwise in the southern hemisphere. They obtain their energy from the warm waters where the sea-surface temperature is greater than 26.5 °C. Once formed, they can persist over cooler seas and for many days (sometimes weeks). Their paths may be quite erratic, but they usually lose their energy over land or colder oceans.

Tropical cyclones occur along the tropical north coast of Australia but may travel several hundred kilometres inland before dissipating. Figure 1.57 shows a simplified model of how tropical cyclones form. The diameter of the eye is around 40 km and is surrounded by a dense ring of cloud about 16 km high. The diameter of the cyclone averages 1000 km.

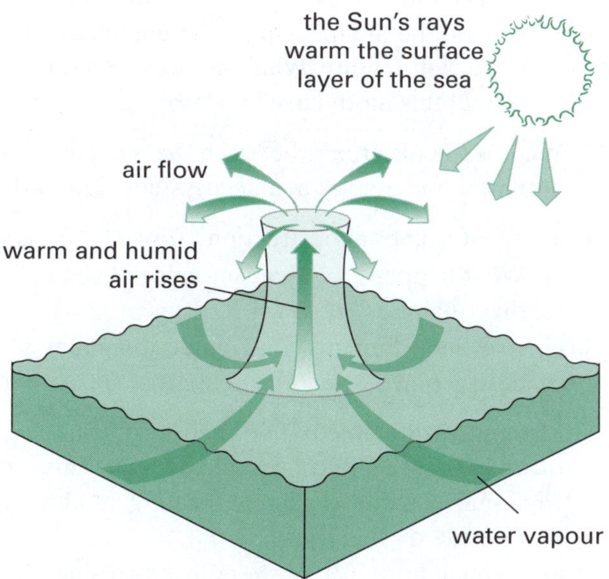

Figure 1.57 Modelling tropical cyclone formation

Cyclones are dangerous because they produce the following.

- Destructive winds can cause extensive property damage, create potentially lethal missiles from airborne debris, uproot trees and damage animal habitats, and whip up seas causing damage to ships and to marine life in harbours and shores.
- Heavy rainfall occurs, persisting as the cyclone moves inland.
- Flooding can be brought on by rainfall and can cause further damage and death by drowning.
- Damaging storm surges can cause low-lying coastal areas to be inundated, and are responsible for more deaths than any other feature. A storm surge is a raised dome of water, about 2 to 5 m higher than normal and about 60 to 80 km across, and may flood low-lying areas for kilometres inland. (See Figure 1.58.)

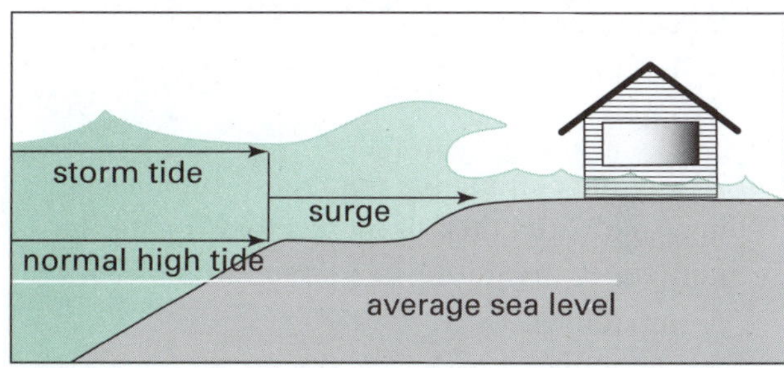

Figure 1.58 A storm surge is a rise above the normal water level along a shore resulting from strong onshore winds. Adding the normal tide to a storm surge produces a storm tide.

Australia's most destructive cyclone was Cyclone Tracy, which struck Darwin on Christmas Eve, 1974. Producing winds around 250 km/h, 71 people were killed, many thousands injured and almost 60% of the population was left homeless. Many were evacuated, but those who remained in Darwin faced the possibility of diseases because much of the city was without water, electricity or basic sanitation. It took years for both the human and natural environment to recover from this cyclone.

Test yourself 2

Part A: Knowledge

1. An organism that creates its own food is a *(1 mark)*

   A decomposer.

   B scavenger.

   C consumer.

   D producer.

2. An example of a biotic component of an ecosystem is *(1 mark)*
   A temperature.
   B rainfall.
   C sunlight.
   D omnivores.

3. Which of the following does *not* recycle in an ecosystem? *(1 mark)*
   A minerals
   B water
   C energy
   D nitrogen

4. Food, water, shelter and a space in which to live all refer to an organism's *(1 mark)*
   A habitat.
   B community.
   C population.
   D niche.

5. A way of counting the number of organisms in a small area then using this to estimate the number of organisms in a larger area is *(1 mark)*
   A modelling.
   B sampling.
   C tagging.
   D culling.

6. Complete the following restricted-response questions using the appropriate word. *(1 mark for each part)*
   a) Animals that have tasted unpalatable plants tend to .... them afterwards on the basis of their most conspicuous features, such as their flowers.
   b) A food .... is a network of food chains representing the feeding relationships among organisms in an ecosystem.
   c) A .... is a group of individuals of the same species that live in the same ecosystem.
   d) An .... pyramid is a diagram that shows the relative amounts of energy in different trophic levels in an ecosystem.
   e) .... deals with the relationship and interactions between organisms and their environment.

7. Use the code letters to match the terms or phrases in each column. *(1 mark for each part)*

Column 1	Column 2
A decomposers	F tropical storm
B biomass	G food, behaviour, non-living conditions
C plankton	H aquatic
D cyclone	I total organic matter
E niche	J bacteria or fungi

Part B: Skills

8. Poison oak, poison ivy and poison sumac produce urushiol, which is an oily organic allergen. In humans these chemicals produce an allergic skin rash. Suggest a reason these plants produce urushiol. *(1 mark)*

9. *Angraecum sesquipedale* is a night-flowering orchid native to Madagascar. It is pollinated at night by moths and switches on a sweet, rich scent just at sunset, flowering all night long.
   a) Why does this plant flower at night and not during the day? *(1 mark)*
   b) What is the purpose of the scent? Why a scent? *(2 marks)*
   c) The flower has an extremely long and narrow neck with nectar at the end of it. When Charles Darwin first saw it 150 years ago, he didn't know what the moth looked like, but he did make a prediction about an important feature. What important feature should this moth have? *(1 mark)*

10. What is the difference between the movement of matter and energy in an ecosystem? *(2 marks)*

11. **a)** What is habitat destruction? *(1 mark)*
    b) What happens to organisms which lived in these habitats? *(2 marks)*
    c) How have humans contributed to habitat destruction? *(3 marks)*

12. Mosquitoes are able to find their host using three sensors: chemical, visual and heat. This is what makes them so good at locating and biting their host.
    a) Explain how each of these three sensing techniques may be of benefit to the mosquito in finding its host. *(6 marks)*

b) Can a mosquito be called a parasite? Why? *(1 mark)*

c) With only short mouth parts, male mosquitoes feed on plant nectar. A female mosquito, on the other hand, has a long proboscis (an elongated feeding tube) that she uses to bite animals and humans and feed on their blood. Shortly afterwards she lays her eggs. Why do you think a female mosquito needs to feed on a meal of blood? *(1 mark)*

d) Mosquito saliva contains a blood anti-clotting agent that it secretes into the bite before feeding. What do you think the purpose of this anti-clotting agent is? *(1 mark)*

13. a) In estimating the number of fish in a lake, biologists captured 232 fish, marked their fins, then returned them to the lake. Several weeks later, another sample of 329 fish were captured. Of this sample, 16 had marks on their fins. Use the capture–recapture technique to estimate the number of fish in the lake. *(2 marks)*

b) List three factors which are important in a capture–recapture method. *(3 marks)*

14. Figure 1.59 shows part of a food web.

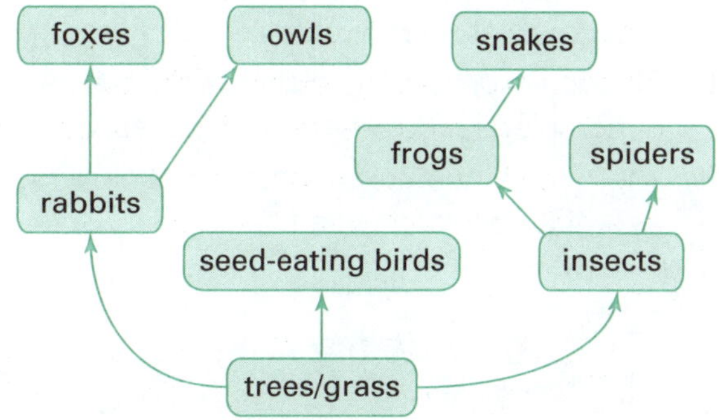

Figure 1.59 Extract of a food web

a) What would happen to the snake and insect populations if the frogs were to die out? *(2 marks)*

b) If the foxes were to be removed from an environment, how might this affect rabbit numbers? *(2 marks)*

Go to pp. 225–226 to check your answers.

Summary

1. The human body consists of a number of organ systems which have specific functions to maintain a functioning body.
2. The complex nature of organ systems requires coordination by the nervous and endocrine systems.
3. The nervous system consists of a network of nerve tissues which transmit electrical information from one site to another.
4. The endocrine system consists of various glands that release hormones directly into the bloodstream and bodily fluids in order to control the various body systems.
5. Microbes that cause disease are called pathogens.
6. The body's immune system attacks the microbes in order to return the body to a healthy state.
7. Vaccination and immunisation protect the body against pathogens.
8. The discovery of antiseptic and antibiotics has played an important role in combating diseases caused by pathogens.
9. Various bands of radiation from the electromagnetic spectrum can be used to diagnose and treat disease.
10. Various imaging technologies have been developed to diagnose and treat disease.
11. Spray-on Skin technology has been developed in Australia to treat burns to the skin.
12. Organisms in ecosystems are connected in different ways.
13. Interdependent organisms form communities.
14. Predator–prey relationships are formed in communities.
15. A parasite is an organism that benefits at the expense of its host.
16. Competition occurs naturally between living organisms that co-exist in the same environment.

Summary cont.

17. Pollination leads to the creation of new seeds that grow into new plants.
18. All populations of organisms are limited partly or completely by diseases in their ecosystems.
19. There are a number of factors that limit the size of a population in an ecosystem
20. In ecosystems, matter is recycled but energy flows through in one direction.
21. Humans and catastrophes have profound impacts on ecosystems.
22. Cyclones affect the biosphere, hydrosphere and atmosphere.

Syllabus checklist

Are you able to answer every syllabus question in this chapter? Tick each question as you go through the list if you are able to answer it. If you cannot answer it, turn to the appropriate page in the guide as is listed in the column to find the answer.

For a complete understanding of this topic		Page no.	✓
1	Can I name the common organ systems of the body and state their specific functions to maintain a functioning body?	2–4	
2	Can I explain why the complex nature of organ systems requires coordination by the nervous and endocrine systems?	4–5	
3	Can I state the purpose of the network of nerve tissues in the body?	5–7	
4	Can I name the important glands of the endocrine system and why they release hormones directly into the bloodstream and bodily fluids?	7–8	
5	Can I recall what the microbes that cause disease are called?	9–10	
6	Can I describe how the body's immune system attacks the microbes and describe how the body is returned to a healthy state?	10–11	
7	Can I explain how vaccination and immunisation protect the body against pathogens?	11	
8	Can I explain how the discovery of antiseptic and antibiotics has played an important role in combating diseases caused by pathogens?	14–15	
9	Can I explain how various bands of radiation from the electromagnetic spectrum can be used to diagnose and treat disease?	16–17	

For a complete understanding of this topic		Page no.	✓
10	Can I describe the use of various imaging technologies that have been developed to diagnose and treat disease?	17–18	
11	Can I explain the importance of Spray-on Skin technology which has been developed in Australia?	18	
12	Can I identify predator–prey relationships formed in communities?	22–23	
13	Can I name the type of organisms that benefit at the expense of their hosts?	24	
14	Can I explain the consequence of competition between organisms in the same environment?	26	
15	Can I explain how pollination is important in the formation of new seeds and new plants?	26–27	
16	Can I explain how diseases influence the population of organisms in an ecosystem?	27–29	
17	Can I name the factors that limit the size of a population in an ecosystem?	29–33	
18	Can I describe examples of different ways in which organisms in ecosystems are connected?	33	
19	Can I describe examples of communities and the interdependence of organisms in it?	33	

	For a complete understanding of this topic	Page no.	✓
20	Can I compare the directions in which matter and energy move in ecosystems?	33–34	
21	Can I state the impact that human activity and catastrophes have on ecosystems?	34–39	

	For a complete understanding of this topic	Page no.	✓
22	Can I identify the ways in which cyclones affect the biosphere, hydrosphere and atmosphere?	39	

Chapter test

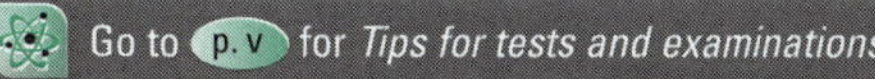
Go to p. v for *Tips for tests and examinations*

Part A: Multiple-choice questions

(1 mark for each)

1. Name the site in the digestive system where digestion is completed and nutrients are absorbed into the bloodstream.
   - **A** large intestine
   - **B** small intestine
   - **C** stomach
   - **D** liver
2. The kidneys are a major organ of the
   - **A** excretory system.
   - **B** digestive system.
   - **C** respiratory system.
   - **D** circulatory system.
3. An important function of the cerebellum is to control
   - **A** muscles involved in involuntary movements and fine motor control.
   - **B** memory and intelligence.
   - **C** behaviour.
   - **D** breathing, temperature and thirst.
4. Involuntary nerves regulate many bodily functions including
   - **A** heart muscle contraction.
   - **B** raising an arm.
   - **C** bending at the waist.
   - **D** chewing food.
5. In a reflex arc the effector is normally
   - **A** skin.
   - **B** the spinal cord.
   - **C** the brain.
   - **D** a muscle or gland.
6. Which of the following are abiotic components of the ecosystem?
   - **A** producers
   - **B** carnivores
   - **C** herbivores
   - **D** water
7. Tapeworms live inside the intestines of humans and feed on the nutrients of the food they eat. A tapeworm can be called a
   - **A** host.
   - **B** parasite.
   - **C** mutualist.
   - **D** cannibal.
8. Predators such as dingoes in Australia, lions in Africa or tigers in Asia are moving closer to highly populated areas. As this is posing an increased danger to people, predator populations are being culled. This can have a devastating effect on the ecosystem because without this natural control
   - **A** the numbers of prey will decrease.
   - **B** the numbers of prey will remain the same.
   - **C** the numbers of prey will increase.
   - **D** plant species will increase.
9. Producers and consumers are organisms in an ecosystem. An organism that eats both producers and other consumers is called a
   - **A** carnivore.
   - **B** omnivore.
   - **C** herbivore.
   - **D** prey.
10. Biological control can be used to control pests. However, there are risks involved if a new

species is introduced to an area. This is because the new species

A could be killed off more easily than native organisms.

B may not find enough food to survive.

C might overpopulate an area as it has no natural predators.

D could possibly eliminate the pest altogether.

Part B: Short-answer questions

11. A person is involved in a motor vehicle accident that leads to a severing of his spinal cord in the middle of his back. A doctor conducts a reflex test on his knee.
    a) Will the knee produce a reflex arc when tested? *(1 mark)*
    b) Will the patient feel the pressure on his knee? *(1 mark)*
    c) Why can't the patient walk following this accident? *(1 mark)*

12. Cause and effect statements are presented in the following table but they are jumbled. Match the Cause code letters with the correct Effect code letters. *(4 marks)*

Cause		Effect	
A	Pancreas detects low blood glucose.	E	More glycogen is broken down to glucose.
B	Glucagon stimulates the liver.	F	Pancreas secretes insulin into the blood stream.
C	Blood glucose levels are too high.	G	Glucose is converted to glycogen for storage.
D	Insulin stimulates the liver.	H	Pancreas secretes glucagon into the blood.

13. The flow chart in Figure 1.60 shows a model of the way a room air conditioner cycles between fixed limits. The air conditioner is to be set to cooling so the room temperature remains between 19 and 21 °C.
    a) Identify the stages (by code letter) of the loop where information is transferred via electrical signals through a wire. *(1 mark)*
    b) Identify the stage (by code letter) where the room temperature has fallen below the set range and the switch is turned off. *(1 mark)*
    c) In this flow chart a temperature sensor in the air conditioner monitors the air temperature. Our brain has a control centre that also monitors the temperature of the blood. Name this control centre in the brain. *(1 mark)*
    d) Stage C in the diagram involves the cooling of the air in the room. This is achieved by the air conditioner extracting the heat from air flowing into it and releasing that heat to the outside of the house. Explain how our body removes excess heat from our blood. *(3 marks)*

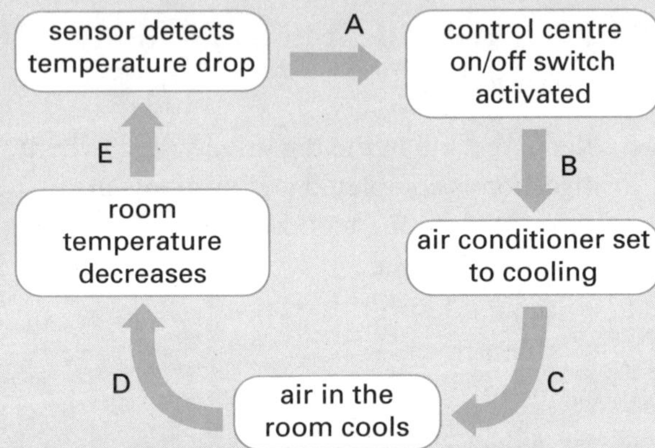

Figure 1.60 Flow chart

14. Urination is an involuntary reflex which we learn to control when we are children. By learning to control the bladder sphincter (muscle), we can wait to urinate at a more convenient time. The flow chart in Figure 1.61 shows the reflex arc.

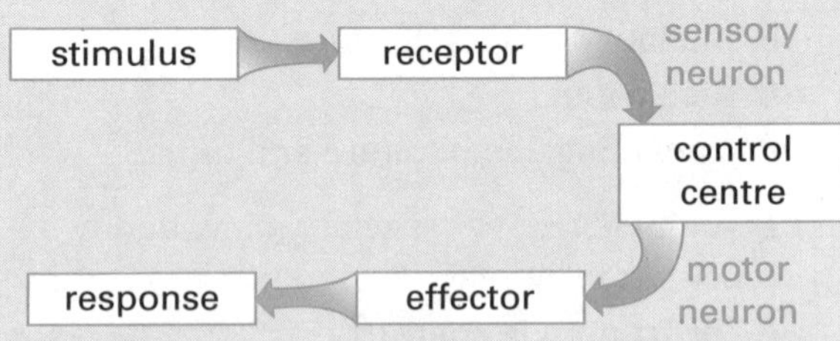

Figure 1.61 Stimulus response

Match the five components of the reflex arc to the following random list: *(5 marks)*

A. Brain
B. Urination
C. Large urine volume
D. Bladder sphincter opens
E. Stretch receptors in bladder activate

15. Cortisol is an important hormone produced by the adrenal glands which are located on the top of the kidneys. Cortisol is sometimes called the 'anti-sleep' hormone. There is a daily rise and fall in the concentration of cortisol in the blood. The production of cortisol gradually wakes us up. As night approaches, the level of cortisol goes down and helps to relax us and prepare us for sleep.

Figure 1.62 shows information about the control of cortisol production. The hypothalamus is at the top of the brainstem.

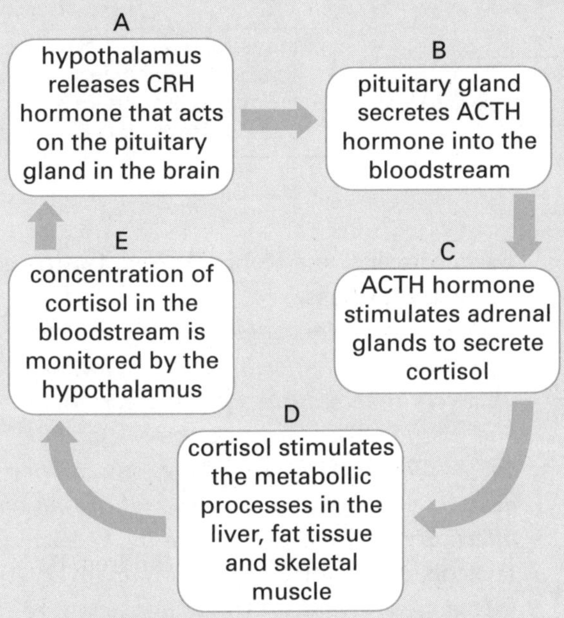

Figure 1.62 Flow chart for cortisol

a) Answer true or false to each of the following statements.
   i) The pituitary gland monitors the level of cortisol in the blood. *(1 mark)*
   ii) The adrenal glands secrete cortisol. *(1 mark)*
   iii) The concentration of cortisol in the blood will be lower than normal following stage D. *(1 mark)*
   iv) The concentration of cortisol in the blood will be higher than normal following stage C. *(1 mark)*

b) Cortisol levels rise in times of stress. Suggest a reason for this change. *(1 mark)*

16. Four components of the digestive system are listed. Some words are missing. Insert the correct words. *(4 marks)*

a) Oesophagus: muscular movements help the .... move through this tube from the mouth to the stomach.

b) Stomach: acids and .... continue to break down the food to form a semisolid material (chyme).

c) Liver: produces bile to aid in .... emulsification (breaking down into small particles).

d) Gall bladder: stores .... until it is released into the small intestine

17. Figure 1.63 shows a scale drawing of a microbe. It has two flagella to help it move. Use the scale to determine the length (L) of the body in micrometres. *(1 mark)*

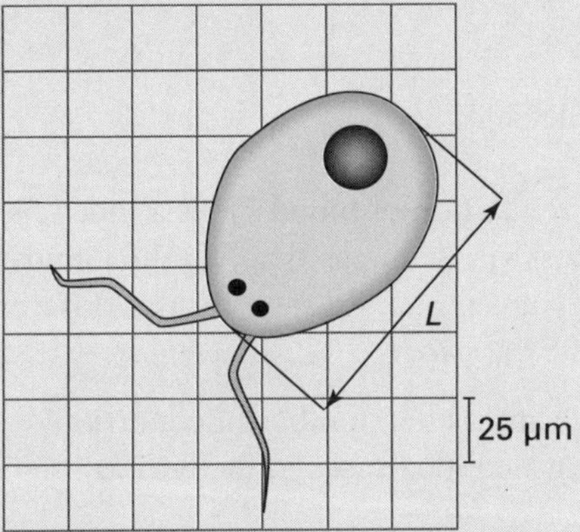

Figure 1.63 Microbe

18. The lungs and kidneys are involved in keeping the pH of your blood in a state of balance. The blood's pH is normally between 7.35 and 7.45. This is called the set point. Imbalances lead to acidosis (blood pH < 7.35) or alkalosis (blood pH > 7.45). The process of maintaining a balanced pH is called buffering. The medulla oblongata in the brainstem is involved.

Figure 1.64 shows an example of how blood pH can be maintained in the set point range.

a) Does this figure show an example of recovery from acidosis or alkalosis? *(1 mark)*

b) Where is the respiratory centre located? *(1 mark)*

c) How does the respiratory centre send messages to the diaphragm and rib muscles? *(1 mark)*

d) What happens to the pH of the blood if hydrogen ions are removed? *(1 mark)*

e) What basic substance is released from the kidney that neutralises excess acid? *(1 mark)*

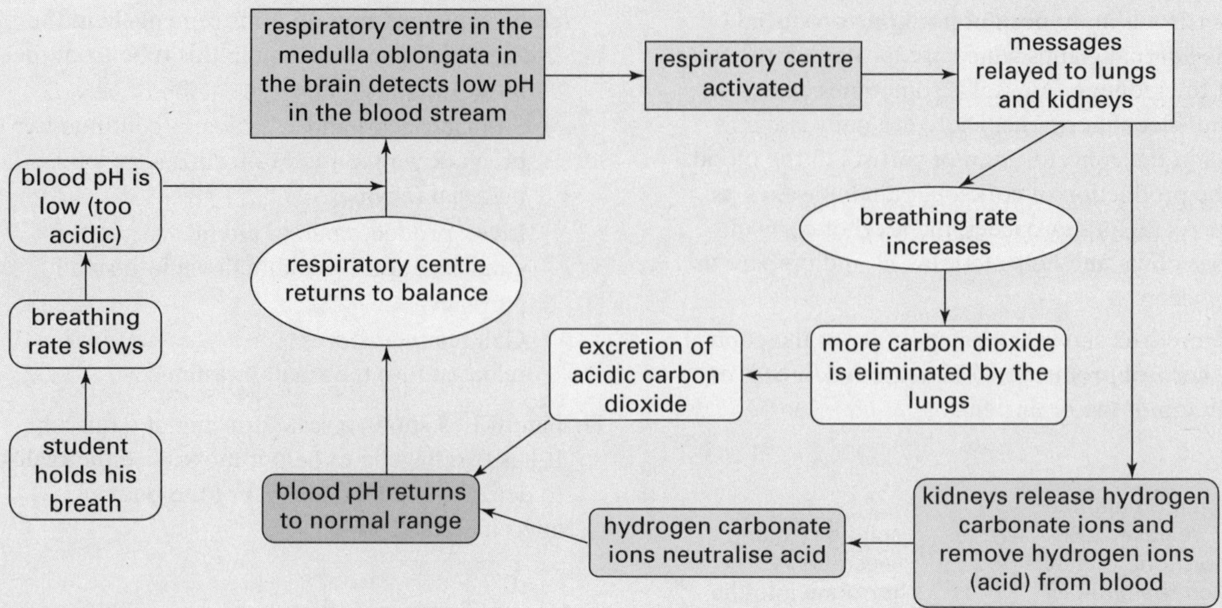

Figure 1.64 Blood pH

f) The regulation of blood pH is a much faster process via the lungs than via the kidneys. Can you suggest a reason for this difference in rate? *(2 marks)*

19. Table 1.6 shows the number of countries reporting smallpox cases in the period 1950 to 1980.

Table 1.6 Countries reporting smallpox cases, 1950–80

Year	Number
1950	82
1953	71
1956	60
1959	51
1962	56
1965	46
1968	45
1971	16
1974	9
1977	0
1980	0

a) Plot a line graph of this data. *(5 marks)*

b) Account for the sudden decline in reported smallpox cases after 1968. *(2 marks)*

20. Read the following passage and answer the questions.

Barry Marshall and Robin Warren: Gastric and peptic ulcer disease

In 2005 Barry Marshall and Robin Warren shared the Nobel prize in Medicine for their discovery of the cause of peptic ulcer disease and gastritis. They began their work in 1981 and by 1983 had discovered the answer. They discovered that these diseases are the result of an infection of the stomach caused by the bacterium Helicobacter pylori. *Prior to this discovery it was thought that ulcers of the stomach were caused by excess acid. They succeeded in culturing and studying the bacteria and showed that these microbes were responsible for chronic ulcerations of the stomach. Antibiotics are now used to cure the disease. By 1994 their research was well accepted by the scientific community around the world.*

a) Name the microbe that causes peptic ulcers. *(1 mark)*

b) What type of medication is now used to cure stomach ulcers? *(1 mark)*

c) Suggest why a decade passed after their successful research before Marshall and Warren received the Nobel prize. *(1 mark)*

21. In 1882 Koch isolated tuberculosis bacteria from infected cattle. How was he able to show that the isolated bacteria were the cause of the disease? *(2 marks)*

22. Read the information below and answer the questions that follow.

Climate change and disease

Scientists have collected evidence that suggests that human activity is causing climate change. The term 'global warming' is used to describe this change. Human activities such as the burning of fossil fuels have been proposed as the major cause of global warming.

Global warming will have an impact on the spread of disease. Alteration of rainfall patterns and changes in temperature and humidity could lead to new areas becoming infested with mosquitoes and ticks. Many viral diseases are transmitted through the bites of these insects. West Nile virus and Rift Valley fever, for example, are transmitted by mosquitoes. Bluetongue virus, which causes disease in cattle and sheep, is carried by midges. Warmer and moister climates in Europe in the last few decades have seen this disease move from northern Africa into southern Europe. Cholera bacteria thrive in warmer waters and without improved sanitation in developing countries, global warming could cause cholera to spread. Many scientists and non-scientists have speculated about the potential of climate change to increase the incidences of malaria. Many media articles predict that transmission will extend to higher latitudes and even higher altitudes. There has been criticism of the simplistic models used to make these predictions. Dengue fever is also considered to be a disease that could increase in frequency due to global warming. Dengue fever is a viral disease spread by mosquitoes; no vaccine has been developed to deal with this disease.

a) What human activity has been proposed as the major cause of global warming? *(1 mark)*

b) What has been the effect of increased rainfall on disease spread in Europe? *(2 marks)*

c) Name a disease caused by a i) protist and ii) virus, that could increase due to global warming. *(2 marks)*

23. Complete Table 1.7 using the following list of words or phrases: measles; eating contaminated food; virus; fungus; malaria; bacterium. *(6 marks)*

24. There is an ancient belief in China that the universe is composed of two opposite forces called Yin and Yang. These two forces merge to form a balance called tai chi. Yin and Yang are complementary. Consider the following example.

Your body gets too hot due to strenuous exercise. Your blood now has an excess of Yang and a deficit of Yin. To feel more comfortable and achieve tai chi you must rid yourself of some Yang and gain more Yin.

Use one example to relate these tai chi concepts to the way our bodies are kept in a state of balance. *(2 marks)*

25. Diseases also affect other animals. Australia has strict quarantine laws and as a result many infectious diseases are prevented from entering the country. Table 1.8 lists some infectious diseases of animals caused by different pathogens.

Table 1.7 Pathogens and diseases

Disease	Pathogen	Transmission	Symptoms of the disease
salmonella food poisoning	a)	b)	diarrhoea; inflammation of the intestine; nausea
tuberculosis	bacterium	inhaling infected droplets in the air; drinking contaminated milk	lesions form in lungs, producing fever and coughing
c)	protozoan	protozoan enters human during a mosquito bite	severe fevers; shivering; sweating; nausea; reoccurs periodically
mumps	d)	contact; inhalation of infected droplets	swollen salivary glands in cheek/neck; fever
e)	virus	contact with infected people; inhalation of infected droplets	red spots (rash) on face and body; fever
tinea	f)	contact of the foot with infected areas (e.g. communal bathing)	itching and blistering of skin between toes; cracking of skin

Table 1.8 Some examples of infectious diseases caused by animals

Animal	Disease	Pathogen	Other information
horse	equine influenza	a)	This disease is spread by contact between horses or via contact with b)......................... equipment. Horses experience muscle pain and stiffness as well as laboured breathing. Older horses may die.
c)	rabies	virus	Rabies is a fatal viral disease that infected dogs can d)......................... to other animals, including humans. Infected animals suffer seizures. Eventually paralysis and death result.
bird	psittacosis	bacterium	Infected birds have trouble breathing, and have e)......................... eyes and watery droppings. The microbe is transmitted in the droppings of infected birds.
bird	aspergillosis	f)	Spores may be present in the bird feed. Symptoms include loss of appetite and breathing difficulties due to formation of lung nodules.
g)	trypanosomiasis	protozoa	Cattle suffer h)......................... and intermittent fevers. Weight loss, breathing difficulties and weakness also occur.

Complete Table 1.8 using the following list of words or phrases: anaemia; fungus; cows; virus; contaminated; dog; transmit; inflamed. *(8 marks)*

26. A predator and its prey evolve together, as they are part of each other's environment.
    a) The predator will die if it does not get food, so it evolves whatever is necessary in order to eat the prey. List at least six different advantages a predator could develop to help it catch its prey. *(6 marks)*
    b) The prey dies if the predator eats it, so it evolves whatever is necessary to avoid being eaten. List at least six different advantages a prey could develop to help it avoid its predator. *(6 marks)*

27. The neem tree (*Azadirachta indica*) produces over 100 different compounds with pesticide properties, which act especially against invertebrates. One of these chemicals is found in all parts of the tree, but is mainly concentrated in its fruit, especially in its seeds.
    a) What is the function of this chemical? *(1 mark)*
    b) Describe at least three different ways this chemical might work. *(3 marks)*
    c) Some of the chemicals produced by the neem tree can be taken up by other plants through their roots and leaves, spreading the material throughout plant tissues. State an advantage of growing neem trees near other agricultural crops. *(1 mark)*

28. The prickly pear cactus (mainly the species *Opuntia stricta*, shown in Figure 1.65) is an introduced pest from South America, and makes large areas of Australian farmland unusable. In 1920 prickly pear had infested some 23 million hectares of land in New South Wales and Queensland and was spreading at a rate of 400 000 hectares a year.

Figure 1.65 Prickly pear

    a) Name two strategies that would have initially been tried to eradicate this pest. *(2 marks)*

A biological control was eventually tried. When the moth cactoblastis (*Cactoblastis cactorum*), whose caterpillar stage feeds on this cactus, was introduced (also from South America) it rapidly brought the cactus under control.

b) Why was cactoblastis sourced from the same continent as the prickly pear? *(1 mark)*

c) State one important factor that needed to be considered before cactoblastis was released into the Australian environment. *(1 mark)*

d) Several years after introduction both moth and cactus were rare. Explain why this might be. *(2 marks)*

e) Suggest a reason why both the moth and prickly pear are not completely extinct in Australia. *(2 marks)*

29. Plants that are pollinated by insects are often brightly coloured and have a strong smell.

a) Why might this be? *(2 marks)*

b) Another way plants are pollinated is by the wind. Describe three features flowers of these plants might have that contrast them with insect-pollinated plants. *(4 marks)*

30. Figure 1.66 shows part of a food web along a rocky shoreline.

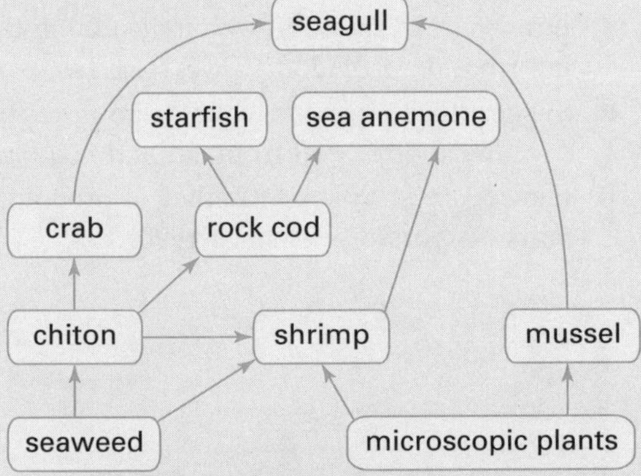

Figure 1.66 Food web on a rocky shoreline

a) What do sea anemones eat? *(2 marks)*

b) Name a producer and a herbivore. *(2 marks)*

c) Draw the food chain beginning at seaweed and ending at seagull. *(2 marks)*

d) The seagull can be described as both a second-order and third-order consumer. How can this be? *(2 marks)*

e) Suppose all the rock cod died out. What might be the consequences to the ecosystem? *(3 marks)*

f) If the shrimp were to die out, the rock cod numbers might increase or decrease. Suggest a reason to support each of these views. *(4 marks)*

31. Figure 1.67 shows laboratory experiments where two different species of paramecium were grown under identical conditions.

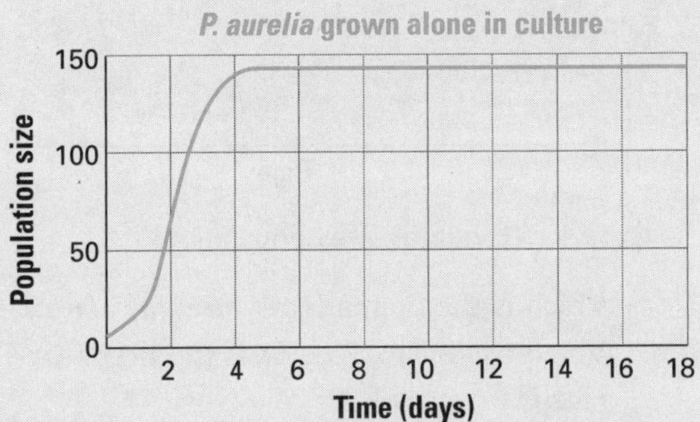

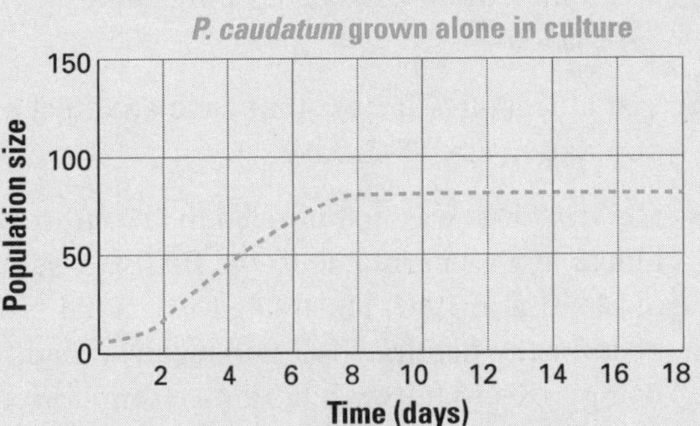

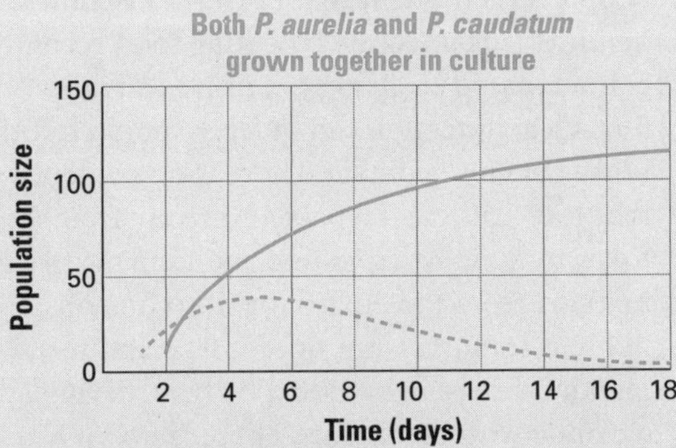

Figure 1.67 Growth of paramecium

a) Which species grows the fastest? How do you know? *(2 marks)*

b) At time $t = 0$ days, the first two graphs do not begin at 0 on the vertical axis, but at some identical small value above it. Why is this? *(2 marks)*

c) Explain why the first two graphs flatten out after several days. *(1 mark)*

d) Describe and account for the shape of the graphs when both species of paramecium are grown together. *(3 marks)*

32. Figure 1.68 shows population sizes of predator and prey over time.

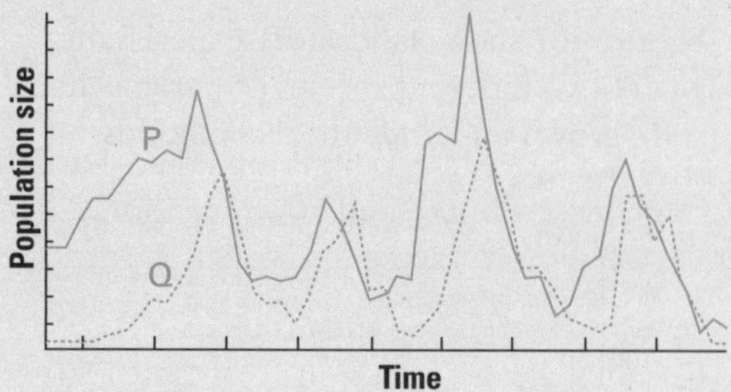

Figure 1.68 Predator–prey populations

a) Which is the dependent variable? *(1 mark)*

b) Which organism, P or Q, is the predator? *(1 mark)*

c) Comment on the shape of the curves. *(3 marks)*

33. List at least five factors that have an effect on population size. *(5 marks)*

34. Hendra virus was first noticed in an outbreak of illness in a racing stable in the Brisbane suburb of Hendra in 1994. The flying fox is a natural host for the hendra virus, and it can spread from flying foxes to horses but, rarely, from horses to people. It is thought that horses may contract hendra virus infection by eating food recently contaminated by flying fox urine, saliva or other fluids. The infection can quickly spread to other horses in close contact. There is no evidence that the virus can be passed directly from flying foxes to humans, from infected humans to other humans, from the environment to humans, from humans to horses, nor does it float in the air. Hendra virus can be killed by heat, drying and cleaning with detergents. There is no known specific treatment for Hendra virus infection, and antiviral medications are not effective.

a) List two steps that can be taken to prevent hendra virus in horses. *(2 marks)*

b) If a horse is suspected of having hendra virus, list two steps humans can take to avoid the possibility of contracting the virus. *(2 marks)*

35. Read the information below and answer the questions that follow.

*The spotted owl (*Strix occidentalis*) lives in the lush, 'old-growth' forests of the Pacific Northwest of the United States. It prefers to eat wood rats and flying squirrels, but is also partial to small mammals, reptiles and birds. In 1990 it was listed as a threatened species. Threats to its survival come from two main areas.*

- *Logging. As a result of heavy logging, these ancient forests have dwindled over the last 150 years. Only about 10% of the forests now remain, mostly on federally owned lands.*
- *The barred owl (*Strix varia*). This close relative has migrated westward and is competing with the spotted owl. Barred owls are more aggressive, have a broader diet and occur in more varied habitats.*

Northern goshawks and crows may prey on juvenile spotted owls, while great horned owls, red-tailed hawks and golden eagles are likely predators of both juveniles and adults.

a) Why are barred owls causing a decrease in the numbers of spotted owls? *(2 marks)*

b) How is logging threatening the survival of the spotted owl? *(2 marks)*

c) i) What are two steps that can be taken to prevent the spotted owl from becoming extinct? *(2 marks)*

ii) What are some issues involved in these solutions? *(4 marks)*

d) Draw part of the food web indicated in the information provided above. *(4 marks)*

e) Suggest three reasons why it is important to save the spotted owl from extinction. *(3 marks)*

f) Figure 1.69 shows the number of breeding pairs of spotted owls since 1990.

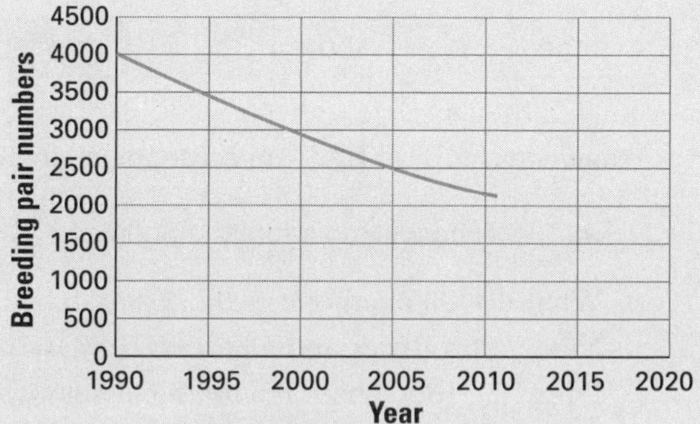

Figure 1.69 Spotted owls

i) By what percentage did the breeding pair number of owls decline between the estimates in 1990 and 2005? *(1 mark)*

ii) What is an estimate of the breeding pair numbers in 2020? What assumption are you making in obtaining this value? *(2 marks)*

36. As the tide recedes on a rocky seashore, not all of the coast is left dry. Depending on the rock type, there are usually shallow or deep rock pools containing a variety of organisms (see Figure 1.70). The sea does not always cover a rock pool, and so these organisms have adapted to changing conditions.

Figure 1.70 Rock pool

Describe how each of the following abiotic features would change between tides.

a) temperature *(2 marks)*
b) salinity *(2 marks)*
c) oxygen *(2 marks)*
d) light *(2 marks)*

37. List some impacts tropical cyclones have on the environment. *(7 marks)*

38. Some species of the frog *Cyclorana* burrow underground and can remain dormant for more than 5 years. They can store large amounts of water in their bladder, forming a type of cocoon around themselves. They are often called water-holding frogs. These frogs only return to the surface to breed and eat, and normally only after heavy summer rains.

a) In what areas of Australia is this frog most suited? *(1 mark)*
b) Why do they burrow underground and remain dormant? *(2 marks)*
c) How do these frogs prevent water loss? *(2 marks)*
d) Eggs are normally laid in temporary water puddles on the surface with tadpoles developing faster than other frog species. Why does this occur? *(2 marks)*

39. a) 'Authorities say it has become harder to control rabbits since the drought has broken.' Comment on this statement. *(6 marks)*

b) 'Whether rabbits are controlled by disease, poisoning, warren destruction, exclusion or a combination of these methods, the point is not how many rabbits are killed, but how many are left behind.' Comment on this statement. *(2 marks)*

40. The southern bluefin tuna is a highly prized fish caught in the southern Pacific, Atlantic and Indian Oceans. Australian scientists have made a study of the size of the spawning stock. The results are given in Table 1.9.

Table 1.9 Southern bluefin tuna spawning stock biomass, 1930–2005

Year	Spawning stock biomass (millions tonnes)
1930	810
1935	810
1940	810
1945	800
1950	790
1955	700
1960	610
1965	380
1970	300
1975	200
1980	180
1985	100
1990	90
1995	50
2000	50
2005	60

a) Draw a labelled line graph of this data. *(4 marks)*
b) In which decade did fishing the bluefin tuna become widespread? How do you know? *(2 marks)*
c) Describe the shape of the curve. *(3 marks)*
d) Suggest a reason for the curve not continuing to decline after about 1995. *(1 mark)*
e) Suggest two things Australia can do to allow the fish stocks to recover. *(2 marks)*

Go to pp. 226–231 to check your answers.

CHAPTER 2

Atomic theory and chemical reactions

Overview

In this chapter you will learn about:

- the historical development of the atomic theory
- the structure of the atom and the subatomic components
- electron shells and electron configurations
- isotopes
- radioactivity and types of radiation
- uses of radioactivity
- indicators of chemical change
- word equations and symbolic equations
- conservation of mass in chemical reactions
- types of chemical reactions including reaction of acids on metals, oxides and carbonates
- neutralisation reactions and indicators
- endothermic and exothermic reactions
- combustion reactions
- respiration and photosynthesis
- pollution of the environment

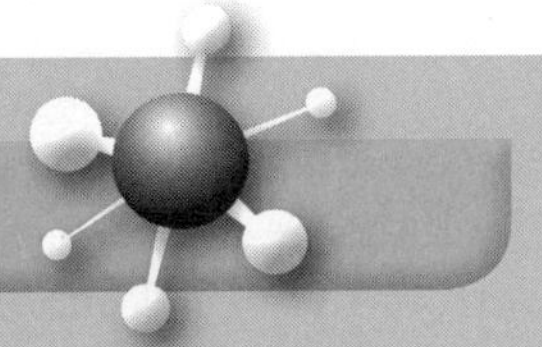

Glossary

Atom—the smallest unit of an element; composed of protons, neutrons and electrons

Atomic mass unit—(symbol 'u' or 'Da' for dalton) a unit that is used for indicating mass on an atomic or molecular scale (1 atomic mass unit = 1.66×10^{-27} kg)

Atomic number (Z)—the number of protons in the nucleus of an atom

Combustion—the reaction of a fuel and oxidiser (oxygen) to release energy

Effervescence—gas bubbles in a liquid

Electron—a negatively charged subatomic particle located outside and moving around the nucleus

Electron configuration—the arrangement of electrons in their shells

Endothermic—a reaction where heat energy is absorbed

Exothermic—a reaction where heat energy is released

Indicators—dye molecules that change colour in the presence of an acid or base

Isotope—variations of atoms of a particular chemical element, which have differing numbers of neutrons but the same number of protons

Half-life—the time needed for half of any given amount of a radioactive substance to decay. The rate of decay of radioactive substances is measured in terms of their half-life

Mass number (A)—the number of protons plus neutrons in the nucleus of an atom

Neutralisation—the destruction of the properties of acids or bases when they react together

Neutron—a neutral subatomic particle

Nucleus—the central positive core of an atom

Proton—a positively charged subatomic particle located in the nucleus

Radioactive decay—the spontaneous disintegration of an unstable atomic nucleus into a lighter one, where radiation is released in the form of alpha particles, beta particles, gamma rays and/or other particles

Shells—energy levels (or orbits) around the nucleus occupied by electrons

Salt—a compound formed when an acid neutralises a base

Transmute—convert one physical substance to another, such as a base metal into a valuable metal

2.1 Historical development of the atomic theory

Over the centuries many people thought about what substances were made from. Many of these ideas were based on very little evidence and abstract philosophical reasoning as these early 'scientists' had no way of seeing atoms, and they did very little experimentation. Atoms are minuscule objects with tiny masses. A scanning tunnelling microscope can now be used to observe atoms, but this instrument was only invented a little over 30 years ago. Almost all of an atom's mass is concentrated in the nucleus, with the mass shared almost equally between protons and neutrons. The modern atomic theory was developed from John Dalton's simple atomic model. Dalton believed, like some philosophers in ancient Greece, that all matter was composed of small, indivisible particles called atoms. Dalton proposed that each element was composed of unique atoms with different atomic weights. It wasn't until the end of the 20th century that subcomponents of the atom were identified.

Early Greek philosophers

Around 600 BC Thales of Miletus found that if he rubbed a piece of amber with fur, it attracted light objects such as hair, fluff or feathers. However, he did not connect this with any atomic particle, suggesting instead that this mysterious force came from the amber.

Then in 460 BC the Greek philosopher, Democritus, followed ideas from his teacher Leucippus to develop the notion of atoms. He reasoned that if you break a piece of matter into two, and then break it again and again, how many breaks will you have to make before you can break it no further? Democritus thought that it ended at some point, a smallest possible bit of matter. He called these basic matter particles atoms (from the Greek *atomos*, meaning 'indivisible'). But he was largely ignored by other thinkers at the time.

In the 5th century BC, Empedocles proposed that there were four 'elements' (which he called roots) from which everything was made. These were air, earth, water and fire, and everything else was some combination of these (see Figure 2.1). This belief became very popular in the medieval ages, giving rise to alchemy. Alchemists believed that since everything was made of only four elements, you could transmute one substance into another substance of the same type by tinkering with how much of each element it had. For example, it was believed that lead could be made into gold. Being fundamental, these four elements were infinitely divisible but, at the same time, eternal and unalterable.

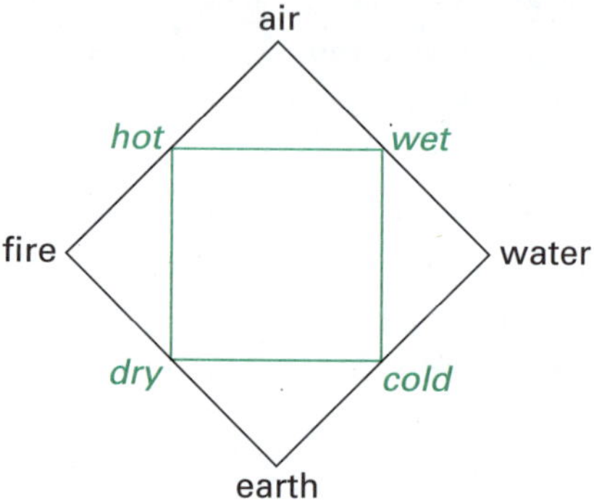

Figure 2.1 Empedocles' theory of the four elements became the standard established belief for the next 2000 years as alchemists tried to turn worthless metals into precious metals.

John Dalton

It was not until the early 1800s that people began to again question what matter is made of. The English chemist John Dalton (1766–1844) performed experiments with different chemicals that showed that matter seems to be made of elementary and tiny particles (atoms). He came up with an atomic theory that stated the following.

- Elements are made of tiny particles called atoms.
- All atoms of an element are identical and they have the same mass.
- Atoms of a given element are different from those of any other element.
- Atoms of one element can combine with atoms of other elements to form chemical compounds.
- Any particular compound always has the same relative numbers of types of atoms.
- Atoms cannot be created, divided into smaller particles, nor destroyed by any chemical means.
- A chemical reaction simply changes the way atoms are grouped together.

Dalton used his own symbols, such as those shown in Figure 2.2, to visually represent the different atoms.

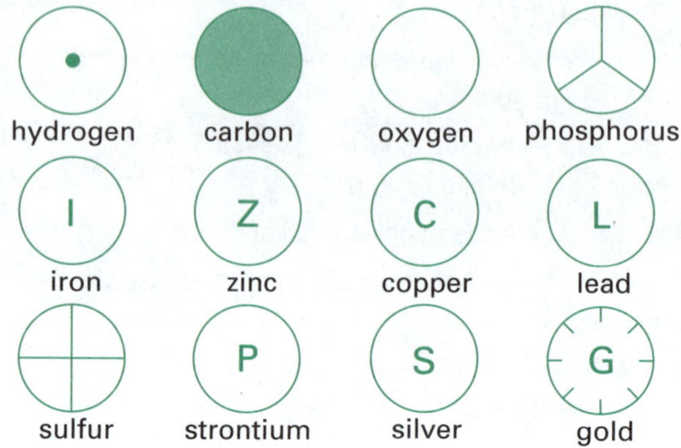

Figure 2.2 An example of the types of symbols used by Dalton to represent different atoms

J. J. Thomson

In 1897, the English physicist Joseph John Thomson (1856–1940) discovered the electron and in 1906 he suggested a model of the atom. This he assumed was a sphere of positive matter in which electrons are stationed by electrostatic forces. Thomson knew that electrons had a negative charge and thought that matter must have a positive charge. He therefore assumed the basic shape of an atom to be a spherical object containing a number of electrons confined in a homogeneous jelly-like, but relatively massive, positive charge distribution (see Figure 2.3). The total positive and negative charges would cancel each other out, since atoms are normally electrically neutral.

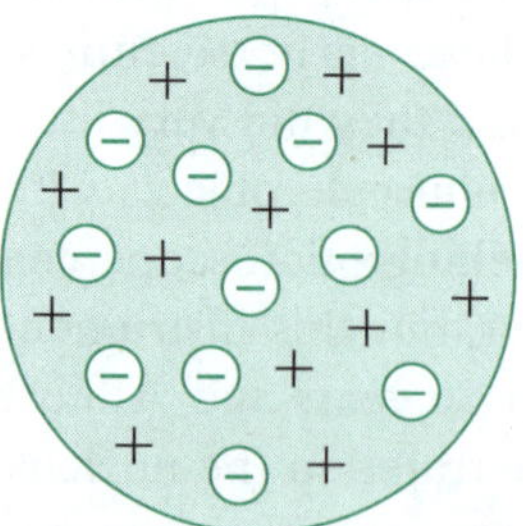

Figure 2.3 Thomson's atomic model looked like raisins stuck in and on the surface of a lump of pudding. This is sometimes known as the 'plum pudding' or 'raisin bread' model.

Ernest Rutherford

By the beginning of the 20th century, scientists had found that certain elements emitted fast-moving particles. One of these was called the alpha particle (α particle), which was heavy and positively charged.

In 1911 the New Zealand-born chemist and physicist Ernest Rutherford (1871–1937) bombarded atoms with these alpha particles, hoping that this experiment would uncover more details of the insides of the atom. Using radium as the source of the alpha particles, he directed them onto a gold foil. Surrounding the foil was a fluorescent screen where he could observe the impact of these particles. This is shown in Figure 2.4.

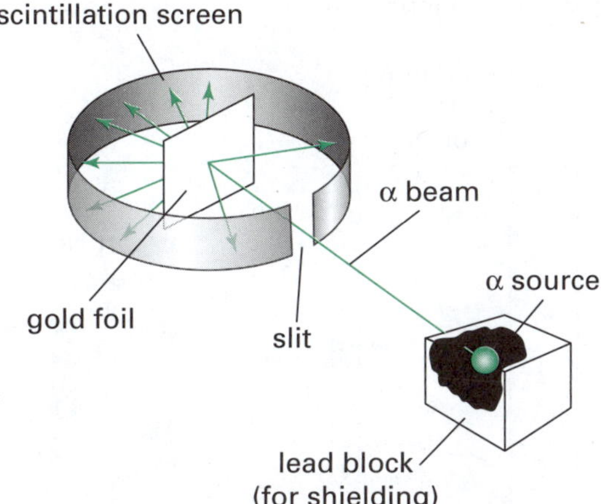

Figure 2.4 Rutherford beamed alpha particles through a gold foil, only 0.0004 mm thick. The alpha particles are detected as light flashes (scintillations) on a zinc sulfide screen.

He did not expect the results of the experiments he got. Most of the alpha particles went right through the foil with very little deviation. Only the occasional alpha particle veered sharply from its original course, sometimes bouncing straight back from the foil! Rutherford realised that the Thomson model can't be correct. The alpha scattering must be due to tiny concentrations of positively charged matter and surrounding these centres must be empty space (see Figure 2.5). He argued that electrons must exist somewhere within this space. Rutherford concluded that the negative electrons orbited these positive centres somewhat like the planets orbit the Sun in our solar system.

In 1919 Rutherford finally concluded that the nucleus of an atom contained discrete positive charges of matter. In an experiment with alpha particles, he knocked positive particles out of six different atoms: boron, fluorine, sodium, aluminium, phosphorus and nitrogen. He called these positive particles protons (from the Greek *protos* meaning 'first') as they were the first identified building blocks of the nuclei of all elements. He also discovered these protons had a mass 1836 times as great as the mass of the electron. Figure 2.6 shows the differences between Thomson's model and Rutherford's.

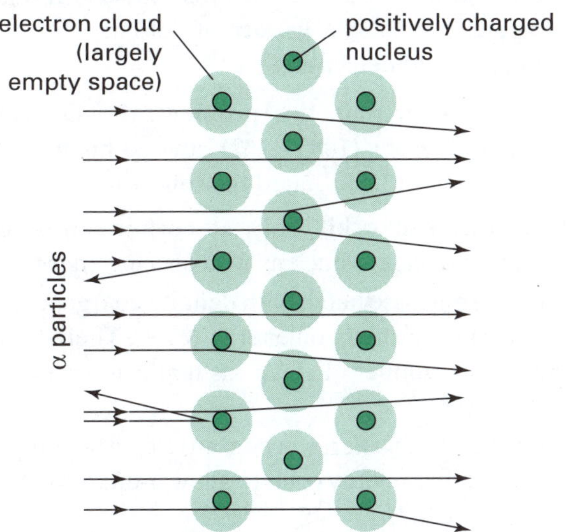

Figure 2.5 Rutherford's concept of the atom

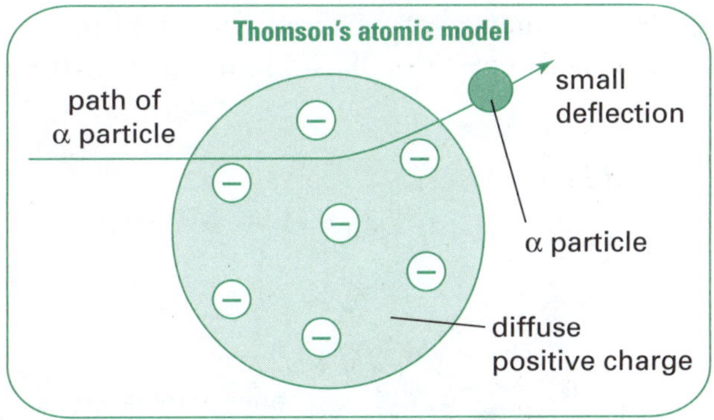

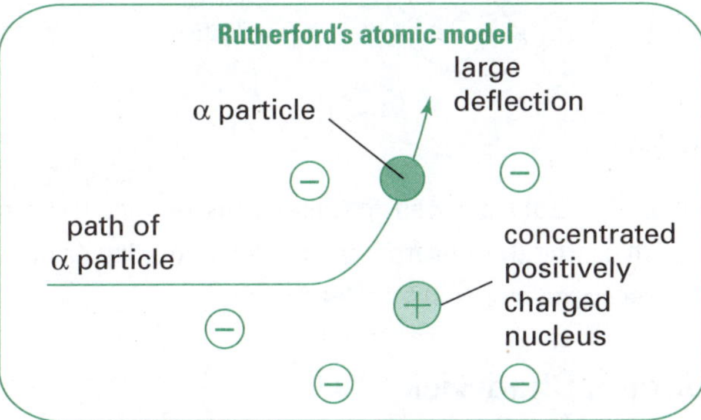

Figure 2.6 With the positive charge spread over a large volume, as in Thomson's model, the alpha particles would not have been deflected as greatly as Rutherford observed.

Niels Bohr

While Rutherford's model of the atom was an improvement, there seemed to be something wrong. Scientists had by now worked out that opposite charges attract each other. This would mean the negatively charged electrons should gradually lose energy and spiral inward towards the positive nucleus. And as these electrons lost energy, atoms should release this as a variety of colours. But no experiment could detect this.

In order to overcome this problem, the Danish physicist Niels Bohr (1885–1962) devised his model of the atom in 1913. He stated the following.

- Electrons can orbit only at certain allowed distances from the nucleus with specific energies.
- These orbits, associated with definite energies, are called energy shells or energy levels. There is a maximum number of electrons that can occupy a given orbit.
- Atoms radiate energy when an electron jumps from a higher-energy orbit to a lower-energy orbit (see Figure 2.7).
- Atoms absorb energy when an electron jumps from a low-energy orbit to a high-energy orbit.
- The frequency of this radiation emitted or absorbed depends on the orbits it is moving from and to.

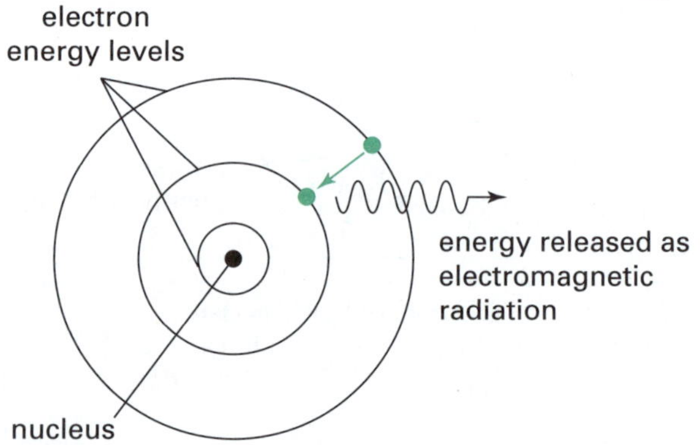

electron energy levels

energy released as electromagnetic radiation

nucleus

Figure 2.7 Bohr showed that electrons can move from a higher orbit (further from the nucleus) to a lower orbit (closer to the nucleus) by releasing energy.

James Chadwick

By 1931 scientists had found out that if very energetic alpha particles produced by polonium was aimed at certain light elements (such as beryllium, boron or lithium), an unusually penetrating radiation was produced. While it was thought this radiation was gamma radiation, it was more penetrating. Then in 1932, Irène Joliot-Curie and Frédéric Joliot showed that if this unknown radiation fell on paraffin, or some other hydrogen-containing compound, it ejected very high-energy protons.

In 1932, the British physicist James Chadwick (1891–1974) performed a series of experiments that showed these unknown rays were not gamma rays as they had the wrong properties. For example, they were not deflected in a magnetic field and could penetrate lead for several centimetres. He suggested that the new radiation consisted of uncharged particles of approximately the mass of the proton, and went on to prove this (see Figure 2.8).

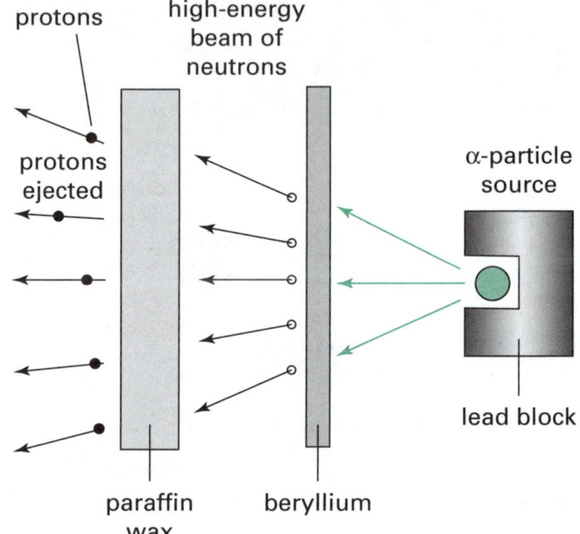

Figure 2.8 Chadwick showed that all the observations could be explained if the unknown radiation between beryllium and paraffin consisted of neutral particles of a mass about that of a proton.

This particle became known as the neutron because it didn't have an electric charge. The neutron is a subatomic particle with no net electric charge and a mass slightly larger than that of a proton. Except for hydrogen, atomic nuclei consist of protons and neutrons.

2.2 Modelling subatomic structure

Atoms are so small it takes a high-powered electron microscope to see just a fuzzy outline of them. A very small unit, the picometre (pm), is used to measure their diameters. One million million picometres equals one metre (1 000 000 000 000 pm = 1 m).

In drawing atoms, we often show them in two dimensions with the nucleus large, compared to where the electrons are. For example, if the proton in the nucleus of a hydrogen atom was enlarged to the size of a soccer ball, the electron would be around 30 km away! The size of an atom is difficult to describe because atoms have no definite outer boundary. But one thing is certain: the nucleus is much, much smaller than the size of the atom (see Figure 2.9).

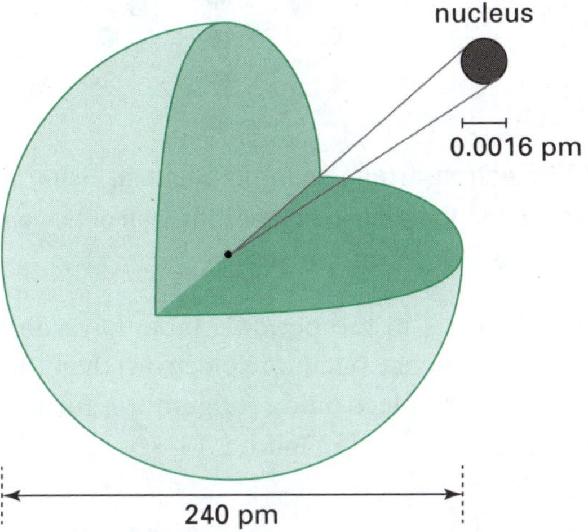

Figure 2.9 Comparing an atom to the size of its nucleus

The proton and neutron are over 1800 times heavier than the electron. The weight of an atom is therefore concentrated in the tiny volume of the nucleus. These subatomic particles have very small masses; so instead of grams they are measured in atomic mass units ('u'). Figure 2.10 shows a model of an atom of carbon.

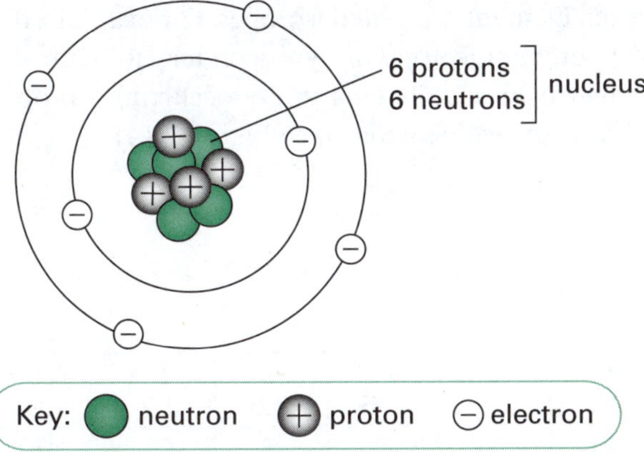

Figure 2.10 Both protons (positively charged) and neutrons (no charge) reside in the nucleus, with electrons (negatively charged) whizzing around the nucleus in distinct orbits.

For example, the approximate mass of a carbon atom can be calculated knowing that protons and neutrons have masses close to 1 u and that electrons have very small masses (~0.0005 u).

The carbon atom consists of 6 protons, 6 neutrons and 6 electrons so its total mass can be calculated:

6 protons = 6 u
6 neutrons = 6 u
6 electrons = 0.003 u

So the mass of one carbon atom is approximately 12 u.

Table 2.1 provides a comparison of the three main subatomic particles.

Table 2.1 The particles making up atoms

Particle	Charge	Mass (compared to electron)
proton	+1	1836
neutron	0	1838
electron	–1	1

Electron shells

Electrons are arranged around the nucleus of an atom in shells. While you can think of them like planets orbiting a central sun, they are more like 'clouds' of electric charge surrounding the nucleus.

The shells are numbered outwards from the nucleus ($n = 1$, $n = 2$ and so on; sometimes they are labelled K, L, M and so on.) The greatest number of electrons found in each shell can be calculated by $2n^2$. For example, the maximum number of electrons that can occupy the second shell ($n = 2$) is $2 \times 2^2 = 8$.

In the titanium atom shown in Figure 2.11, there are two electrons in the first shell, eight electrons in the second shell, 10 in the third shell and two in the fourth shell. This gives a total of 22 electrons, so there must be 22 protons in the nucleus to keep the atom neutral.

Electrons in outer shells travel farther from the nucleus and have higher average energy than those of shells closer in. The outermost electrons are more important in determining how the atom reacts chemically and whether or not the element is a conductor. This is because there is a weaker pull by the atom's nucleus and so the attraction holding the electrons can be broken more easily. Hence, the reactivity of a given element depends on its electron configuration.

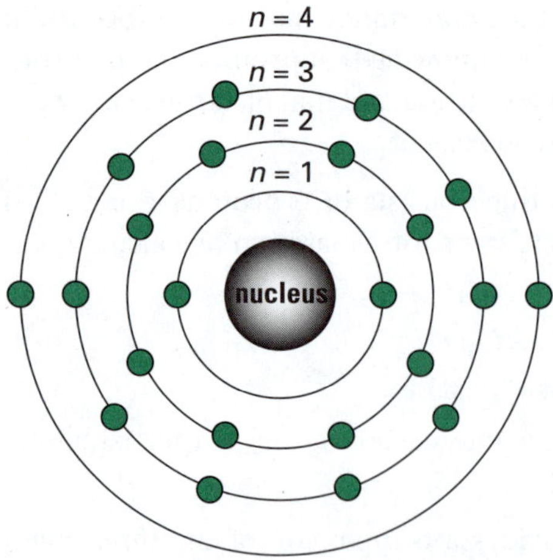

Figure 2.11 Electron diagram for titanium

Electron configuration

The number of protons in the nucleus of an atom of a given element is given by the atomic number (Z). This is also equal to the number of electrons in an atom. So, given the atomic number for an element, it is an easy matter to determine the number of electrons each atom has.

The electron configuration shows the number of electrons in each energy level. For example, looking on a periodic table, calcium has an atomic number of 20. This means there are 20 protons in its nucleus. It also means there are 20 electrons whizzing around in orbits. From the $2n^2$ rule given earlier, only two electrons can occupy the first (innermost shell), and only eight can be in the second shell. For the third shell, there is a maximum of $2 \times 3^2 = 18$ electrons possible, but in calcium only eight occupy this shell. This means that the remaining two electrons occupy the fourth (outermost shell).

So why doesn't the third shell completely fill before electrons are added to the next shell? The reason is that eight electrons (an octet) provides additional stability. This electron arrangement is shown in Figure 2.12. The electron configuration for calcium can be given as 2.8.8.2 where each number indicates the number of electrons in each shell, beginning from the first (innermost) shell.

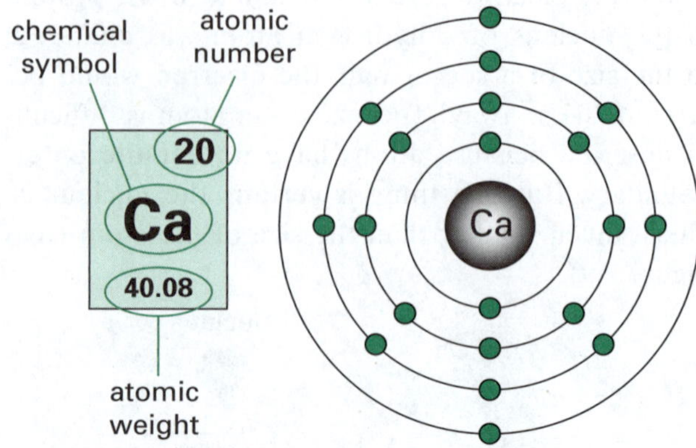

Figure 2.12 Electron arrangement in calcium. Using a periodic table and the atomic number for elements, you can work out how electrons are arranged in an atom.

Successive elements in the periodic table have one more proton (and hence one more electron) than the element before. The electronic configuration for the first 20 elements is given in Figure 2.13.

2.3 Isotopes and the atomic weight of elements

All the atoms of calcium have 20 protons in each of their nuclei. Similarly, all atoms of carbon contain six protons in each of their nuclei. But the number of neutrons may vary. The different possible varieties of each element are called isotopes. For example, the most common isotope of hydrogen has no neutrons, but hydrogen also has naturally occurring isotopes with one or two neutrons (see Figure 2.14).

H 1							He 2
Li 2.1	Be 2.2		B 2.3	C 2.4	N 2.5	O 2.6	F 2.7	Ne 2.8
Na 2.8.1	Mg 2.8.2		Al 2.8.3	Si 2.8.4	P 2.8.5	S 2.8.6	Cl 2.8.7	Ar 2.8.8
K 2.8.8.1	Ca 2.8.8.2	transition metals						

Figure 2.13 The electronic configuration for the first 20 elements on the periodic table

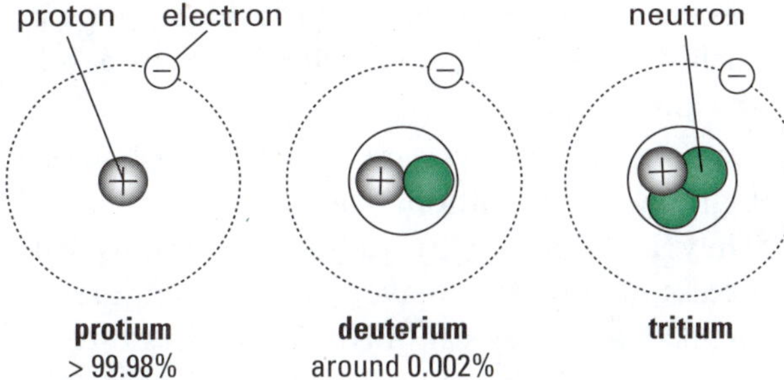

Figure 2.14 There are three naturally occurring isotopes of hydrogen: H-1, H-2 and H-3. Hydrogen is the only element that has different names for its isotopes in common use.

For any element there are given combinations of neutrons and protons, where the forces holding the nuclei together seem to work best. Light elements tend to have about the same number of neutrons as protons; heavy elements have a few more neutrons than protons in order to keep the nucleus together. Atoms with a few too many or too few neutrons can exist for a while, but they are unstable. For example, while H-1 and H-2 are stable, H-3 is radioactive and decays. So H-3 (tritium) can be called a radioisotope.

Some terms will now be defined.

- Atomic number (Z): the number of protons in the nucleus of an atom. (This was defined earlier.)
- Mass number (A): the number of protons + neutrons in the nucleus.
- The number of neutrons in the nucleus of an atom is therefore:

  neutron number = mass number – atomic number
  $= A - Z$
- Each element can be represented symbolically using the atomic number and mass number. For element X, the symbol is written as ${}^{A}_{Z}X$.

For example, the element sodium has an atomic number of 11 and a mass number of 23. This information can be represented as ${}^{23}_{11}Na$. It can also be written as Na-23 where the atomic number (11) is understood since all sodium atoms contain 11 protons (see the periodic table at the front of this book). The number of neutrons in a sodium atom is therefore 23 – 11 = 12.

Atomic weight

The weights of individual atoms of an element are very small. Consequently, chemists refer to the relative atomic weight of an element. This weight is compared to a standard which is the carbon-12 isotope. By setting its atomic weight as 12 atomic mass units (exactly), the relative atomic weights of all other elements can be compared. Because elements consist of mixtures of isotopes, the relative atomic weight (which is an average) is not a whole number. For example, sulfur: 32.06, chlorine: 35.45, calcium: 40.08.

2.4 Radioactivity

Radioactivity refers to the particles and radiation that are emitted from nuclei as a result of instability. Carbon, for instance, has a number of isotopes, many of which are radioactive. Three of the most important ones are C-12, C-13 and C-14. C-12 and C-13 are stable. The longest-lived radioisotope is C-14 with a half-life of 5730 years. All other radioisotopes of carbon have half-lives under half an hour, most less than a minute.

The half-life is the time it takes for a substance undergoing decay to decrease by half (see Figure 2.15). For example, beryllium-11 has a half-life of 13.81 seconds. Suppose you started with 12 g of Be-11. After 13.81 seconds, ½ × 12 = 6 g is left. After another 13.81 seconds you are left with ½ × ½ × 12 = 3 g of Be-11; after a further 13.81 seconds you have ½ × ½ × ½ × 12 = 1.5 g ... and so on.

Half-lives can range from more than a billion years for some nuclei to less than a thousandth of a second for others. Much of the decay occurs in the early stages when there are many atoms, with less decaying when there aren't so many. The half-life always remains the same, but the half gets smaller and smaller.

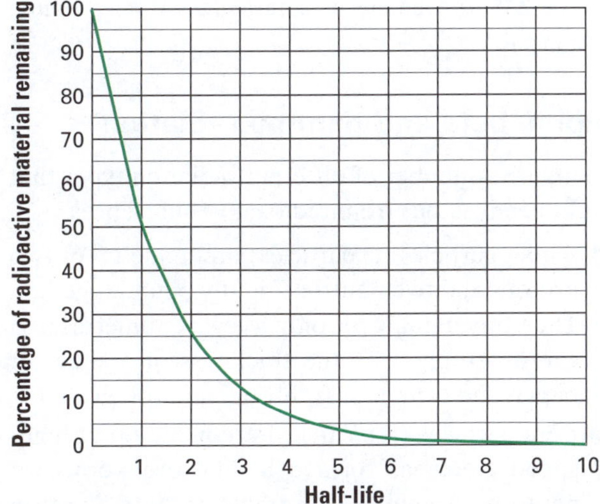

Figure 2.15 The half-life is the amount of time it takes for half of the atoms in a sample to decay. The half-life for a given isotope is always the same.

Unstable nuclei

But what makes an atom radioactive? Not all nuclei are stable. Radioactive elements have unstable nuclei. An unstable nucleus will break down, or decay, into a more stable nucleus. As the nucleus breaks down, it will emit radiation in the form of particles or rays.

The nucleus contains protons (positively charged) and neutrons (no charge). Electromagnetic forces push the protons apart because they have the same charge. But there are strong nuclear forces that pull the protons and neutrons together. So a nucleus is only stable if it has the right balance of protons and neutrons. An unstable nucleus will sooner or later change its state by undergoing radioactive decay.

Americium-241 is radioactive as it has too many protons. It decays producing neptunium, with the release of an alpha particle. It has a half-life of 432.7 years. This is shown in Figure 2.16. How quickly a radioactive element decays is measured in terms of its half-life.

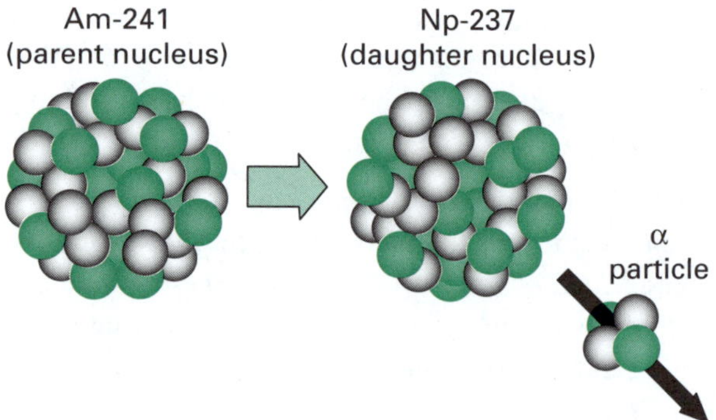

Figure 2.16 The decay of americium-241 to produce neptunium

Alpha, beta and gamma radiation

There are a number of different radiation types that could be given out from radioactive substances.

- Alpha particles (α particles) are made up of two protons and two neutrons tightly bound together. They travel in air for only a few centimetres and can be stopped by the thickness of a sheet of paper, and cannot pass through unbroken skin.
- Beta particles (β particles) are made up of high-speed electrons formed by the breakdown of neutrons. A neutron can transform into a proton and an electron. While the proton stays in the nucleus, the electron (beta particle) is ejected with a lot of energy. Beta particles can travel a few metres through air and can be stopped by such objects as a sheet of aluminium or a 1-cm thickness of wood.
- Gamma rays (γ rays) are not particles but very high-energy radiation. They are not charged. They are emitted by radioactive atoms with excess energy. They have energies even greater than X-rays. They can travel kilometres through air. They can be stopped by such objects as thick concrete blocks, or at least 2-cm to 3-cm thicknesses of lead. Gamma rays are often released together with alpha or beta particles.

Figure 2.17 shows the differences in the penetrating abilities of alpha, beta and gamma rays.

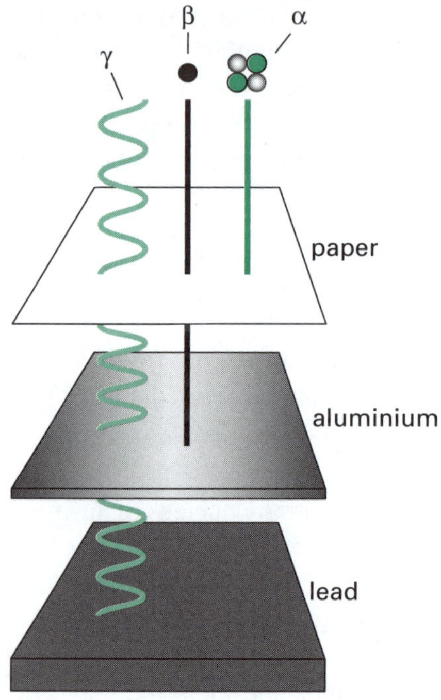

Figure 2.17 The penetrating ability of radiation

Figure 2.18 shows how alpha, beta and gamma rays behave in electric and magnetic fields.

Ernest Rutherford and Marie and Pierre Curie investigate radioactivity

Ernest Rutherford, together with Frederick Soddy, performed a series of experiments which showed that elements such as uranium and thorium became different elements (i.e. they were transmuted) during radioactive decay. This was an incredible idea at the time since scientists had by now established the alchemists' notion of changing cheap metals into gold or other precious metals was fake. He further went on to find the following.

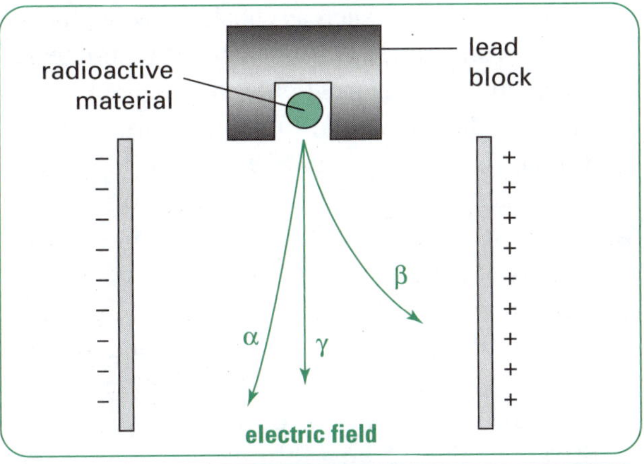

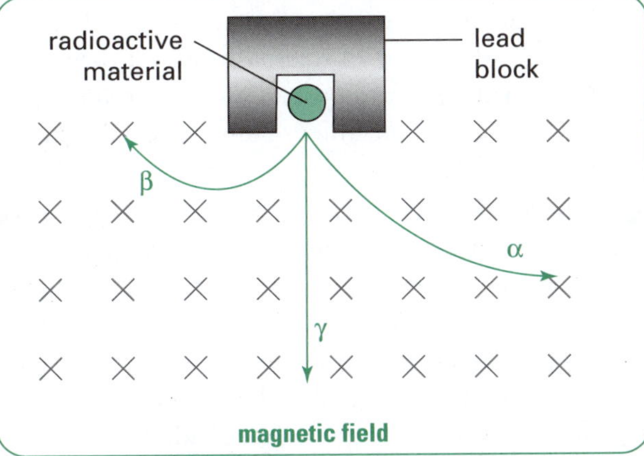

Figure 2.18 Alpha particles are positively charged, and beta particles are negatively charged, so they are affected by electric and magnetic fields. Gamma rays are not charged and so are unaffected by these fields.

- In 1899 he described two distinct types of radiation emitted by these nuclei on the basis of their penetrating power: alpha rays and beta rays, terms that he invented.
- A sample of radioactive material took the same amount of time for half the sample to decay; its 'half-life'.
- In 1903 he realised that a new type of radiation emitted from radium because it had a very much greater penetrating power. He called this third type of radiation the gamma ray.
- Together with Hans Geiger, he developed ionisation chambers and zinc sulfide scintillation screens to count alpha particles. He established that the alpha particle has two positive charges.
- In 1917 he was the first person to deliberately transmute one element into another, when he changed nitrogen into oxygen using a nuclear reaction:

  $$\text{N-14} + \alpha \rightarrow \text{O-17} + \text{proton}$$
- In 1921 he came up with the idea of a new particle, the neutron, which could somehow compensate for the repelling effect of the proton's positive charge. (Neutrons were discovered later by James Chadwick, in 1932.)

Rutherford won the 1908 Nobel prize in chemistry 'for his investigations into the disintegration of the elements, and the chemistry of radioactive substances'. However, he did much of his important work after this date.

Marie Curie (1867–1934) was a Polish-born (maiden name Sklodowska) French physicist and chemist who was famous for her pioneering research on radioactivity, a term that she coined. Pierre Curie (1859–1906) had already established himself as a scientist when he turned his attention to radioactive substances. These studies were made together with his wife, Marie, whom he married in 1895. Their achievements include the following.

- A theory of radioactivity. Using an electrometer (a sensitive device to measure electric charge invented by Pierre), they discovered that the air around a sample of uranium conducted electricity. They worked out that the activity of the uranium compounds depended only on how much uranium was present. They figured out that the radiation came from the atom itself and was not some molecular interaction. In 1903 Marie and Pierre, along with Henri Becquerel, were awarded the Nobel prize in Physics.
- They developed techniques for isolating radioactive isotopes.
- They discovered of two elements, polonium and radium, for which Marie received the 1911 Nobel prize in Chemistry.

Since these discoveries, many other radioactive elements have been discovered or artificially produced. A number have been put to use in various ways ranging from medical to industrial.

Uses of radioactivity in medicine and industry

Since their discovery over a century ago, a number of uses have been found for radioactive isotopes. These uses include the following.

- Medical instruments are sterilised using radioactive isotopes. Penetrating radiation, such as electron beams, X-rays, gamma rays or subatomic particles are used to kill all forms of life, including fungi, bacteria, viruses and spores that might exist on the surface.

- Gamma rays can be used to kill insects and microorganisms such as bacteria and mould in food. While this process prolongs the shelf life of the food, it sometimes changes its taste but it does not make food radioactive. Food irradiation acts by damaging the target organism's DNA beyond repair.
- Radioactive isotopes are used to kill cancer cells inside the body (radiotherapy).
- X-rays are used to look at bone structure and for detecting some diseases in soft tissue (see Figure 2.19). Examples include chest X-rays for lung diseases such as pneumonia, lung cancer or pulmonary oedema; abdominal X-rays for detecting intestinal obstruction, and sometimes gallstones and kidney stones. In dentistry X-rays are used to diagnose common oral problems, such as cavities.

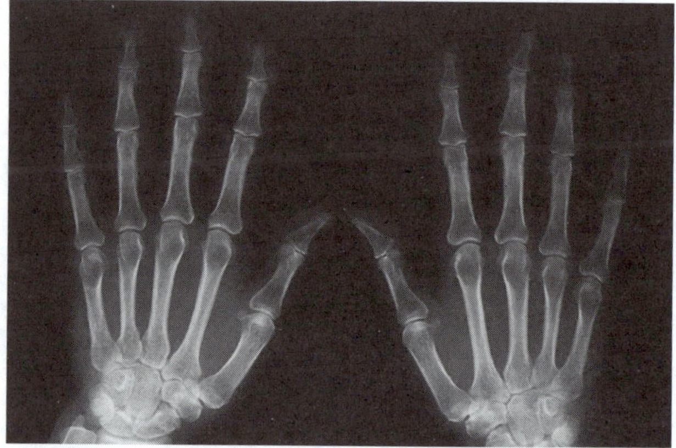

Figure 2.19 X-rays are a form of invisible, high-energy electromagnetic radiation. Their most obvious use is to image bones.

- X-ray imaging is used to investigate paintings which have been painted over. Satellites with X-ray detectors on them perform X-ray astronomy. X-ray technology is used by customs agents to detect both the movement of prohibited and restricted goods from large shipping containers to small envelopes.
- Nuclear power stations use uranium in fission reactions as a fuel. Heat released during the fission process generates steam, which turns a turbine to produce electrical energy.
- A tracer is a radioactive element whose pathway can be followed. Radioisotopes can be used for medical purposes, such as checking for a blocked or damaged kidney. A small amount of iodine-123 is injected into the patient, and its path traced through the kidneys using a radiation counter. Similarly, radioactive iodine-131 can be used to study the function of the thyroid gland.
- Tracers can be used to detect leaks in underground water and gas pipes.
- Radioactive isotopes are used in carbon dating, determining the age of ancient documents, wooden objects or other substances containing carbon. Animals and plants have a known proportion of radioactive carbon-14 in their tissues. When they die they stop taking in carbon, and so the amount of carbon-14 decreases at a known rate. By measuring the amount of carbon-14 that is left, the age of the ancient organic materials can be determined.
- Smoke alarms contain a weak source of americium-241 that emits alpha particles. Any smoke entering the device is detected and an alarm sounds. (While smoke detectors provide some radiation exposure, it is very low. The health risk reduction from the fire protection vastly outweighs the health risk from radiation.)
- In paper mills, the thickness of paper can be controlled by measuring how much beta radiation passes through the paper to a Geiger counter (see Figure 2.20).

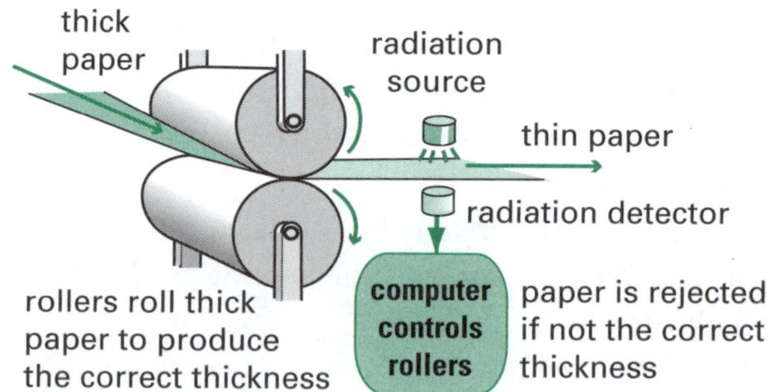

Figure 2.20 Measuring paper thickness

While useful in industry and medicine, radioactive isotopes can also be dangerous.

- Too much radiation can cause cancer in living tissue.
- Radioactive wastes causes problems, as these waste materials remain radioactive and dangerous to humans and animals for hundreds, and sometimes thousands, of years.

Test yourself 1

Part A: Knowledge

1. Which of the following scientists would have made the following statement?

   'The atoms of the elements consist of a number of negatively electrified corpuscles enclosed in a sphere of uniform positive electrification.' *(1 mark)*

   A Dalton
   B Thomson
   C Rutherford
   D Chadwick

2. In an atom of oxygen, there are eight protons and eight electrons. How many electrons does a neutral oxygen atom have in its second shell? *(1 mark)*

   A 2
   B 4
   C 6
   D 8

3. Which shell in the atom shown in Figure 2.21 is *not* full? *(1 mark)*

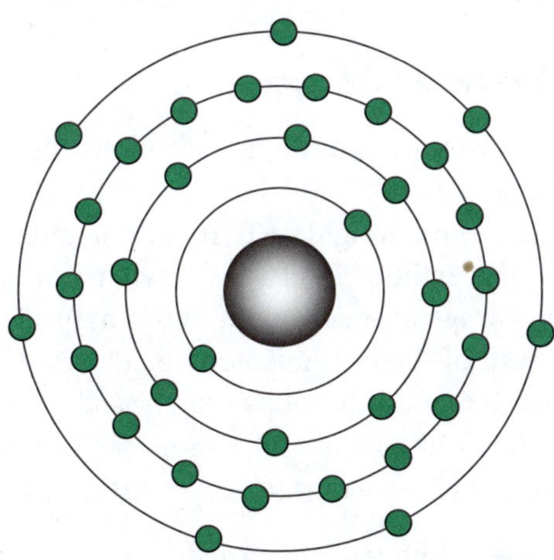

Figure 2.21 Shell structure of an atom

   A $n = 1$
   B $n = 2$
   C $n = 3$
   D $n = 4$

4. This scientist bombarded gold foil with alpha-particles then stated: 'It was as incredible as if you had fired a 15-inch shell at a piece of tissue paper and it came back and hit you'. The scientist is *(1 mark)*

   A Dalton.
   B Thomson.
   C Rutherford.
   D Chadwick.

5. Select the correct statement about the substructure of the atom. *(1 mark)*

   A An electron is a negative particle of similar mass to a neutron.
   B Neutrons and protons are located inside the nucleus of an atom.
   C The number of neutrons equals the number of protons in all atoms.
   D Protons and neutrons are nuclear particles with identical masses.

6. Complete the following restricted-response questions using the appropriate word. *(1 mark for each part)*

   a) James Chadwick discovered a particle in the atom called the ....
   b) .... was the first person to propose a set of symbols to represent the elements.
   c) All elements, apart from hydrogen, have .... electrons in their first (or K shell).
   d) Carbon-14 is a .... isotope of carbon-12.
   e) The ancient Greek thinker, ..., maintained the impossibility of dividing things *ad infinitum* (forever).

7. Use the code letters to match the terms or phrases in each column. *(1 mark for each part)*

Column 1	Column 2
A alpha particle	F negative
B Curie	G neutral
C beta particle	H positive
D gamma ray	I polonium
E neutron	J electromagnetic

Part B: Skills

8. Figure 2.22 shows the nucleus of an atom.

   a) How do you know whether or not the nucleus shown is radioactive? *(2 marks)*

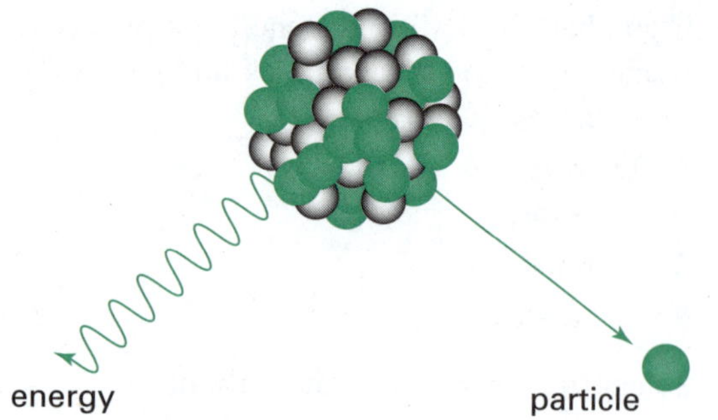

Figure 2.22 Nucleus

b) What might this energy be? Give one feature of it. *(2 marks)*

c) What might this particle be? Give one feature of it. *(2 marks)*

9. Write the electron configuration for the following elements.

a) Nitrogen (Z = 7) *(1 mark)*

b) Magnesium (Z = 12) *(1 mark)*

c) Argon (Z = 18) *(1 mark)*

10. Describe how radioactivity can be used to determine the thickness of a stack of paper (see Figure 2.23). *(2 marks)*

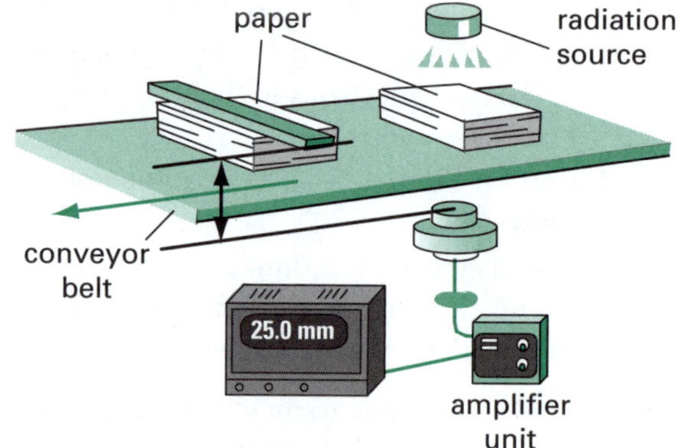

Figure 2.23 Thickness of paper

11. Copy and complete Table 2.2. *(12 marks)*

12. For each of the following elements, use the information provided to determine the number of protons and the number of neutrons. *(2 marks each)*

a) Isotope X-31; electron configuration: 2.8.5

b) Isotope Y-60; electron configuration: 2.8.15.2

13. Use the following symbols to determine the number of protons, neutrons and electrons in each element. Then use a periodic table to name each element. *(4 marks each)*

a) $^{207}_{82}\text{Pb}$

b) $^{222}_{86}\text{Rn}$

c) $^{99}_{43}\text{Tc}$

14. Figure 2.24 shows a sample of atoms of the element boron (B). Boron consists of a mixture of isotopes of atomic weight 10 u and mass 11 u.

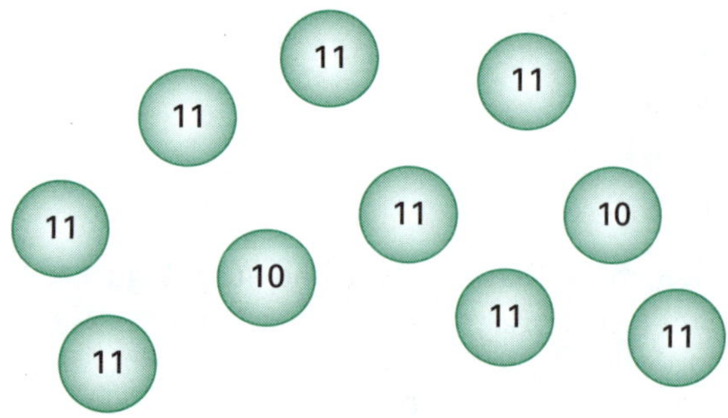

Figure 2.24 Isotopes of boron

a) Which isotope is in the greater abundance? *(1 mark)*

b) If the atomic weight of B-10 is 10 u and the atomic weight of B-11 is 11 u, what would the average atomic weight of the natural mixture of the boron isotope be closest to? Choose one of the following. *(1 mark)*

i) 10.0 u **ii)** 10.5 u **iii)** 10.8 u **iv)** 11.0 u

c) Why did you make that choice? *(2 marks)*

Go to pp. 231–232 to check your answers.

Table 2.2 Properties of some elements and subatomic particles

Element	Symbol	Atomic number	Mass number	Number of neutrons	Number of protons	Number of electrons
Oxygen		8	16			
	C		12	6		
	F			10		9

2.5 Reactions and chemical equations

Making accurate observations is an important skill when performing chemical reactions. Some events during a reaction cannot be detected using our senses, while others are easy to observe. The following are some common observations:

- a substance dissolves
- effervescence occurs (i.e. gas bubbles form)
- the reaction mixture changes colour
- the reaction mixture gets hot or cold
- an insoluble substance (or precipitate) forms when solutions are mixed
- flames are produced or an explosion occurs.

Observations should be consistent with the chemical word equation for the reaction.

Word equations

A chemical equation summarises the events of a chemical reaction.

- The reacting chemicals (reactants) are written first.
- The new substances that form (products) are written last.
- An arrow separates the reactants and products. The arrow stands for the phrase 'reacts to form'.

Chemical equations can be written as a word equation.

Let's examine some simple cases.

Example 1: Burning magnesium in air

Observation: Silvery magnesium is heated in a Bunsen burner flame. The magnesium burns with a dazzling bright white light and a crumbly white powder is formed.

Explanation: The hot magnesium atoms react with oxygen molecules in the air to form a white compound called magnesium oxide.

Word equation:

magnesium + oxygen → magnesium oxide

Example 2: Reaction of sodium and chlorine

Observation: Silvery sodium metal is heated until it forms a pool of molten sodium. The hot sodium is placed in a gas jar of yellow-green chlorine gas. A white smoke of very fine crystals is seen to form.

Explanation: The hot atoms of sodium react with chlorine molecules to form white crystals of the compound called sodium chloride.

Word equation:

sodium + chlorine → sodium chloride

Naming products

In each of these examples the name of the compound can be determined using naming rules.

The rules for reactions between metals and non-metals are as follows.

- Name the metal first.
- Name the non-metal second.
- Delete the last few letters of the non-metal's name and substitute '-ide'.

For example:

1. reactants: zinc; sulfur
   product: zinc sulfide
2. reactants: calcium; bromine
   product: calcium bromide.

The rules for reactions between non-metals are as follows.

- Name the non-metal with the lower periodic group number first. Periodic groups are the 18 vertical columns of the periodic table. They are numbered left to right on the periodic table.
- Name the non-metal with the higher periodic group number second.
- Use Greek prefixes to indicate the number of each type of atom (mono = 1; di = 2; tri = 3; tetra = 4; penta = 5).
- Delete the last few letters of the second non-metal's name and substitute '-ide'.

When non-metals react together, often one or more products can form. For example:

1. reactants: carbon; oxygen
   products: carbon monoxide (CO) or carbon dioxide (CO_2)
2. reactants: phosphorus; chlorine
   products: phosphorus trichloride (PCl_3) or phosphorus pentachloride (PCl_5)

Chemical reactions between metals do not normally occur. Metals can combine to form mixtures called alloys.

Modelling chemical reactions

Chemical reactions can be modelled using atomic model kits. These consist of plastic or wooden balls of different colours and sizes which represent the different types of atoms. Atoms can be joined together with plastic or metal connectors to show the structure of the compound formed. These connectors represent chemical bonds linking atoms.

Example 1: Hydrogen gas + oxygen gas → water

Hydrogen gas consists of diatomic molecules (i.e. two atoms of hydrogen joined together). Oxygen gas is also diatomic. Water molecules are triatomic with two hydrogen atoms joined to one oxygen atom (see Figure 2.25). In a chemical reaction the number of atoms of each element in the reactants must equal the number of these atoms in the products.

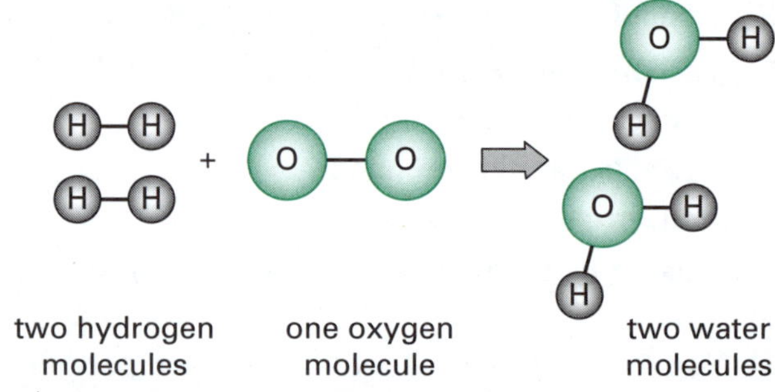

Figure 2.25 Hydrogen gas reacts with oxygen gas to form water.

In this example, we need to check that there is a balance of atoms on each side of the arrow.

Reactants: 4 atoms of H + 2 atoms of O
Products: 4 atoms of H + 2 atoms of O

Thus, the atoms are conserved.

Example 2: Nitrogen gas + hydrogen gas → ammonia

Nitrogen gas and hydrogen gas are both diatomic molecules. Ammonia molecules are tetra-atomic with three hydrogen atoms joined to a nitrogen atom (see Figure 2.26). The name of the product molecule is named ammonia rather than nitrogen trihydride for historical reasons.

In this example, we need to check that there is an atom balance.

Reactants: 6 atoms of H + 2 atoms of N
Products: 6 atoms of H + 2 atoms of N

Thus, the atoms are conserved.

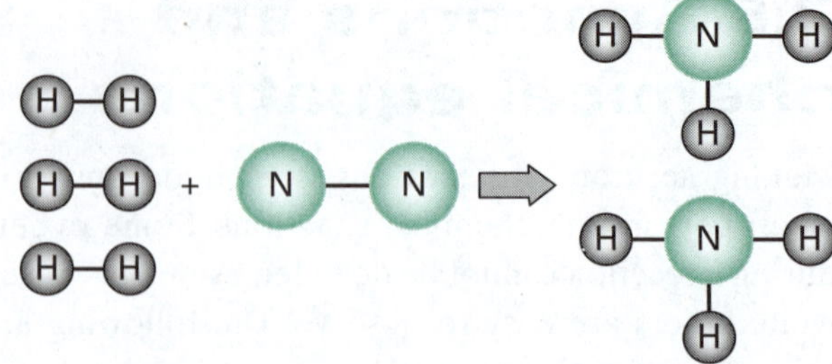

Figure 2.26 Hydrogen gas reacts with nitrogen gas to form ammonia.

Symbolic equations

The models of the reactions presented in the previous section allow us to write balanced equations using chemical symbols.

Example 1: Reaction of hydrogen and oxygen

The model of this reaction in Figure 2.25 showed that two molecules of hydrogen combine with one molecule of oxygen to form two molecules of water. The balanced symbolic equation is:

$$2H_2 + O_2 \rightarrow 2H_2O$$

The coefficients in front of each formula represent the number of each of these molecules.

Example 2: Reaction of nitrogen gas and hydrogen gas

The model of this reaction in Figure 2.26 showed that one molecule of nitrogen combined with three molecules of hydrogen to form two molecules of ammonia. The balanced symbolic equation is:

$$N_2 + 3H_2 \rightarrow 2NH_3$$

2.6 Mass conservation

The French chemist Antoine Lavoisier (1743–1794) performed the following experiment in the late 18th century. He placed some red mercury oxide in an evacuated glass vessel. The tap was closed and the vessel heated. The red substance broke down to form mercury and a colourless gas that Lavoisier named oxygen. He weighed the vessels before and after the reaction and found that there was no change in mass. This experiment is shown in Figure 2.27.

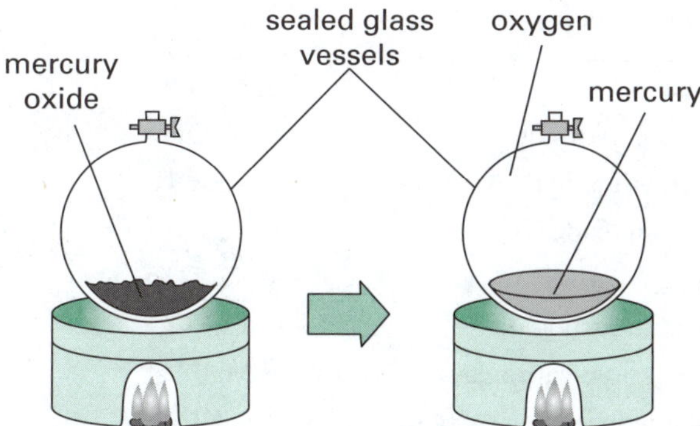

Figure 2.27 Lavoisier's experiment

Lavoisier found that 217 g of the red oxide of mercury produced 201 g of mercury (Hg). He concluded that 16 g of oxygen had been formed as the mass of the apparatus had not changed. This observation led to the law of mass conservation in a chemical reaction.

Law of mass conservation

In a chemical reaction, the mass of reactants equals the mass of products.

In the early 19th century, John Dalton used the law of mass conservation to develop his atomic theory. Dalton knew that hydrogen was the lightest element and he assigned the atomic weight of 1 unit to this element. He then set about performing various chemical reactions to measure the combining weight of elements. Some of these experiments produced inaccurate results for the atomic weights. It was not until much later in the 19th century that more accurate atomic weights were measured. These atomic weights are listed in the periodic table on the inside cover of this book.

Experiment 1

Combining weights of elements

Aim

- To determine the combining weights of carbon and oxygen and deduce the chemical formula of the product

Method

1. Twelve grams of carbon is burnt in oxygen in a closed container. The mass of oxygen is increased until all the carbon has combusted and no oxygen remains.
2. The product gas is collected and weighed.

Results

Mass of carbon = 12 g

Mass of gaseous product = 28 g

Analysis

Analyse these results and predict the possible formula of the gaseous product. Use your prediction to determine the atomic weight of oxygen if the atomic weight of carbon is 12 units.

Go to p. 231 to check your answer.

Conclusion

Write a suitable conclusion.

Go to p. 231 to check your answer.

Modelling mass conservation

The known atomic weights of elements together with atomic models can be used to demonstrate the law of mass conservation.

Example: Decomposition of copper carbonate

Copper carbonate is a green solid that will decompose when heated strongly to form carbon dioxide gas and black copper oxide powder. Figure 2.28 shows an atomic model of this reaction.

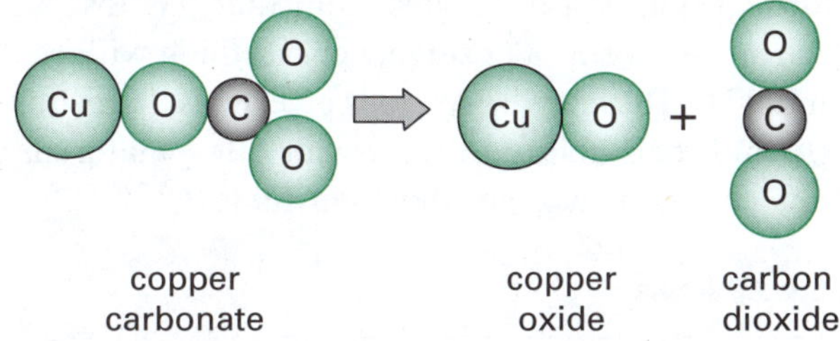

Figure 2.28 Decomposition of copper carbonate

The atomic weights of each element are listed below:

- copper, Cu = 63.5
- carbon, C = 12.0
- oxygen, O = 16.0.

The mass of the copper carbonate reactant is:

$$63.5 + 12.0 + 16.0 + 16.0 + 16.0 = 123.5$$

The total mass of the two products is:

$$(63.5 + 16.0) + (12.0 + 16.0 + 16.0) = 123.5$$

Thus, mass is conserved in this decomposition reaction.

2.7 Reactions of acids

The ancient Egyptians, Greeks and Romans recognised that some common substances, including lemons, limes, sour milk and vinegar, tasted sharp or sour. According to legend, Cleopatra dissolved pearls in wine vinegar and drank the solution that resulted. The eighth-century Arabian chemist Jabir prepared some liquids that we use in industry and the school laboratory today. These included nitric acid and sulfuric acid.

Today we classify compounds as being acidic not by their taste but by the presence of replaceable hydrogen atoms in their structures. Hydrochloric acid and sulfuric acid are two common acids used in the laboratory. The structure of these compounds is shown in Figure 2.29.

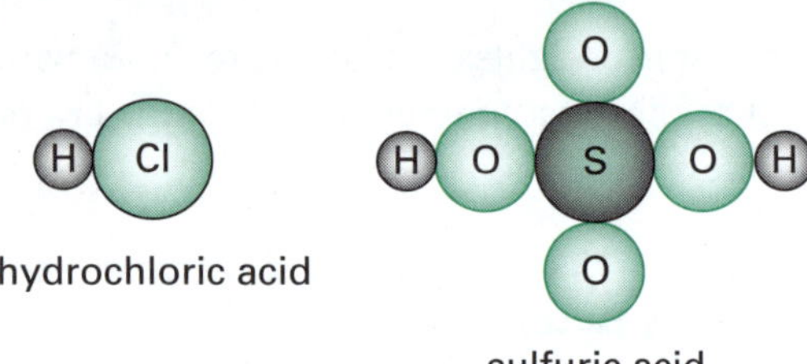

Figure 2.29 Hydrochloric acid and sulfuric acid

When these acids are added to water, the hydrogen atoms break free and form hydrogen ions. The hydrogen ion is a hydrogen atom that has lost an electron and forms a positively charged ion with the symbol H^+. These hydrogen ions give acids their sour taste and their ability to react with other materials. Some common acids are shown in Table 2.3.

Table 2.3 Acids

Acid	Formula
hydrochloric acid	HCl
sulfuric acid	H_2SO_4
nitric acid	HNO_3
carbonic acid	H_2CO_3
acetic acid	CH_3COOH

Bases are substances that can react with acids and destroy their properties. Many soluble bases dissolve in water to release hydroxide ions (OH^-) and such solutions are referred to as alkaline. Some common bases are shown in Table 2.4. Table 2.5 shows some common acidic and basic substances.

Table 2.4 Bases

Basic	Formula
sodium hydroxide	$NaOH$
ammonium hydroxide	NH_4OH
calcium oxide	CaO
sodium carbonate	Na_2CO_3
calcium carbonate	$CaCO_3$
magnesium carbonate	$MgCO_3$
sodium hydrogen carbonate	$NaHCO_3$

Table 2.5 Common acidic and basic substances

Common acidic substances	Common basic substances
car battery acid	caustic soda (drain cleaner)
vinegar (acetic acid)	ammonia solution (window cleaner)
lemon juice	laundry detergents
soft drinks	baking soda
soda water	limestone
urine	lime

Neutralisation

Neutralisation reactions involve acids and bases. When a base neutralises an acid, a solution of a 'salt' is formed. In most neutralisation reactions, water is formed and in some cases gases are also released.

- A 'salt' is an ionic compound formed during the neutralisation reaction. Many salts are water soluble but some are not.
- Soluble salts can be recovered from the solution by evaporating the water.
- Neutralisation reactions can be used to relieve indigestion. Weak bases such as sodium hydrogen carbonate or magnesium hydroxide in antacid tablets neutralise excess stomach acid.
- Salts are named after the acid that is used:
  - hydrochloric acid forms chlorides
  - sulfuric acid forms sulfates
  - nitric acid forms nitrates
  - carbonic acid forms carbonates
  - phosphoric acid forms phosphates.

The general word equation for neutralisation reactions involving oxide and hydroxide bases is:

$$\text{acid} + \text{base} \rightarrow \text{salt} + \text{water}$$

Figure 2.30 models a typical neutralisation reaction.

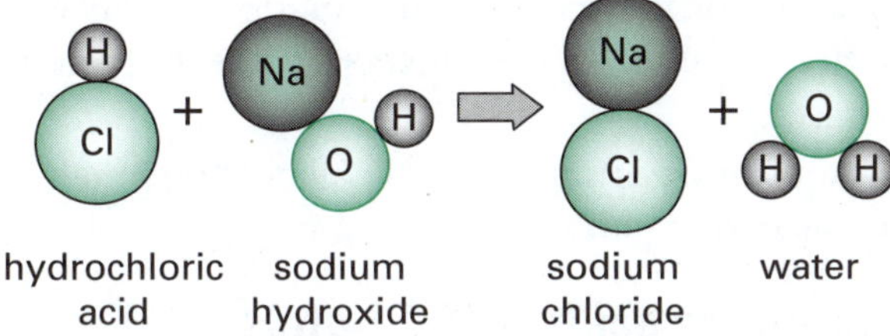

Figure 2.30 Neutralisation model

Example 1: Neutralisation of an oxide

Hydrochloric acid was added to solid calcium oxide in a beaker. The calcium oxide dissolved to form a solution of calcium chloride. The solution becomes hot due to the neutralisation reaction. The word equation for this neutralisation reaction is:

hydrochloric acid + calcium oxide
→ calcium chloride + water

In this example the calcium chloride salt is dissolved in the water and it can be recovered by evaporation or crystallisation.

Example 2: Neutralisation of a hydroxide

If sulfuric acid is added to potassium hydroxide solution, no visible change occurs as reactants and products are all colourless. There is a chemical reaction and a significant rise in temperature occurs.

The products of the reaction are potassium sulfate (a soluble salt) and water.

sulfuric acid + potassium hydroxide
→ potassium sulfate + water

Example 3: Neutralisation of a carbonate

If nitric acid is added to green copper carbonate powder, there is considerable fizzing as a gas is evolved. The final solution is a blue colour. The products of this neutralisation reaction are copper nitrate (a blue, soluble salt), carbon dioxide gas and water. The solution also becomes quite hot. The blue salt, copper nitrate, can be recovered from the solution by crystallisation.

nitric acid + copper carbonate
→ copper nitrate + water + carbon dioxide

The carbon dioxide released can be tested for using the limewater test. Limewater is a colourless solution of calcium hydroxide in water. If carbon dioxide is bubbled through the limewater, it turns a milky white.

Using indicators to observe neutralisation

Indicators are substances that change colour in the presence of an acid or base. They can be prepared easily from plant material such as flower petals. The petals can be soaked in water to dissolve the dye. The dye extract changes colour if acids or bases are added. These types of indicators are not stable for long-term use.

Indicators can often be used to determine whether a solution is acidic, neutral or basic. This is useful in the laboratory as well as domestically. Indicators are used to detect whether the water in a swimming pool is too acidic or basic.

Indicators can often be used in neutralisation reactions to determine the point where exact neutralisation occurs. This is important, especially when the reacting solutions are colourless.

Table 2.6 shows examples of some common indicators used in the school laboratory.

Table 2.6 Useful indicators

Indicator	Colour in acidic solution	Colour in water (neutral)	Colour in basic solution
litmus	red	purple	blue
methyl orange	red	orange	yellow
bromothymol blue	yellow	green	blue
phenolphthalein	colourless	colourless	crimson

Example: Testing solutions with indicators

Table 2.7 shows the results of testing various solutions with three indicators.

Table 2.7 Testing solutions with indicators

Solution	Methyl orange	Phenolphthalein	Bromothymol blue
X	orange	colourless	green
Y	yellow	pink	blue
Z	orange-red	colourless	yellow-green

From these results we can conclude that:

- X is a neutral solution
- Y is a slightly basic solution (the phenolphthalein is pink rather than crimson)

- Z is a slightly acidic solution (the bromothymol blue is yellow-green rather than yellow; the methyl orange is orange-red rather than red).

How acidic or basic a solution is can also be represented by a number on the pH scale (see Figure 2.31).

- pH is a measure of the acidity or basicity of a solution.
- Neutral solutions have a pH of 7.
- Acidic solutions have a pH less than 7.
- The lower the pH the more acidic is the solution.
- Strongly acidic solutions have a pH around 0 to 2.
- Basic solutions have a pH greater than 7. The higher the pH, the more basic is the solution.
- Strongly basic solutions have pH around 12 to 14.

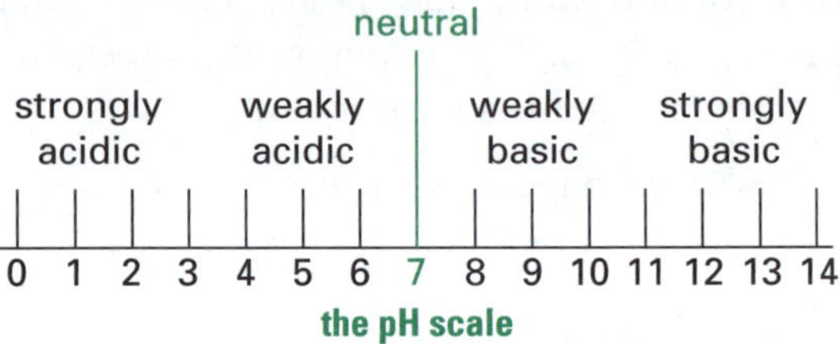

Figure 2.31 pH scale

The pH of a solution can also be related to the colour of indicators. Universal indicator is a useful mixed indicator that shows a large range of colours over the pH scale. Table 2.8 shows the correlation of the colour of universal indicator and the pH scale.

Note: The exact colours and ranges may vary in some different brands of universal indicator. Always use the supplied colour card.

Example: Using universal indicator

A student dissolved some baking soda (sodium hydrogen carbonate) in water and tested the solution with universal indicator and phenolphthalein. The universal indicator turned green-blue and the phenolphthalein was very faintly pink. From these observations the student made the following conclusions.

- The pH of the baking soda solution is between 8 and 9 based on the universal indicator colour chart.
- The solution is slightly basic as phenolphthalein turns faintly pink.

Reaction of acids with metals

The reactions of dilute acids with a variety of different metals can be investigated in the school laboratory. Small, fresh samples of metals are placed in 2-mL volumes of dilute hydrochloric or sulfuric acid. The rate of reaction depends on the following.

- Concentration of the acid: if the acid is too dilute the reaction is very slow.
- Type of metal: some metals (e.g. magnesium) react faster than other metals (e.g. tin).
- Surface area of the metal: powdered metals react faster than large lumps.
- Temperature: warm acid solutions attack metals faster than cold solutions.

For acids such as dilute hydrochloric acid and dilute sulfuric acid, the general equation for reaction with reactive metals is:

acid + metal
→ ionic compound (salt) + hydrogen gas

Example: Zinc + sulfuric acid → zinc sulfate + hydrogen

The hydrogen gas can be collected and identified using the 'pop' test. (A small sample of hydrogen in the presence of air and a spark/flame will explode.)

The reaction above is for hydrochloric and sulfuric acids. While nitric acid is strong, it behaves differently to the other acids. Its reactions do not produce hydrogen gas. Nitrogen monoxide or nitrogen dioxide is produced instead.

The results of such tests with a large variety of metals show that some metals such as potassium and calcium are very reactive and that metals such as zinc, iron and tin are moderately reactive. Metals such as lead, copper and silver are quite unreactive.

Table 2.8 pH and universal indicator

pH range	0–4	4–5	5–6	7–8	8–9	9–10	10–14
universal indicator colour	red	orange	yellow	green	blue-green	blue-violet	violet

Experiment 2

Activity of metals

Aim

- To use the reactions of metals with dilute hydrochloric acid to determine the relative activity of the metal

Method

1. Clean samples of selected metals (zinc, magnesium, copper, calcium and iron) are placed in separate test tubes.
2. Measure out 5 mL of dilute hydrochloric acid into each tube.
3. Determine the relative rate at which the metals react with the acid by observing the speed at which gas bubbles are released from each metal's surface.

Result

Figure 2.32 shows the results of this experiment

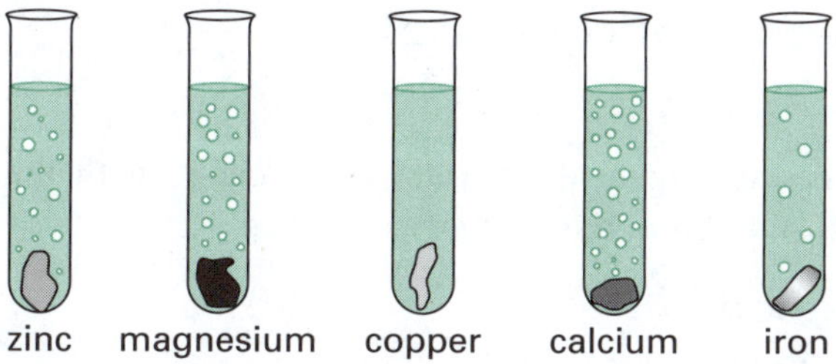

Figure 2.32 Activity of metals in dilute hydrochloric acid

Analysis

Analyse these results to determine the relative activity of these metals in dilute acid. Go to p. 231 to check your answer.

Conclusion

Write a suitable conclusion.

Go to p. 231 to check your answer.

2.8 Combustion reactions and pollution

When you see a candle flame or a burning log of wood, you are seeing light emitted from hot gases rising up either from the site of combustion or from the source of combustion. The colour of the emitted light depends on the gases and particles in the flame as well as the temperature of the flame.

Combustion is a chemical reaction in which a fuel combines with an oxidiser (usually oxygen from the air) to release energy.

$$\text{fuel} + \text{oxidiser} \rightarrow \text{products} + \text{energy}$$

Examples of combustion reactions are:

- burning a wax candle
- burning natural gas in a Bunsen burner
- combustion of petrol in a car's engine
- a bushfire (wood cellulose and other fuels burn in oxygen)
- explosion of hydrogen in air
- burning sulfur in air.

Each of these combustion reactions do not happen spontaneously when the reactants are combined. An ignition source is needed. We often use a spark or burning match to ignite other fuels. Once the combustion starts, the heat energy produced is sufficient to keep the reaction going.

The three requirements for combustion are called the fire triangle, as shown in Figure 2.33.

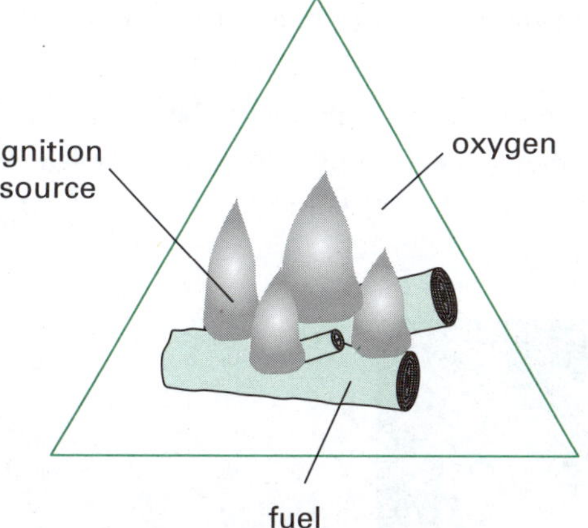

Figure 2.33 Fire triangle

Common fuels and ignition sources

Fuels can be classified according to their state at room temperature. Vapours and gaseous fuels more readily form combustible mixtures with air so they usually ignite more readily than liquid and solid fuels. Ignition sources include:

- frictional heat
- matches, lighters and cigarettes

- sparks from electrical equipment or from lightning
- cooking and heating appliances.

Table 2.9 lists some common fuels.

Table 2.9 Common fuels

Solid fuels	Liquid fuels	Gaseous fuels
wood	petrol	natural gas (methane)
coal	kerosene	propane
coke	oils/waxes	hydrogen
paper	alcohol	LPG
plastic	paint/solvents	acetylene (ethyne)

Extinguishing fires

If you can remove one or more of the components of the fire triangle, then the fire can be extinguished. Here are some ways of achieving this.

- Remove or separate unburnt fuel. In the case of a gas fire, for example, the gas mains can be turned off.
- Remove the ignition (heat) source; for example, cool down the fuel by spraying with water or dumping water from an aircraft.
- Remove the oxygen supply. Carbon dioxide fire extinguishers can smother a fire by cutting off the oxygen supply. If clothing is on fire, the flames can be put out by using a fire blanket to exclude the oxygen.

Figure 2.34 shows one of the many symbols indicating where a fire blanket is located. Do you know where the emergency fire blanket is located in your school laboratory?

Figure 2.34 Example of a fire blanket symbol

Example: Combustion of natural gas in a plentiful supply of oxygen

In this example we assume that natural gas consists only of methane (CH_4). (Actually, natural gas is a mixture consisting mainly of methane but with up to 20% of ethane and other hydrocarbons.)

Observations: the gas burns with a blue flame; heat and light energy are produced.

Explanation: the hot molecules of methane and oxygen combine to form carbon dioxide molecules and water molecules.

Word equation:

methane + oxygen → carbon dioxide + water

If the supply of oxygen is limited, then the methane can produce a variety of combustion products including carbon dioxide (CO_2), carbon monoxide (CO), carbon (C) solid and water (H_2O).

Experiment 3

Investigating candles

Aim

- To investigate the variables associated with the formation of flames in wax candles

Method

This is an experiment that can be done at home or at school. You will need candles of different lengths and diameters. Investigate the effect of each of the following variables on the appearance of the flame of a wax candle and the rate at which the wax of the candle melts:

- candle diameter
- length of exposed wick
- conducting heat away from the flame using metal gauze.

Figure 2.35 provides information about the set-up of the candle experiment.

Analysis

Analyse these results to determine the effect of each change on the candle flame. Go to p. 231 to check your answer.

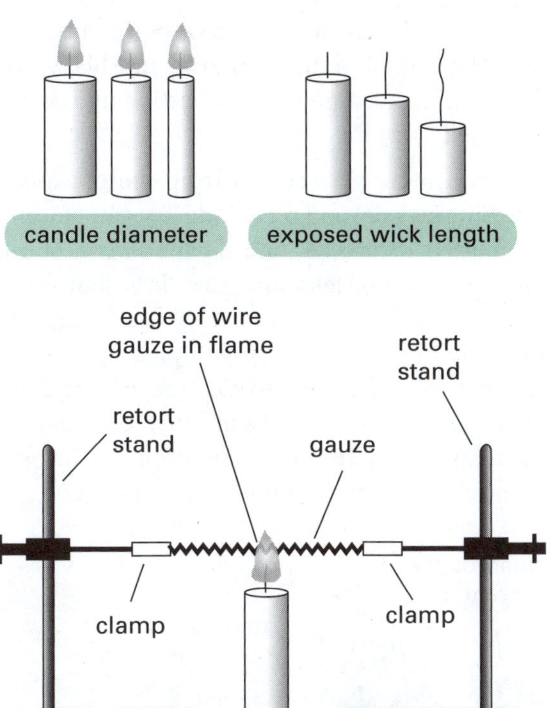

Figure 2.35 Investigating candle flames

Conclusion

Write a suitable conclusion.

Go to p. 231 to check your answer.

Exothermic and endothermic reactions

Physical or chemical changes that release energy are classified as exothermic. The energy is often released in the form of heat but other forms such as light and sound may be involved. The following are some examples of exothermic reactions.

- Condensation of water vapour. This is a physical process in which gaseous water molecules turn into droplets of liquid water. The kinetic energy of the moving water molecules in the water vapour is released as heat as the water condenses.
- Neutralisation of a base with an acid. This is a chemical change in which a salt and water are formed as the acid neutralises the base. The final mixture is warmer than the reactants.
- Combustion of kerosene. This is a chemical change in which some of the chemical potential energy of the kerosene is released as heat as the combustion occurs. This reaction is used in a kerosene heater to warm a room.

Physical or chemical changes that absorb energy are classified as endothermic. In these cases the surroundings cool down as some energy is absorbed by the reactants. The following are some examples of endothermic reactions.

- Melting of ice. This is a physical process in which solid water or ice absorbs energy from the surroundings and is converted to liquid water. Energy is required to separate the water molecules in the ice so they can move about in the liquid state.
- Electrolysis of water. This is an electrochemical process in which electrical energy is passed into water via a battery and metal electrodes. The energy is used to break down the chemical bonds in the water and produce hydrogen and oxygen gas.
- Thermal decomposition of a compound. This is a chemical change in which the heat energy is used to break down the chemical bonds between the elements of a compound to produce new substances. For example, mercury oxide will decompose when heated to form mercury and oxygen.

Pollution

The Earth is virtually a closed system. Any changes we make to any part of the environment eventually affect the whole system. Pollutants are constantly being released into the environment. A pollutant is any substance that does not naturally occur in a particular environment or one that has a concentration outside the normal limits.

Pollution of the environment occurs in many ways. Vehicle exhaust and the emissions from factories and bushfires add smoke, unburned fuels, carbon dioxide and black soot particles, as well as noise, to the atmosphere. Natural pollution can come from bushfires, volcanoes, pollen and dust blowing in the wind, gases from decaying plants and animals, and marsh gas or methane (CH_4) from swamps and animal wastes.

High-quality air is mostly composed of nitrogen, oxygen, argon, carbon dioxide and water vapour. Let's examine some common pollutants that are present in the atmosphere.

Sulfur dioxide

Sulfur dioxide is usually generated by the combustion of coal or oil that contains small amounts of sulfur minerals. The sulfur combines with oxygen during the combustion process and sulfur dioxide forms. Sulfur dioxide can also be released from the smelting of sulfide ores such as copper sulfide.

sulfur + oxygen → sulfur dioxide

Sulfur dioxide is a poisonous gas. It also dissolves in rainwater, producing acid rain. Acid rain can damage built structures such as statues and stone buildings. Acid rain can also damage forests and lakes. If acid rain accumulates in lakes, the aquatic life can be severely affected. For example, fish eggs will die if the pH is less than 5.5.

Nitrogen oxides

Nitrogen oxides include nitrogen monoxide (NO) and nitrogen dioxide (NO_2). These pollutants form inside high-temperature furnaces and engines. As the fuel burns, the high temperatures cause the nitrogen and oxygen in the air to combine to form nitrogen monoxide and nitrogen dioxide. A mixture of oxygen and nitrogen dioxide can absorb ultraviolet light from the Sun and produce ozone gas. This poisonous gas is a component of photochemical smog. Nitrogen gas can also dissolve in rain water to produce acid rain which can cause considerable damage to the natural and built environment. For example, acid rain can attack steel structures and cause corrosion.

Particulates and aerosols

Particulates (tiny particles of solid matter) and aerosols also contribute to pollution. Aerosols consist of tiny solid particles or liquid droplets that remain suspended in the air for long periods. Ash and soot from fires and metal particles (such as lead and arsenic) from exhausts and mining can pollute the air for long periods of time under certain environmental conditions. People with emphysema and asthma should not be outdoors when there are high levels of particulates in the air.

Carbon monoxide

Carbon monoxide (CO) is a toxic pollutant released in the combustion of hydrocarbon fuels. When hydrocarbons are burned in a plentiful supply of oxygen, carbon dioxide and water are formed. This process is called complete combustion.

Carbon monoxide is formed when carbon compounds are burned in a limited supply of oxygen. This type of burning is called incomplete combustion and soot is often formed as well.

Most carbon monoxide comes from motor vehicle exhaust, burning coal and oil in industry, smouldering leaves, lighted cigarettes and cigars. It is a subtle poison, being odourless and tasteless. Poisoning occurs because the red blood cells that normally swap CO_2 for O_2 in the lungs will pick up any CO in the lungs in preference to O_2 (see Figure 2.36). This means the red blood cells are then carrying less oxygen than is required for respiration. In this way, carbon monoxide deprives body cells of oxygen.

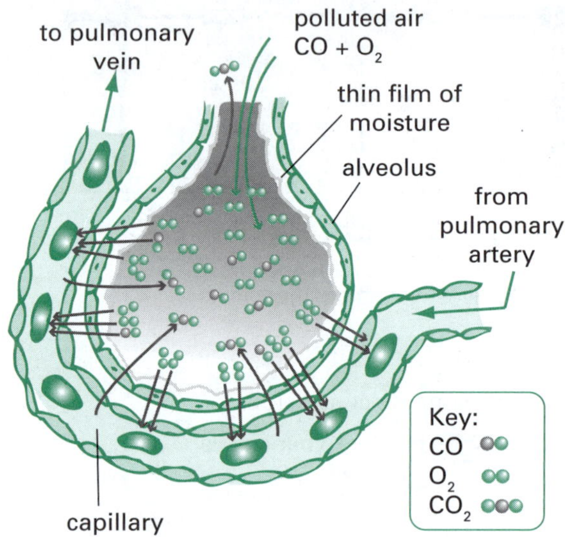

Figure 2.36 Inhaling carbon monoxide

Monitoring and minimising pollution of the atmosphere

The Environmental Protection Agency (EPA) monitors air quality and the levels of pollutant gases. In major cities in Australia, the level of nitrogen oxides has remained fairly stable despite the increasing numbers of cars on the road. This is largely due to the requirement that vehicles have fitted a catalytic converter unit in the exhaust system that will minimise the amounts of nitrogen oxides and carbon monoxide released. The catalytic converter contains two chambers in which different catalysts are used to convert nitrogen oxides and carbon monoxide to carbon dioxide and nitrogen as well as oxidise any fuel gases that may not have ignited in the engine (see Figure 2.37). The air quality in Australian cities compares favourably to cities such as Beijing which have less stringent standards with regard to emission of pollutants into the atmosphere.

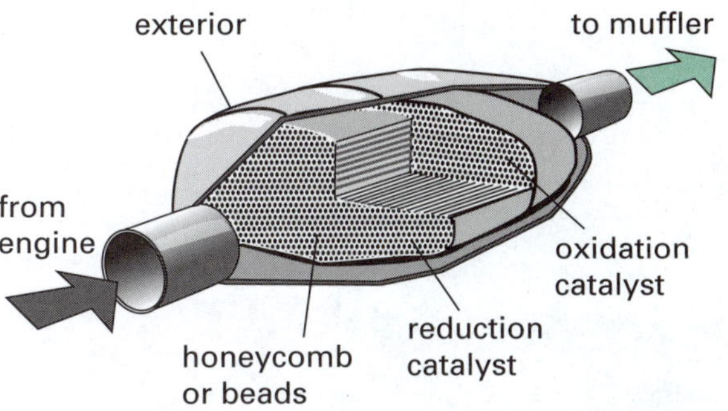

Figure 2.37 Catalytic converter in a car exhaust system

In the case of coal combustion and the generation of electricity, the levels of toxic emissions into the atmosphere can be reduced using electrostatic precipitators and scrubbing technologies. In electrostatic precipitators, small particles are attracted to charged electrodes which capture the particles. In wet scrubbers, acidic flue gases are passed through a scrubbing solution which usually consists of a watery mixture of weak bases (see Figure 2.38). Such mixtures dissolve and neutralise the acidic or toxic gases. The gas stream that emerges from the scrubber is then released into the environment.

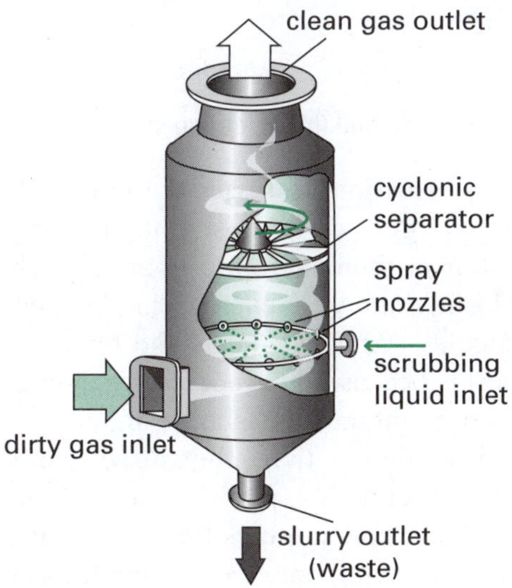

Figure 2.38 Scrubbers remove toxic and acidic gases from flue gases.

Comparing combustion to other oxidation reactions

Combustion is one example of a large class of chemical reactions called oxidation reactions. Oxidation reactions always involve the loss of electrons from an element or compound. In a combustion reaction involving a hydrocarbon fuel, the oxygen molecules are responsible for removing electrons from the fuel. This removal leads to a chemical change and new products form.

Here are some other examples of oxidation reactions which do not involve combustion or fires.

Rusting of iron

When iron metal is exposed to air and water, the iron oxidises to form rust. Rust is a hydrated iron oxide compound. In the absence of water, the iron will still be oxidised by the oxygen to form a different oxide which is not rust.

iron + oxygen + water → rust

Dissolving metals in dilute acids

When magnesium metal is placed in a beaker of dilute sulfuric acid, the magnesium dissolves as it is oxidised by the acid. The metal loses electrons to the acid and forms magnesium ions that dissolve in the water. The sulfate ions of the acid are also present and if the solution is evaporated, magnesium sulfate crystals are obtained.

magnesium + sulfuric acid
→ magnesium sulfate + hydrogen

Acid sulfate soils

Some waterlogged soils contain iron sulfide minerals. If the water evaporates in times of drought or restricted water supplies, then these iron sulfide minerals can become oxidised by contact with oxygen in the air and sulfuric acid is formed. The acid sulfate soils that form can kill plants and aquatic life. In this example, the iron sulfide minerals have lost electrons to the oxygen and water molecules.

iron sulfide + oxygen + water
→ sulfuric acid + iron hydroxide

2.9 Respiration and photosynthesis

As green plants absorb carbon dioxide, they take up water and release oxygen. During this process, they manufacture organic matter such as sugar (glucose) which is later used by the plant. This endothermic process is called photosynthesis. The process can be summarised by this word equation:

$$\text{carbon dioxide + water} \xrightarrow{\text{light}} \begin{array}{c}\text{sugar}\\ \text{(glucose)}\end{array} + \text{oxygen}$$

Because only the green parts of plants carry out photosynthesis, the green pigment called chlorophyll must be involved. The chlorophyll molecule is able to absorb light energy and use that energy to power the photosynthetic process.

Chlorophyll is present only in certain cells containing chloroplasts. The chlorophyll molecules are arranged in layers inside structures called grana (see Figure 2.39). The grana are suspended in the stroma (the supportive framework of little tube-like strands connecting one granum to another) of the chloroplasts. The chloroplast is the place where light energy is converted into chemical energy in the form of glucose. Chloroplasts also contain starch granules formed from glucose. The granules store the chemical energy until it is needed. That chemical energy has to be made available to all the cells of the plant. In order to do this, plants must transport the glucose to each cell where the glucose is recombined with oxygen to produce carbon dioxide, water and useable energy. This process is called cellular respiration.

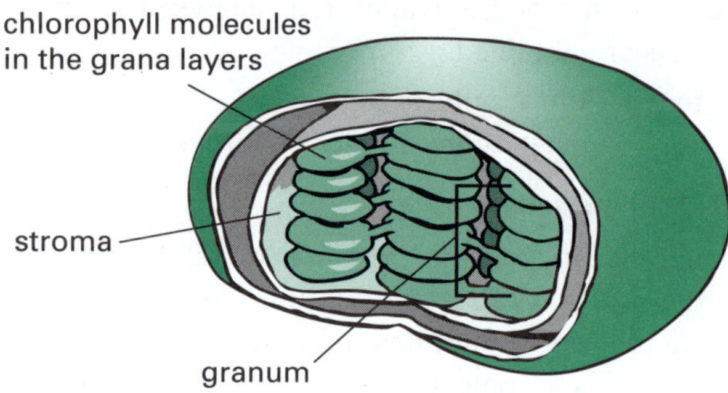

Figure 2.39 Chloroplast

Cellular respiration occurs in both plants and animals. Animal and plant cells use molecules like glucose to supply their energy needs. In the cytoplasm of the plant or animal cell, some energy can be obtained by breaking the bonds of the glucose molecule. Oxygen is not required for this first step. Most energy is generated in the cell by aerobic respiration, which requires oxygen. The overall aerobic respiration process is exothermic and can be summarised by the following word equation:

$$\text{glucose} + \text{oxygen} \rightarrow \begin{matrix}\text{carbon}\\\text{dioxide}\end{matrix} + \text{water} + \begin{matrix}\text{useable}\\\text{energy}\end{matrix}$$

What is the site of cellular respiration in plant or animal cells? Scientists have found that a cell organelle called the mitochondrion is involved. Muscle cells (especially heart muscle cells) are packed with thousands of mitochondria. In sperm cells, the mitochondria are clustered near the tail of the sperm where the energy is required (see Figure 2.40).

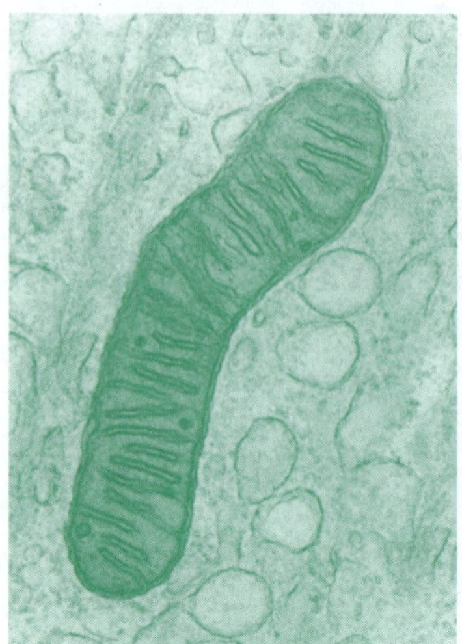

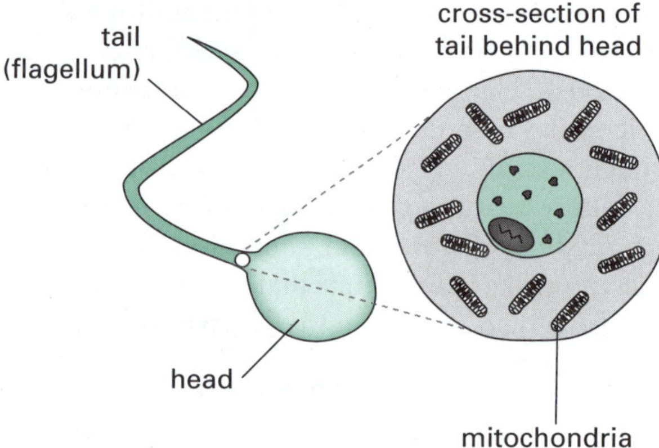

Figure 2.40 Mitochondria and their location in sperm cells

All energy transformation processes are not 100% efficient. Some energy is always lost along the way. Even while resting or running, your cells are actively respiring and losing some heat energy. This lost heat keeps your muscles warm. Mammals and birds are called endotherms because their bodies are able to produce heat to maintain a constant temperature which is usually higher than the surrounding environment. Ectotherms (e.g. reptiles) are less able to control their body temperatures through cellular respiration and must sun themselves during the day or hibernate in the winter to keep their body heat.

Test yourself 2

Part A: Knowledge

1. Sodium carbonate and nitric acid are allowed to react. A gas is evolved. The name of this gas is *(1 mark)*

A water.
B carbon dioxide.
C nitrogen dioxide.
D hydrogen.

2. Which of the following substances would produce carbon dioxide on combustion in oxygen? *(1 mark)*
   A phosphorus
   B sulfur
   C propane
   D hydrogen

3. Which of the following substances contains an acid? *(1 mark)*
   A baking powder
   B oven cleaner
   C salt water
   D grapefruit

4. During an advancing bushfire, firefighters will often burn the scrub some kilometres ahead of the fire. The reason for doing this is to *(1 mark)*
   A remove the fuel.
   B starve the fire of air.
   C prevent convection currents forming.
   D reduce radiant heat.

5. When an acid is neutralised by a base, what new substance is formed? *(1 mark)*
   A water only
   B sodium chloride
   C a salt
   D carbon dioxide

6. Complete the following restricted-response questions using the appropriate word. *(1 mark for each part)*
   a) Dilute sulfuric acid dissolves magnesium metal and releases .... gas.
   b) The law of mass conservation states that the mass of .... equals the mass of products.
   c) Hydrogen gas reacts with oxygen gas to form ....
   d) Baking soda can be .... by adding vinegar.
   e) The salt formed when .... oxide reacts with hydrochloric acid is barium chloride.

7. Use the code letters to match the terms or phrases in each column. *(1 mark for each part)*

Column 1	Column 2
A acid	F oxygen
B salt	G base
C combustion	H toxic gas
D litmus	I indicator
E sulfur dioxide	J ionic

Part B: Skills

8. A student collected some flower petals and extracted the coloured pigment into water. The solution she collected was yellow. She poured this solution into three test tubes A, B and C. To A she added hydrochloric acid and it turned orange. To B she added sodium hydroxide and it turned green. She did not add anything to tube C.
   a) What colour did the pigment turn in the alkaline (basic) solution? *(1 mark)*
   b) What is the purpose of tube C? *(1 mark)*
   c) The student placed some of the flower petal dye in an unknown liquid (X) and it turned orange. What did the student conclude about X? *(1 mark)*

9. Examine Figure 2.41 which shows models of various reactions.

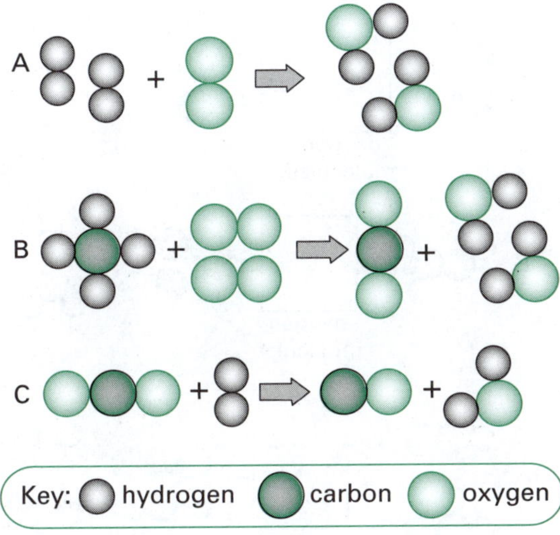

Figure 2.41 Modelling reactions

   a) Use the key to work out which reactions would be classified as combustion reactions. *(1 mark)*

b) Use the periodic table at the front of the book to find the symbols of these elements to convert these model equations into balanced symbolic equations. *(3 marks)*

10. A student placed some sulfuric acid in a test tube and measured its temperature. The initial temperature was 20.8 °C. She then slowly added a slight excess of sodium hydroxide and the temperature increased by 3.6 °C.

a) What was the final temperature of the mixture? *(1 mark)*

b) What energy transformation has occurred in this reaction? *(1 mark)*

c) If some purple litmus solution had been added to the original sulfuric acid, describe the colour changes that the student would observe during this experiment. *(1 mark)*

d) Is the final solution acidic, basic or neutral? *(1 mark)*

e) Name the salt formed in this neutralisation reaction. *(1 mark)*

f) How could the salt be recovered from the reaction mixture? *(1 mark)*

11. The atomic models in Figure 2.42 show combustion reactions. In each case the oxygen molecules are missing. Work out in each case how many oxygen molecules must be needed to balance the atoms on each side of the equation. *(3 marks)*

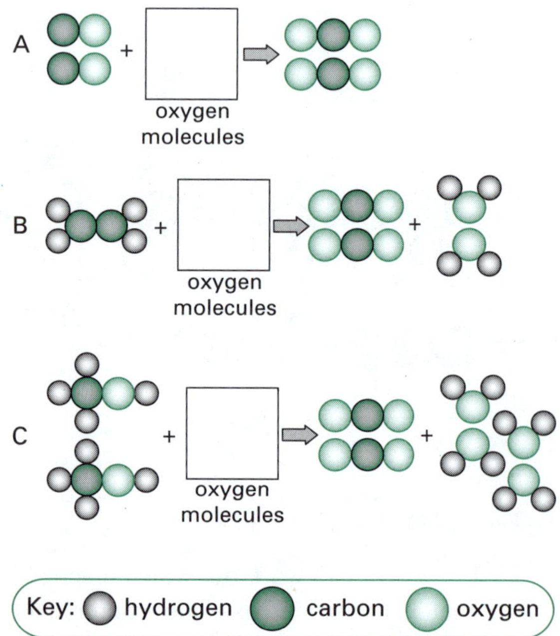

Figure 2.42 Work out the missing coefficients

12. Write word equations for the neutralisation reactions that occur when the following acids and bases are mixed. *(3 marks)*

a) nitric acid and aluminium hydroxide

b) sulfuric acid and copper hydroxide

c) carbonic acid and ammonium hydroxide

13. Two test tubes containing dilute hydrochloric acid and drops of universal indicator were set up in a test-tube rack. Each tube also contained a thermometer. Two white powdered solids (X and Y) were slowly added separately to each tube until no further changes were observed. The results of the experiment are summarised in Table 2.10.

Table 2.10 Properties of two powders

Powder	Observations
X	• The white solid dissolves and the solution turns from red to green and then blue-violet. • The temperature of the mixture increases.
Y	• The white solid dissolves and the solution turns from red to green and then green-blue. • Effervescence occurs. • The temperature of the mixture increases.

a) Classify the type of reaction occurring when X and Y combine with the hydrochloric acid. *(1 mark)*

b) What is the purpose of the universal indicator? *(1 mark)*

c) Why does the temperature rise in each case? *(1 mark)*

d) One substance is magnesium oxide and one is magnesium carbonate. Use the results to determine which substance is magnesium carbonate. Justify your answer. *(2 marks)*

e) What can one conclude about the pH of the final mixture in each tube? *(2 marks)*

14. Jenna lights the wick of an alcohol burner and places it in a sealed glass jar. She notices that the flame goes out after a while, even though there is fuel still left. Explain her observation. *(2 marks)*

Go to pp. 232–233 to check your answers.

Summary

1. Over the centuries people had different ideas of what made up atoms.
2. Democritus was one of the first to recognise atoms as indivisible.
3. Dalton was one of the first scientists to devise an atomic theory.
4. Thomson, Rutherford, Bohr and Chadwick all contributed to what made up atoms.
5. Models of atoms are one way of conceptualising, summarising and predicting.
6. Atoms consist of protons, neutrons and electrons.
7. Electrons are arranged around the nucleus of an atom in shells.
8. The electron configuration shows the number of electrons in each energy level.
9. Elements come in a number of varieties called isotopes.
10. Atomic number, mass number and atomic weight describe features of atoms.
11. Radioactivity refers to the particles and radiation that are emitted from nuclei as a result of instability.
12. The half-life is the time it takes for a substance undergoing decay to decrease by half.
13. Alpha, beta and gamma radiation is given out from unstable nuclei.
14. Rutherford and the Curies did a lot of work to unravel the nature of radiation.
15. There are many uses for radioactivity in medicine and industry.
16. A chemical equation summarises the events of a chemical reaction.
17. Model kits can be used to model chemical reactions.
18. Mass is conserved in a chemical reaction.
19. Acids and bases neutralise each other to form a salt.
20. Dilute acids dissolve reactive metals to form salts and hydrogen gas.
21. Dilute acids react with carbonates to form salts, water and carbon dioxide.
22. Indicators can be use to determine whether a substance is acidic, basic or neutral.
23. Combustion is a chemical reaction in which a fuel combines with an oxidiser such as oxygen.
24. Reactions that release heat are exothermic and reactions that absorb heat are endothermic.
25. The atmosphere can become polluted with gases released from combustion reactions.
26. Combustion is an example of an oxidation reaction.
27. Cells obtain their energy by cellular respiration of nutrients such as sugars.
28. Plants use water and carbon dioxide to produce glucose and oxygen during photosynthesis.

Syllabus checklist

Are you able to answer every syllabus question in this chapter? Tick each question as you go through the list if you are able to answer it. If you cannot answer it, turn to the appropriate page in the guide as is listed in the column to find the answer.

For a complete understanding of this topic		Page no.	✓
1	Do I know some of the different ideas about what made up atoms that people have had over the centuries?	53–56	
2	Can I name the scientist who first proposed that atoms were indivisible?	53–54	
3	Can I name a scientist who first devised an atomic theory and assigned atomic weights to elements ?	53–54	
4	Can I recall the contribution that Thomson, Rutherford, Bohr and Chadwick made to our understanding of atoms?	54–56	

	For a complete understanding of this topic	Page no.	✓
5	Can I explain the importance of developing atomic models?	56–57	
6	Can I recall the three subatomic particles?	57	
7	Can I recall the location of electrons in an atom?	57	
8	Can I recall what the electron configuration tells us about an atom?	58	
9	Can I recall what we call elements that have different neutron numbers and the same proton numbers?	59	
10	Can I distinguish between the terms atomic number, mass number and atomic weight?	59	
11	Can I recall the meaning of the term 'half-life' in relation to radioactive decay?	59	
12	Can I recall the nature of radioactive emissions released from unstable nuclei?	60	
13	Can I recall the names of the three most common forms of radioactive emissions?	60	
14	Can I recall the names of important scientists in the first 20 years of the 20th century, whose work led to an understanding of the nature of radiation?	60–61	
15	Can I identify some common uses for radioactivity in medicine and industry?	61–62	

	For a complete understanding of this topic	Page no.	✓
16	Can I recall what a chemical equation summarises?	65	
17	Can I demonstrate the use of model kits to model chemical reactions?	66	
18	Can I explain why mass is conserved in a chemical reaction?	66–67	
19	Can I recall what happens when an acid and base are mixed together?	68–69	
20	Can I recall the products that form when dilute acids react with carbonates?	69	
21	Can I explain the uses of acid–base indicators?	69–70	
22	Can I recall the products that form when a dilute acid reacts with an active metal?	70	
23	Can I describe the meaning of the term combustion in a chemical reaction?	71	
24	Can I explain the difference between exothermic and endothermic reactions?	73	
25	Can I explain the type of combustion reactions that lead to atmospheric pollution?	73–74	
26	Can I explain why combustion is an example of an oxidation reaction?	75	
27	Can I recall how cells obtain their energy?	75–76	
28	Can I explain how glucose is produced during photosynthesis?	75–76	

Chapter test

Go to p.v for *Tips for tests and examinations*

80 MIN

Part A: Multiple-choice questions

(1 mark for each)

1. He found that electrons move around the nucleus only in certain shells or orbits. The size and shape of these shells depend on the amount of energy the electrons have. Name this scientist.
   - A Bohr
   - B Thomson
   - C Rutherford
   - D Chadwick

2. Who reasoned that matter was made up of tiny particles that could not be split into smaller particles, and was one of the first to use the word 'atom'?
   - A Dalton
   - B Thomson
   - C Rutherford
   - D Democritus

3. An isotope of potassium has the symbol K-40. Select the correct statement about the atom of this isotope.

A Potassium-40 atoms contain 40 protons.
B Each atom contains 19 protons, 19 electrons and 40 neutrons.
C Each atom contains 21 neutrons.
D K-40 atoms have the same number of neutrons as K-39 atoms.

4. The mass of a proton is
   A about the same as the mass of a neutron.
   B similar to the mass of an electron.
   C much smaller than the mass of an electron.
   D positive, whereas the mass of an electron is negative.

5. An element with the electronic configuration 2.8.5
   A is a metal.
   B has an atomic number of 15.
   C has a mass number of 15.
   D is very unreactive.

6. Figure 2.43 shows a particle model of a chemical reaction.

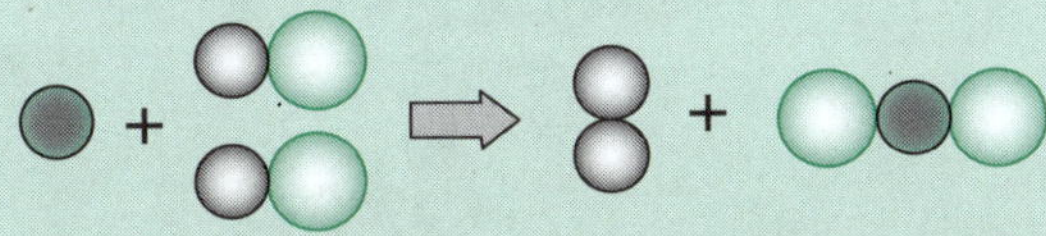

Figure 2.43 Model of a chemical reaction

The type of chemical reaction shown in this particle model is
   A an acid–base reaction.
   B an acid on a metal.
   C an acid on a carbonate.
   D a combustion reaction.

7. Oxidation is a chemical process in which a substance
   A reacts with a base.
   B loses electrons.
   C reacts with hydrogen.
   D decomposes.

8. Acidic flue gases from factories can be cleaned
   A using a wet scrubber containing a weak base.
   B by respiration.
   C using a catalytic converter.
   D by passing them through a fine filter.

9. Which one of the following processes is endothermic?
   A freezing of liquid water into ice
   B neutralisation of an acid and a base
   C combustion of natural gas
   D sublimation of iodine solid to iodine vapour

10. Here is a list of metals: copper; zinc; sodium; silver; iron; magnesium. The least active metals are
   A copper and zinc.
   B silver and iron.
   C silver and copper.
   D sodium and magnesium.

Part B: Short-answer questions

11. Figure 2.44 shows two examples of atomic models. Why do scientists (and teachers) use such models? *(4 marks)*

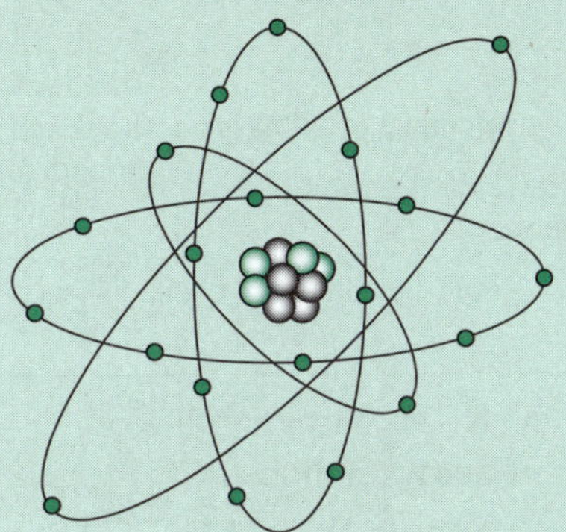

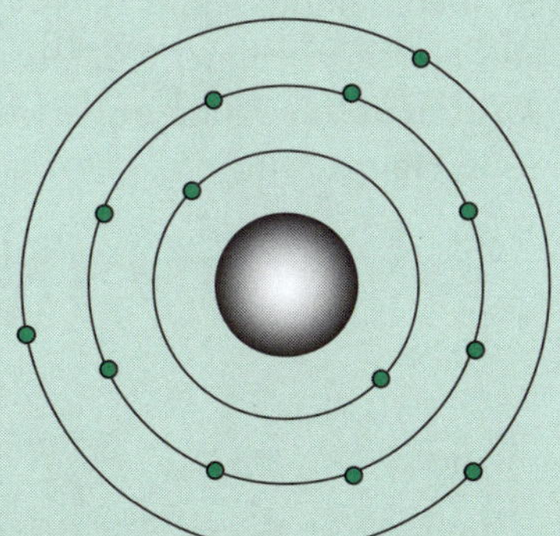

Figure 2.44 Atomic models

12. Figure 2.45 shows the movement of alpha particles, beta particles and gamma rays in an electric field.
   a) Why is the path followed by alpha particles and beta particles in opposite directions? *(3 marks)*
   b) Explain why beta particles bend more than alpha particles. *(3 marks)*
   c) Why do gamma rays show no deviation? *(2 marks)*

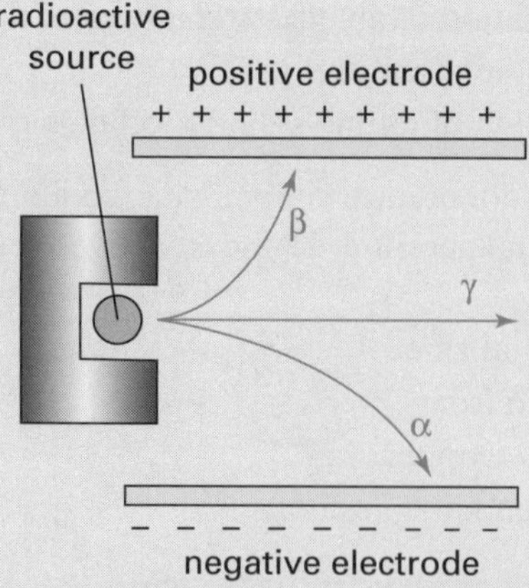

Figure 2.45 Radiation in an electric field

13. Copy and complete Table 2.11. *(9 marks)*

Table 2.11 Properties of radiation

Radiation	Made up of	Charge	Distance travelled through air
alpha particle			
beta particle			
gamma ray			

14. Aluminium-26 (Al-26) has a half-life of 720 000 years. It decays radioactively to magnesium-26 (Mg-26). When a particular rock was formed, it contained 8 units of aluminium-26 and no magnesium. On analysis of a sample, it was found to contain only two units of Al-26. (See Figure 2.46.)

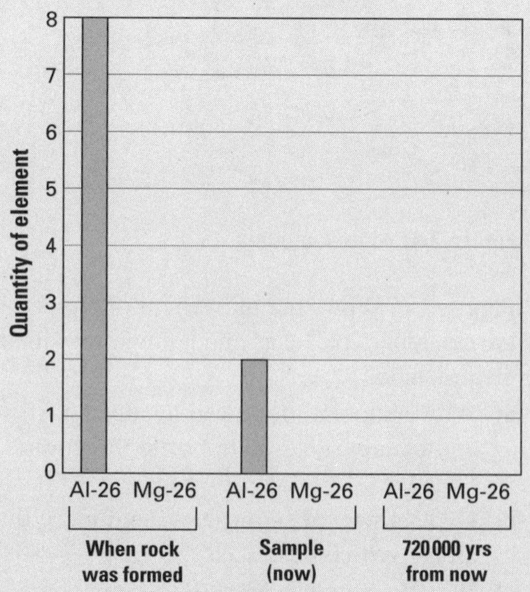

Figure 2.46 Decay of Al-26

a) The quantity of Mg-26 was left off the graph. How many units high should this column be? *(1 mark)*

b) How old is the sample of rock? *(1 mark)*

c) When were the Al-26 and Mg-26 quantities the same? *(1 mark)*

d) In 720 000 years from now, another scientist will analyse this rock sample. Draw the column graphs that scientists would get for both these elements. *(2 marks)*

15. Figure 2.47 shows two separate radioactive nuclei decaying.

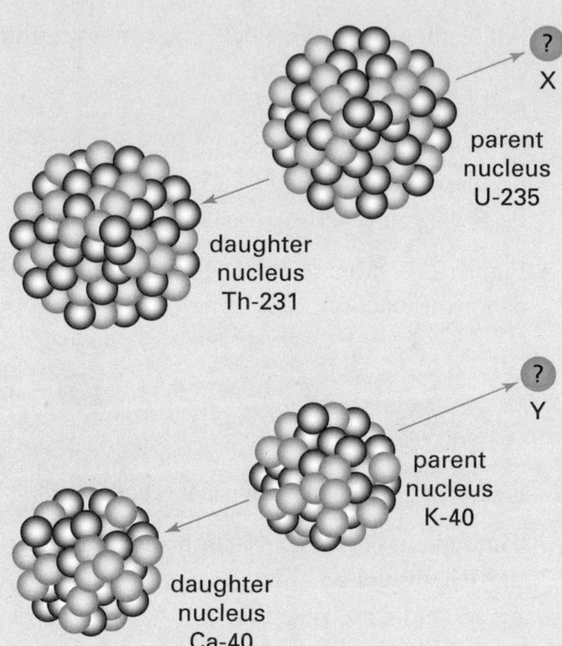

Figure 2.47 Radioactive decay

a) Use a periodic table to complete Table 2.12. *(12 marks)*

Table 2.12 Properties of parent and daughter nuclei in nuclear reactions

	Symbol	Name	Proton number	Neutron number
parent nucleus	U-235	uranium		
daughter nucleus				
parent nucleus	K-40			
daughter nucleus		calcium		

b) Use the information from the table you completed to identify the radiation particles X and Y. *(2 marks)*

16. Figure 2.48 shows how a smoke alarm works. The list that follows explains each step, but the sentences are out of order. Write the letters of each sentence in the correct sequence. *(4 marks)*

A. This causes air particles to ionise (to become charged).
B. This absorption causes less ionisation to take place.
C. Alpha particles move between the two charged metal plates.
D. When a smaller than normal current results, an alarm sounds.
E. As a result, a current flows.
F. Some smoke detectors use americium-241 that produce alpha particles, to detect smoke.
G. These charged particles are attracted to the oppositely charged metal plates.
H. Smoke particles attach to some of the alpha particles.

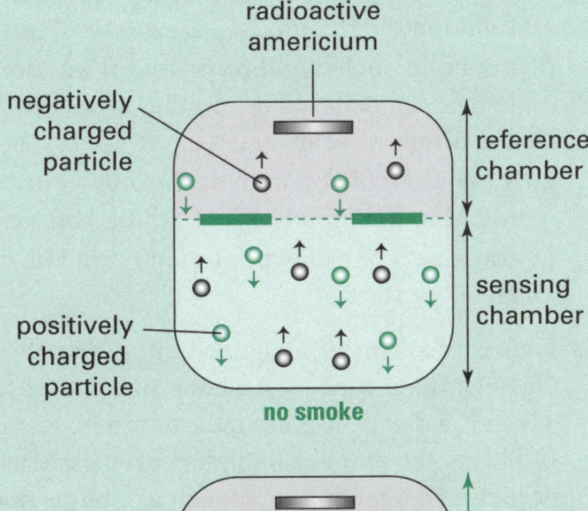

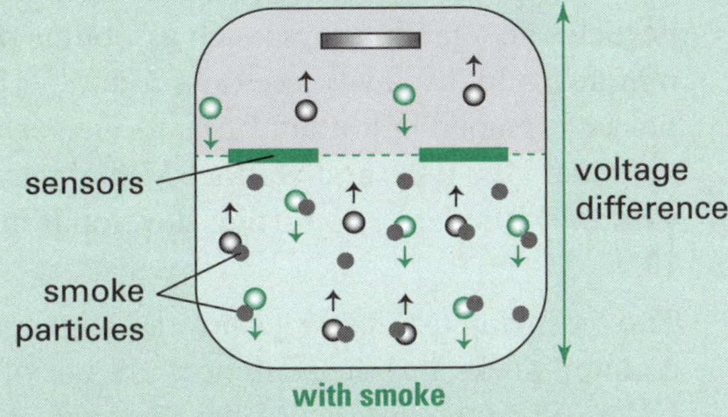

Figure 2.48 Smoke alarm

17. Radiosurgery, using a machine called a gamma knife, is a very precise form of radiation therapy targeting tumours in the brain (see Figure 2.49). In this external radiotherapy, a radioactive source is not injected into the patient.

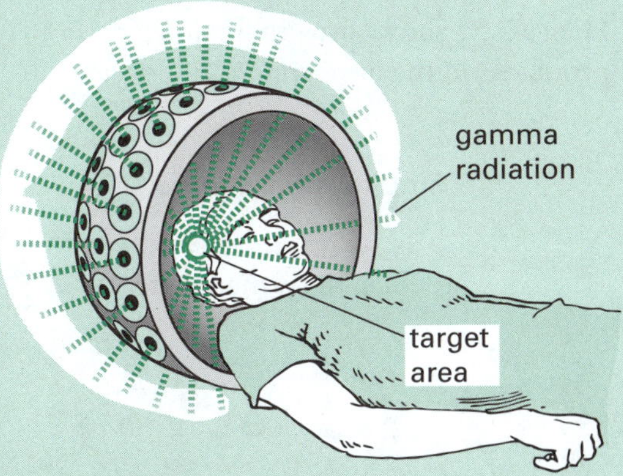

Figure 2.49 Gamma knife

a) Why are gamma rays used and not alpha particles? *(2 marks)*

b) How does radiation destroy tumours? *(2 marks)*

c) Why do you think the gamma radiation comes from different directions and is focused on the target area? *(2 marks)*

18. Figure 2.50 shows a way of detecting underground pipe leaks.

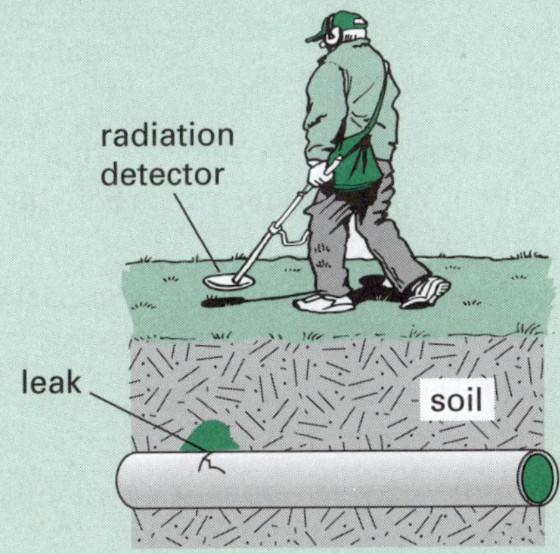

Figure 2.50 Leaking underground pipes

Describe how radiation can be used to detect underground leaks. *(4 marks)*

19. Carbon-14 is a radioactive isotope of carbon. It has two extra neutrons in its nucleus making it unstable. There is a small amount of radioactive carbon-14 in all living organisms. No new carbon-14 is taken in by them when these organisms die. The carbon-14 left in the organism at the time of death decays over

many thousands of years. The approximate time since the organism died can be worked out by measuring the amount of carbon-14 left in the remains of the organic material.

Figure 2.51 shows how radioactive carbon-14 is produced in the atmosphere.

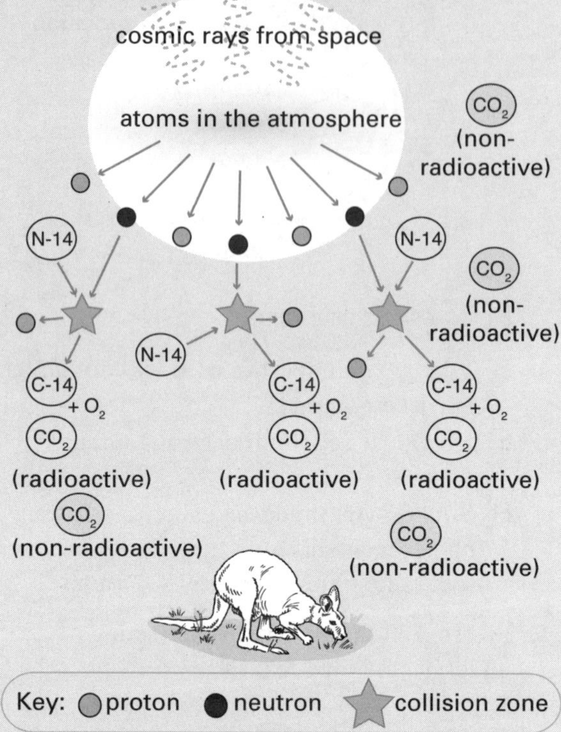

Figure 2.51 Uptake of carbon-14

a) What subatomic particle collides with a nitrogen-14 atom to produce a carbon-14 atom? *(1 mark)*

b) How does a carbon-14 atom become part of a CO_2 molecule? *(1 mark)*

c) How does this radioactive form of carbon enter the food chain? *(2 marks)*

d) Is it only radioactive carbon-14 which enters the food chain? Explain. *(2 marks)*

e) Why doesn't carbon-14 continue to enter the food chain when a piece of wood from a tree has been carved into a boomerang? *(2 marks)*

20. Radon-220 is radioactive and decays over time, as shown in Figure 2.52.

a) What was the original mass of Rn-220? *(1 mark)*

b) What is the half-life of Rn-220? *(1 mark)*

c) After how long will only 5 g of radon-220 remain? *(1 mark)*

d) What mass of radon-220 has been converted after 90 seconds? *(1 mark)*

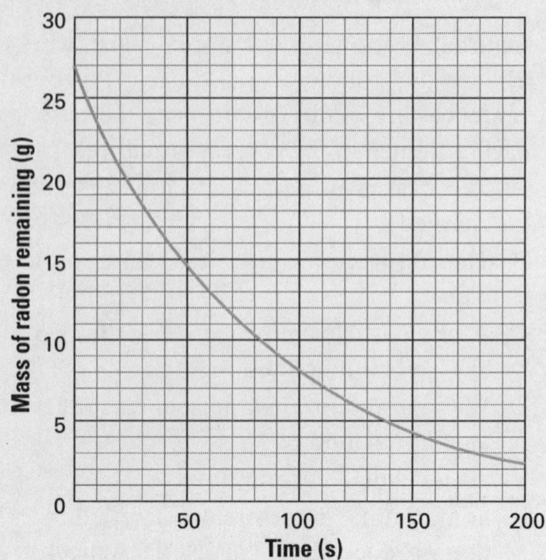

Figure 2.52 Radon-220 decay

21. Which of the following cannot be dated using radioactive carbon dating? Explain why. *(3 marks)*

A. a cloth wrapping from a mummified bull found in the pyramids in Egypt

B. a sample of charcoal recovered from the bed of ash near a volcano that erupted thousands of years ago

C. a bone sample of an extinct primitive tribe found in the mountains of South America

D. fragments of a clay pot found buried in layers of rock

22. Figure 2.53 shows a Geiger counter (actually a Geiger-Müller tube with a counter attached), which is a device used to measure radioactivity. It detects the number of alpha particles, beta particles and gamma rays (such as 'counts per minute') and displays these on a counter. The device is named in honour of Hans Geiger who invented it in 1908, and Walther Müller who worked with Geiger to further develop it in 1928.

The following list describes how a Geiger counter works, but the sentences are out of order. Write the letters of each sentence in the correct sequence. *(5 marks)*

A. This causes an 'avalanche' of electrons.

B. A voltage of some hundreds of volts is applied across this gas between a central thin wire and the casing.
C. A Geiger counter can detect individual particles up to about 10 000 each second.
D. These electrons arriving at the wire create a pulse that is amplified and counted.
E. A Geiger-Müller tube contains a mixture of argon or neon and methane gas at low pressure.
F. Thus one single incoming radioactive particle will cause many electrons to arrive at the wire.
G. The voltage attracts the electron to the central wire.
H. When a charged particle enters the tube, it knocks an electron from a gaseous atom.
I. This device is used widely in medicine and in prospecting for radioactive ores.
J. As the electron rushes towards the wire, it knocks more electrons from other gas atoms.

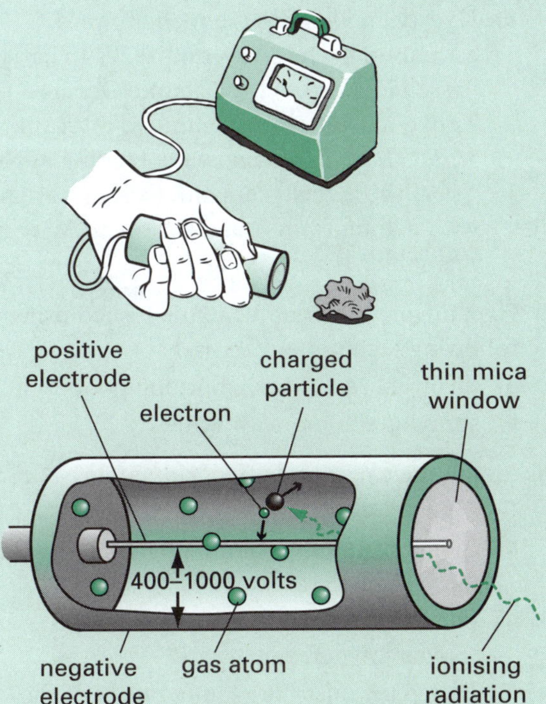

Figure 2.53 Geiger counter

23. An atom has an electronic structure 2.8.10.2.
   a) Draw a diagram showing the arrangement of electrons in the orbits of this atom. *(4 marks)*
   b) What is the atomic number of this atom? *(1 mark)*
   c) Use a periodic table to identify the element. *(1 mark)*
   d) The most abundant isotope of this element has a mass number of 48. How many neutrons does it have? *(1 mark)*

24. Describe, using diagrams as needed, how Thomson's 'plum pudding' atomic model gave way to Rutherford's 'planetary' atomic model. *(6 marks)*

25. The graph in Figure 2.54 shows the penetration of beta particles and gamma rays in a certain material.

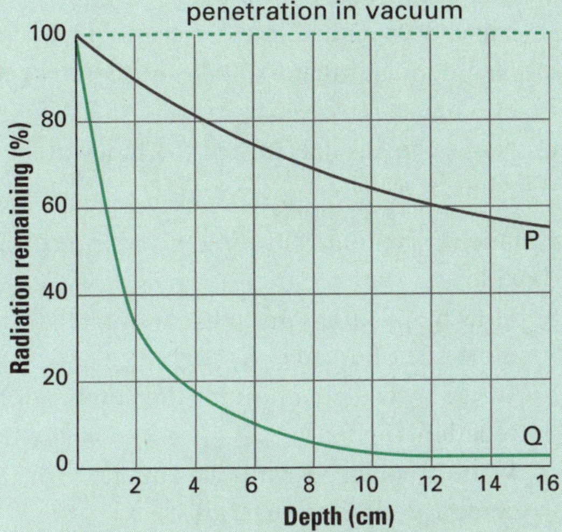

Figure 2.54 Beta and gamma penetration

   a) Identify radiation P and Q. *(2 marks)*
   b) After how many centimetres has radiation Q dropped to half its original amount? *(1 mark)*
   c) Extrapolate the graph and estimate how many centimetres radiation P takes to drop to half its original amount. *(1 mark)*
   d) Comment on the penetration of both radiations after a depth of 8 cm. *(2 marks)*

26. Figure 2.55 shows a ball-and-stick model of a chemical reaction involving molecules.

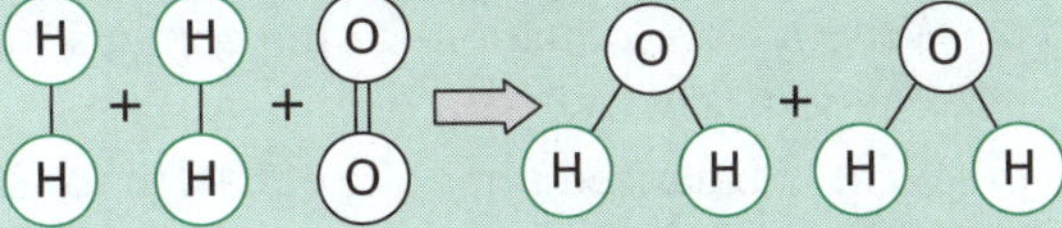

Figure 2.55 Ball-and-stick model

   a) What major scientific idea is illustrated by this equation? *(1 mark)*
   b) Classify the types of molecules according to the number of atoms per molecule. *(2 marks)*
   c) What type of reaction does this model represent? *(1 mark)*

27. Read the following description of substance Q.
- *Q is a colourless liquid.*
- *Q reacts with calcium metal with the release of a colourless gas that explodes in the presence of a flame.*
- *Q reacts with iron carbonate to produce a yellow solution and a colourless gas. The colourless gas turns limewater white.*
- *Q conducts an electrical current.*

a) Suggest two chemical substances that could produce the reactions described. *(2 marks)*
b) Write the chemical formulae for the substances named in a). *(2 marks)*
c) Name the gas that explodes in the presence of a flame. *(1 mark)*
d) Name the gas that turned the limewater white. *(1 mark)*

28. Ethane (C_2H_6) burns in oxygen to form carbon dioxide and water.
a) How many atoms are present in one molecule of ethane? *(1 mark)*
b) Write a word equation for the combustion reaction. *(1 mark)*
c) Write a balanced symbolic equation for the combustion reaction. *(2 marks)*
d) How could a student prove that carbon dioxide is released in the combustion process? *(2 marks)*
e) Why is an ignition source not required after the combustion reaction has begun? *(1 mark)*

29. Yellow sulfur burns in a gas jar of pure oxygen with a pale mauve flame. A colourless gas is formed. The gas dissolves in water and the solution formed turns blue litmus red.
a) What are the indicators of a chemical change in the combustion process? *(2 marks)*
b) Name the gas formed in the combustion reaction. *(1 mark)*
c) Write a word equation for the combustion reaction. *(2 marks)*
d) Write a balanced symbolic equation for the combustion reaction. *(2 marks)*
e) What can one conclude about the acid–base properties of the colourless gas? *(1 mark)*

30. Name the salts produced in each of the following reactions with acids. *(3 marks)*
a) zinc dissolves in sulfuric acid
b) calcium oxide dissolves in nitric acid
c) sodium carbonate dissolves in hydrochloric acid

31. A natural indicator was made by extracting the dye from flower petals into water. The pink extract was tested in a variety of solutions. The results of the experiment were:
- acetic acid turns from pink to deep red
- sodium hydroxide solution turns from pink to deep green.

The extract was then used to test each of the following household substances. Predict the colour change that may occur. *(2 marks)*
a) ammonia window cleaner
b) lemonade

32. a) Hydrogen (a fuel) and oxygen are mixed together in a vessel but they do not react. Explain why they do not burn. *(1 mark)*
b) Once ignition occurs in part of the combustible mixture, why does the combustion reaction continue? *(1 mark)*
c) Hydrogen air ships (zeppelins) were once used in the 1920s and 1930s to carry passengers across the Atlantic Ocean. Zeppelin use was discontinued after the *Hindenburg* exploded while landing at New York. These days, zeppelin balloons are filled with helium rather than hydrogen. Why is this safer? *(1 mark)*

33. Explain why fires can be extinguished using the following techniques. *(2 marks)*
a) throwing sand over a burning pool of oil
b) spraying the fire with water

34. Match the cause and effect statements. *(3 marks)*

Cause:
a) A spark jumped from the landing tower to the zeppelin craft.
b) The burst of heat radiation created a convection current.
c) The spark raises the temperature of the hydrogen and oxygen mixture.

Effect:
i) The flames on the forest floor suddenly became more intense.
ii) Combustion of the fuel and oxygen mixture occurs when the temperature reaches the ignition point.
iii) The hydrogen in the giant balloon exploded.

35. Combustion can be classified as complete or incomplete. When compounds containing carbon are burnt in low concentrations of oxygen, carbon monoxide and carbon (soot) usually forms.

a) What oxide of carbon is produced in complete combustion when the oxygen concentration is high? Write the chemical formula for this oxide. *(2 marks)*

b) Soot is a pollutant formed during poor combustion. What damage can this cause the environment or living things? *(3 marks)*

36. Antoine Lavoisier was convinced that combustion of carbon-containing fuels and respiration in the body are linked.

a) State one similarity between these two processes. *(1 mark)*

b) State one difference between these two processes. *(1 mark)*

c) What simple chemical test shows that carbon dioxide is breathed out from our lungs? *(1 mark)*

37. Hydrogen gas is often referred to as a 'clean' fuel.

a) When hydrogen burns, what combustion product is formed? *(1 mark)*

b) Using the correct formulae for hydrogen molecules and oxygen molecules, write a balanced symbolic equation for this combustion reaction. *(1 mark)*

c) Explain why hydrogen is called a clean fuel. *(1 mark)*

d) Why is hydrogen not more commonly used as a fuel? *(1 mark)*

38. Leif was asked by his teacher to investigate the following problem.

Do small pieces of zinc release hydrogen gas at a faster rate in acid than larger pieces of zinc?

a) Leif first of all considered the possible variables. Here is a list he drew up. Which of these factors do you think are relevant to the investigation? *(6 marks)*

Leif's variables:

- strength (or concentration) of the acid used
- size of the pieces of zinc
- size of the test tubes used in the experiment
- weight of zinc used
- time the investigation was begun
- volume of acid used
- type of acid used
- initial temperature of the acid and the zinc.

b) Leif designed an experiment, taking into account the following rule.

Only one variable should be tested at a time.

i) Which variable did he alter? *(1 mark)*

ii) Leif's friend suggested he use different amounts of acid in each container as part of his investigation. Would you agree with this proposal? Explain. *(2 marks)*

39. Design an experiment to test the following hypothesis.

The total acid content in squeezed citrus juice increases with age and high storage temperatures.

a) Identify the variables that need to be controlled. *(4 marks)*

b) What equipment and chemicals will be needed for this experiment? *(3 marks)*

c) If the hypothesis is supported, what results would you expect to see? *(2 marks)*

d) Write the method for your experiment. *(4 marks)*

40. a) Distinguish between photosynthesis and respiration. *(2 marks)*

b) Identify the site where chlorophyll molecules are located in a plant leaf. *(1 mark)*

c) Identify the cell organelle in which cellular respiration occurs. *(1 mark)*

d) The following equation for respiration is incomplete. What coefficient (number) must be placed in front of the oxygen molecules to balance this equation? *(1 mark)*

$C_6H_{12}O_6 + O_2 \rightarrow 6CO_2 + 6H_2O$

Go to pp. 233–237 to check your answers.

CHAPTER 3 Plate tectonics

Overview

In this chapter you will learn about:

- the structure of the Earth
- tectonic plates and continental drift
- technologies used to measure continental drift
- types of plate boundaries
- mountain building and seafloor spreading
- earthquakes and volcanic activity at plate boundaries
- measuring the magnitude of earthquakes
- stratigraphy and interpreting geologic cross-sections
- the evolution of the Australian continent
- living near tectonically active zones.

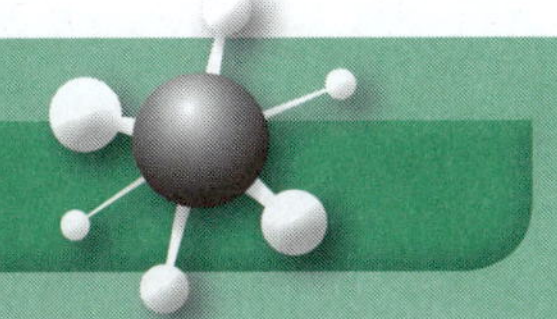

Glossary

Archipelago—a chain or cluster of islands

Asthenosphere—the soft mantle layer below the lithosphere

Convection—a heat current that moves fluids from hot to cooler regions

Correlation—the methods by which the age relationship between various strata of Earth's crust is established

Epicentre—the point on the Earth's surface directly above the focus

Fault—a planar fracture or crack in a volume of rock, along which there has been significant movement

Focus—the point where an earthquake or underground movement originates

Lithosphere—solid outer shell of the Earth which includes the crust and upper mantle

Passive satellite—a satellite that serves only to bounce back a laser beam aimed at it

Plate tectonics—a study of the forces that cause the movement of the crustal plates

Pyroclastic—describes hot rock and ash fragments released in a volcanic explosion

Radiometric dating—determining the age of rocks or fossils using the known half-lives of radioisotopes

Seismology—the study of earthquakes

Strata—layers of rock (singular = stratum)

Subduction—the zone where oceanic crust moves below continental crust

Tectonic—movement and deforming of the Earth's crust

Transform—movement of crustal plates in opposite directions on either side of a fault line

3.1 Tectonic plates

Structure of the Earth

Even though humans have travelled to the Moon, no-one has travelled deeper into the Earth than about 12 km or so (the deepest mine). This is a very small distance compared to the radius of the Earth: 6370 km.

Scientists have discovered from earthquake analysis that the Earth is made up of four layers as shown in Figure 3.1. These four main layers are called:

- crust
- mantle
- outer core
- inner core.

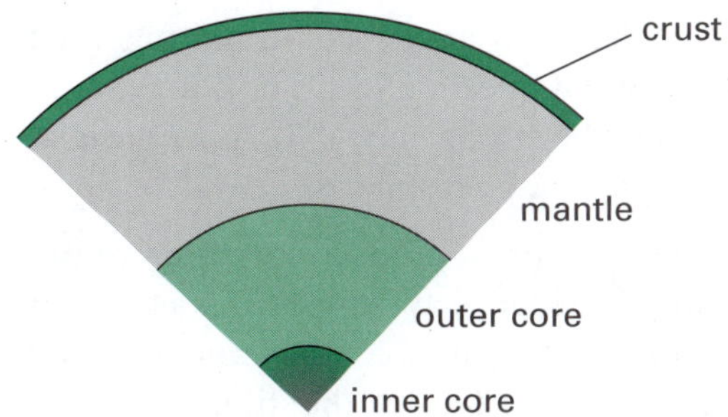

Figure 3.1 Structure of the Earth

Crust

The crust is a very thin outer layer of the Earth. On average it is about 20 km thick. On the continents it is about 35 km thick, but it is thinner under the oceans, roughly only 5 km. Rocks in the Earth's crust have an average density 2.8 times that of water. This is much less dense than other layers. The temperature of the crust gradually increases with depth. It reaches values of about 1000 °C at the base of the crust and in the upper mantle.

Mantle

Over 80% of the Earth's volume lies in the mantle. It extends from the base of the crust to the outer core, a distance of 2885 km. The temperature increase with depth in the mantle is more gradual than in the crust. It rises from about 1000 °C in the upper mantle to 2500 °C at its base. Though solid, under certain conditions of extreme pressure the mantle can flow. In the upper mantle, below about 100 km, lies a zone of weak rocks called the asthenosphere. Rocks here are partially molten and easily deformed. Some of the rocks associated with volcanic activity come from the upper asthenosphee. Above the asthenosphere is the lithosphere. This is the rigid outer layer of the Earth and includes the crust and part of the upper mantle. These rocks are cool enough to behave like a brittle solid. Scientists believe the plastic (able to be moulded by force) material of the asthenosphere moves the rigid lithosphere above it.

Core

From the mantle to the centre of the Earth is a distance of 3485 km. The pressure in the core is about 4 million times the air pressure at the Earth's surface and the temperatures are between 3000 and 5000 °C. The core consists of two parts: a liquid outer layer 2270 km thick; and a solid inner part with a radius of 1215 km. Scientists believe that the core is mainly made of iron and nickel together with small amounts of other elements such as sulfur and oxygen.

Moving continents

Have the continents of the Earth always been in their current location? Many people over the last 400 years have believed that the answer to this question is no but it is only in the last 60 years that the belief in moving continents has been taken seriously. Consider the following observations.

- In 1620 Francis Bacon, writing in his book *Novum Organum*, saw that the west coast of Africa and the east coast of South America could fit together.
- Alexander von Humboldt, the German naturalist and explorer, wrote in 1801 about the similarity between the coasts of West Africa and eastern South America. He saw these as two sides of a huge valley that was once scooped out by the sea.
- In 1858, the Italian Antonio Snider-Pellegrini suggested that after the Earth solidified from a molten state to form one land mass, certain imbalances caused it to be pulled apart.
- In 1912, the German meteorologist Alfred Wegener published two articles where he considered the idea of drifting continents seriously. He argued that about 225 million years ago a super-continent began to split apart. He called this supercontinent Pangaea (see Figure 3.2).

Figure 3.2 Pangaea supercontinent (about 225 million years ago)

- Wegener's views were supported by Alexander du Toit, Professor of Geology at Johannesburg University, and by Arthur Holmes in England. He further suggested that the supercontinent broke first into two large land masses. The northern land mass was called Laurasia and the southern one was called Gondwana (see Figure 3.3). This process was complete by about 170 million years ago. These then broke further apart to eventually form the continents we know today.

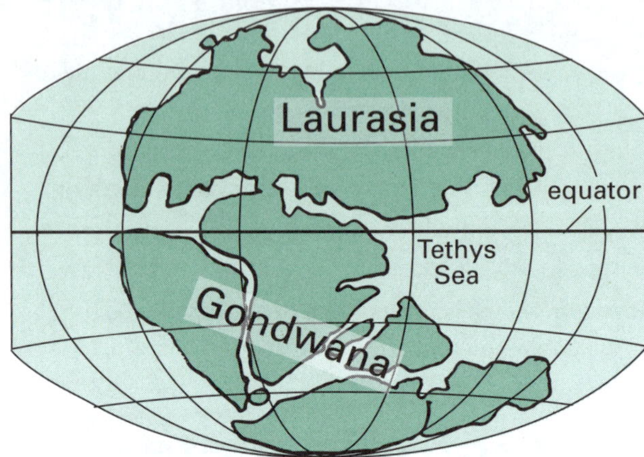

Figure 3.3 Laurasia and Gondwana

- Modern scientific evidence shows that India, Australia and Antarctica split away from the rest of Gondwana about 135 million years ago. Africa and South America remained as one land mass until 105 million years ago.

Throughout much of his life Wegener defended his continental drift theory and worked to find more evidence to support it. But many geologists of the day dismissed him, saying he was meddling in a subject outside his expertise (his training was in astronomy and meteorology). His theory never gained acceptance in his lifetime. He was even turned down for a number of professorships at universities in Germany (his home country) because of his unorthodox interests. In spite of the apparent scientific evidence, Wegener could not answer the most fundamental question of his critics: What kind of forces would be strong enough to move such large masses over such large distances?

While Wegener presented a large amount of circumstantial evidence to support his continental drift theory, he was unable to come up with a convincing mechanism. At the time there were many conservative scientists who were resistant to any change in their current beliefs. One of these arguments was that the oceanic crust was too firm and solid for the continents to simply plough through it. It just goes to show that the 'open-minded' scientist is a myth; scientists, like many others in society, harbour their pet ideas and theories. But, to be fair, solid evidence was lacking at the time.

Before Wegener could answer his critics, he died tragically in a severe storm in Greenland. He was only 50. However, he did set the seed which grew into the theory of plate tectonics some 30 years later. His hypothesis was not accepted until the 1950s when numerous discoveries provided evidence that continents drifted.

Evidence for continental drift

The hypothesis of moving continents is supported by a great deal of evidence. Here is some of that evidence.

Wegener's ideas

During the early part of the 20th century, Alfred Wegener made four main observations.

- Many of the continents 'fit' together, just like a jigsaw puzzle.
- Many rock types and ages are similar between continents.
- There are a number of ancient climatic zones in lands which are very different to their climates today.
- There is a certain distribution of many fossil plants and animals across different continents. These similar or identical fossils occur on continents now far away from each other.

Radioactive dating

The age of rocks can be found by radioactive dating (see Figure 3.4). When the different continents are re-assembled into the ancient southern continent Gondwana, the ages of granite and other rocks on either side of the continental margins coincide. Fossils show this coincidence too.

In the 1950s echo sounding was used to map the ocean floor. A mid-oceanic ridge was found in the Atlantic Ocean. Scientists found that this ridge continued almost all the way around the world; it was about 40 000 km long. Using radioactive dating, it was soon found that the rocks near the mid-ocean ridge were not very old. Close to the ridge their age is less than about a million years (very young geologically). They get older as the distance from the ridge increases. A deep trench was discovered in the middle of the ridge. The temperature in the centre of this ridge was exceptionally high. Further research showed that the trench marked the place where molten material was being pushed up from deep below the surface. This molten rock quickly froze to form new igneous rock, accounting for the young age of the rocks near the ridge. The formation of new crust at these ridges produced huge forces that pushed the continents apart. But where was the old crust going? In 1960 Harry Hess proposed that the old crust was being pushed back down at the margins of the continents. In this way the Earth did not expand but simply recycled its crust.

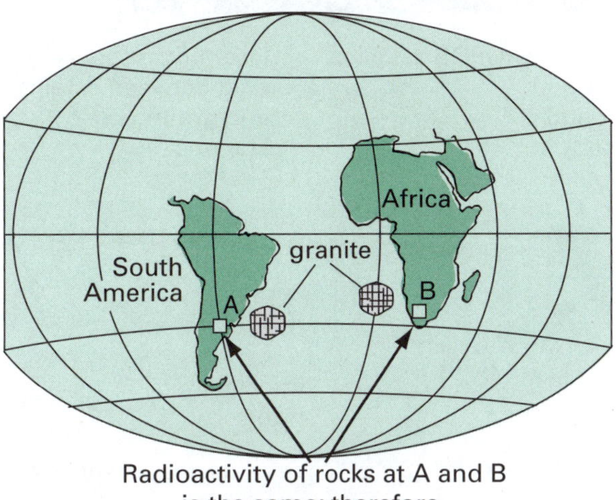

Figure 3.4 Radioactive dating supports continental drift

Tectonic plates

The theory of plate tectonics is based on evidence that the Earth's crust is made up of a number of slabs of rock called tectonic plates, which are able to slowly move over the semi-molten material below. Zones of earthquake and volcanic activity occur at the boundaries of the plates, where they collide. Tectonic refers to the forces which lead to movement of the Earth's crust.

Plate movement can be determined using satellites which can take very sensitive measurements. The first satellites used were the LAGEOS (Laser Geodynamics Satellites) which were launched in 1976 and 1992 (at an altitude of 5900 km) are passive satellites which receive pulses of laser light from ground stations that are located in many countries. The distance can be determined by measuring the time it takes the light to be reflected back to the ground from aluminium-covered brass spheres on the satellites (see Figure 3.5). Distances between ground stations and the satellite can be measured to an accuracy of between 1 and 3 cm.

(a)

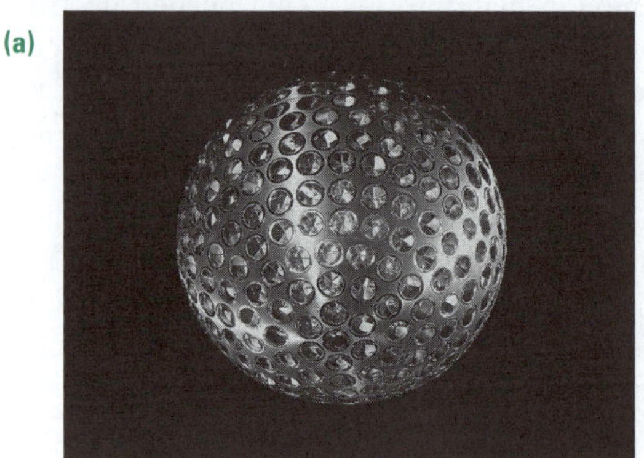

(b)

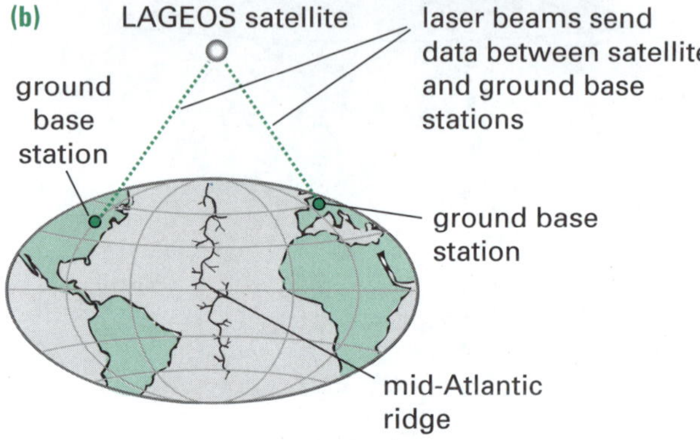

Figure 3.5 (a) LAGEOS satellite reflector; **(b)** Distances are measured by comparing how much the laser beam wave has shifted during the time it took to travel to the LAGEOS reflector and back.

Global Positioning System (GPS) uses satellites to monitor small crustal movements within continental plates. Ground stations are fixed to specific locations, often near fault systems. As the crust moves, the GPS can monitor these changes in position.

The tectonic plates are moving in different directions at different speeds. The tectonic plate on which the Australian continent is located is currently moving 35° east of north at a rate of 67 mm per year (about the same rate your fingernails grow). Geologists suggest that these plates move in response to the combined effects of convection currents in the hot, partly molten region of the upper mantle (called the asthenosphere) that lies below the plates, and gravitational forces that help to pull heavy plate edges downwards at subduction zones. Plate tectonics is the study of the forces leading to plate movement. Figure 3.6 shows these convection currents in the mantle below a mid-ocean ridge.

3.2 Plates and plate boundaries

The crust is divided into numerous major and minor plates. Figure 3.7 shows the major plates only.

Australia and India are located on the Indo-Australian Plate. The eastern side of this plate interacts with the Pacific Plate. The motion of tectonic plates can explain many natural events such as earthquakes and volcanoes.

There are two main types of plate boundaries in which the edges of plates can interact: constructive plate boundaries and destructive plate boundaries.

Constructive plate boundaries

These are sites where two plates are moving apart. Such zones are also called divergent or constructive plate boundaries. They produce mid-ocean ridges

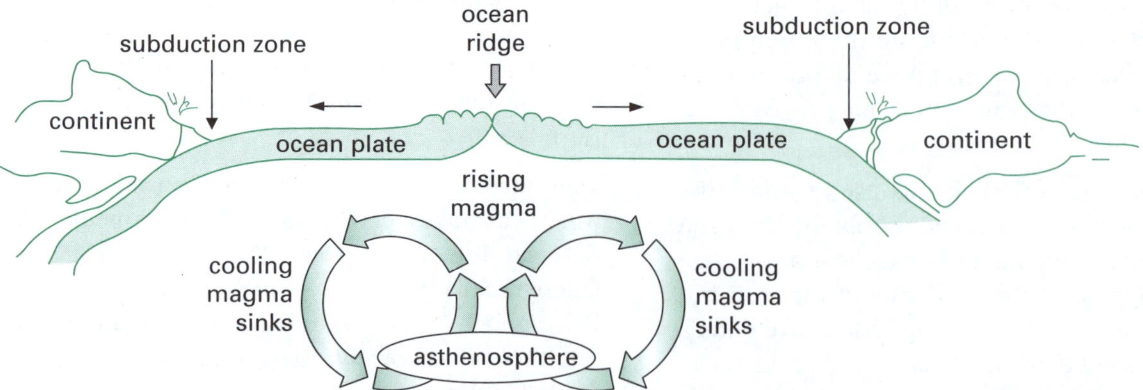

- Ocean plates are dragged away from the ocean ridge by convection currents in the asthenosphere.
- At the subduction zone, gravity pulls the heavy plate edge downwards.

Figure 3.6 Movement of magma in the asthenosphere

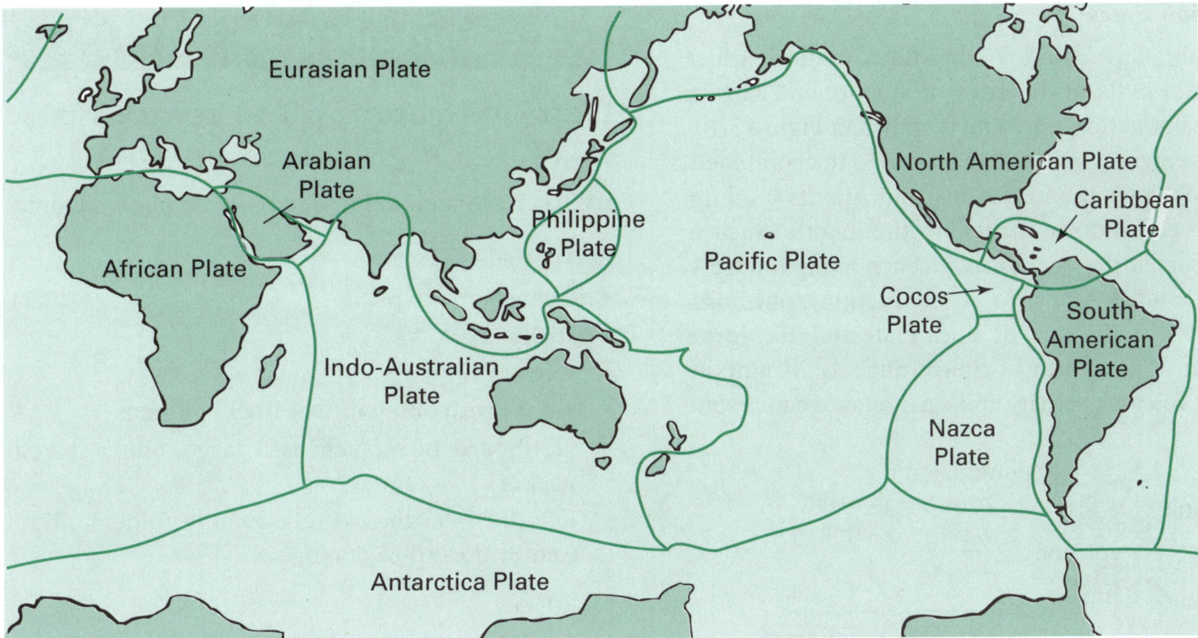

Figure 3.7 Tectonic plates

and rift valleys. Such boundaries usually create large geologic fault systems. A geological fault is a crack or fracture in the crust of the Earth. On each side of the fault zone, there has been movement of the crustal rocks.

Mid-ocean ridges

At mid-ocean ridges, molten rock rises to the surface and a new seabed and chains of volcanoes form as shown in Figure 3.8. The mid-Atlantic trench is an example of a spreading zone. Iceland is a volcanic island formed across this zone. The sea floor spreads as a result of the formation of new oceanic crust. Other smaller spreading zones exist; for example, between the Juan de Fuca Plate and the Pacific Plate.

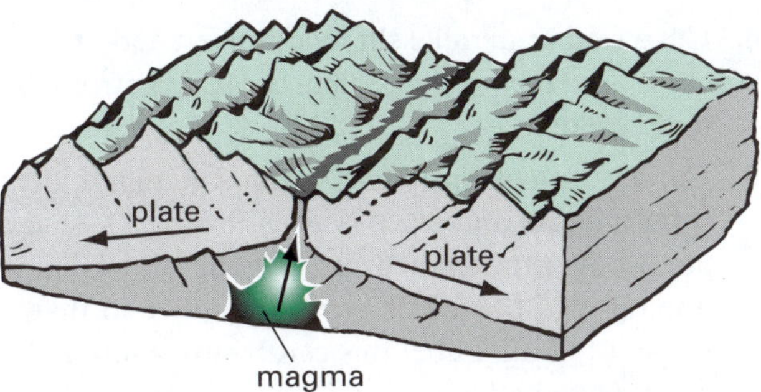

Figure 3.8 Mid-ocean ridge

Rift valleys

Rift valleys result where a divergent boundary exists within a continental mass. The East African Great Rift Valley is an example of a landform produced at a divergent boundary.

Destructive plate boundaries

Collision zones and subduction zones are examples of convergent or destructive plate boundaries. At these sites tectonic plates are moving towards each other.

Collision zones

Continental plates can push into each other and this leads to the formation of mountain ranges as shown in Figure 3.9. The Himalayas are formed as the Indo-Australian Plate and the Eurasian Plate collide. The mountains are rising about 6 cm per year as the land becomes folded upwards due to the immense forces of the colliding plates.

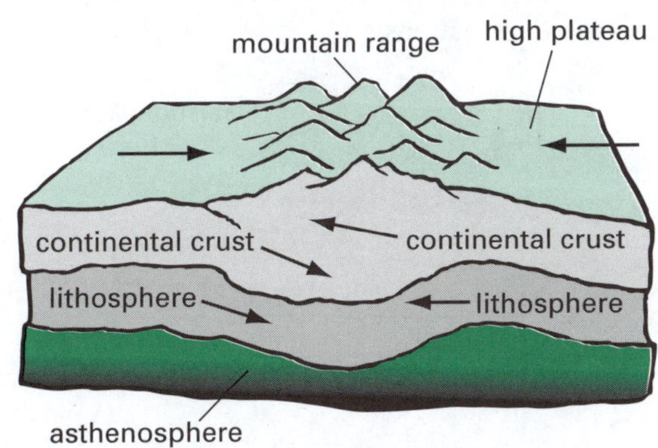

Figure 3.9 Collision zone and mountain building

Subduction zones

An ocean plate collides and slides underneath a continental plate at the edge of a continent. This is called a subduction zone and is shown in Figure 3.10. This movement is believed to be due to the combined action of convection currents and gravity pulling the plate edges downwards into the mantle. Such a zone occurs at the deep ocean trench along the west coast of South America. A subduction zone also exists between the Juan de Fuca Plate and the North American Plate north of San Francisco. Mountain building, volcanic activity and earthquakes can result.

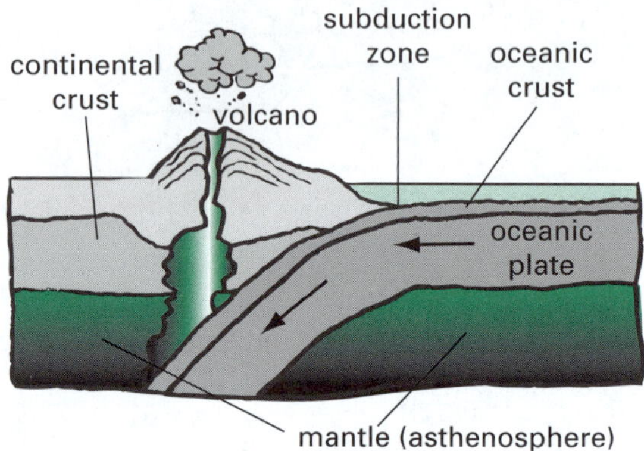

Figure 3.10 Subduction zone

Transform boundary or fault

As well as convergent and divergent plate boundaries, there is a third type of plate interaction called a transform boundary or fault (see Figure 3.11). This interaction neither creates nor destroys lithosphere. It is called a conservative boundary. Most transform faults are located on the ocean floor near spreading zones. The San Andreas fault in California is also an example of a transform fault where the North American Plate slides past the edge of the Pacific Plate. At such boundaries the plates slide past each other either in the same direction at different speeds or in opposite directions.

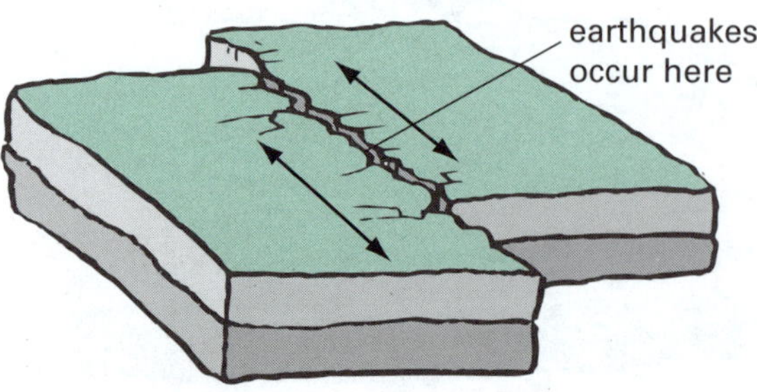

Figure 3.11 Transform fault; plates move relative to each other along an edge

Experiment 1

Demonstrating plate tectonics

Aim

- To create a model to demonstrate plate tectonics

Materials

- 2 sheets of A4 paper
- sticky tape
- scissors
- paint brush and paint, or thick felt pens
- cardboard box (such as a large, empty cereal packet)
- length of cardboard (e.g. manila folder), about 6 cm wide and 23 cm long

Method

1. Tape the two pieces of A4 paper together, end-to-end.
2. Fold the two pieces together along the sticky tape line.
3. Cut a narrow slit (about 1 cm wide) in the middle of one side of your cardboard box (see Figure 3.12).

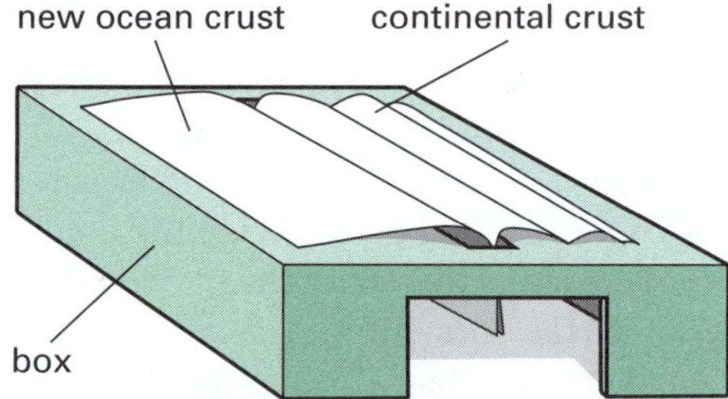

Figure 3.12 Modelling plate tectonics

4. Cut a similar parallel slit near the left end of your cardboard box, no more than about 7 cm from the middle slit.
5. Cut a length of cardboard (from a manila folder) 6 cm × 23 cm and place it as an inverted V-shape almost over the slit near the left edge. Sticky tape one edge of this cardboard to the box so that it doesn't fall off. Label this cardboard 'continental crust' (see the figure).
6. Pass your folded paper, sticky tape end first, through the middle slit. At this stage most of the paper should be in the box.
7. Have one end of the paper pass under the continental crust and slip it down the second slit.

8. Pull the free end of the paper outwards. The gap between the pieces of paper represents the rift valley of the mid-oceanic ridge.
9. Slowly pull out some paper on both sides of your 'rift valley' so that the left-hand sheet slides beneath your cardboard model of the continental crust.

Analysis

Answer the following questions to analyse the results of this activity.

1. What do the two pieces of paper represent?
2. Where is the trench formed?
3. What does the cardboard box represent?
4. How does this activity demonstrate divergent and convergent plate boundaries?
5. Is this a good model to demonstrate plate tectonics? Explain.

Go to p. 237 to check your answers.

Conclusion

Write a suitable conclusion.

Go to p. 237 to check your answer.

Test yourself 1

Part A: Knowledge

1. Alfred Wegener gathered information about rock types, fossils, glaciation and climate zones. He searched for a common explanation for all this data. He eventually decided that the present-day continents were in different positions millions of years ago. Wegner's proposal could best be described as *(1 mark)*
   - **A** a fact.
   - **B** an observation.
   - **C** a prediction.
   - **D** a hypothesis.
2. In the early stages of the break-up from Gondwana, India and Australia were on separate plates with an active spreading zone between them. Now the spreading zone between them has stopped spreading, so some people argue we are now on the same crustal plate. Which of the following statements is true about the movement between India and Australia? *(1 mark)*
   - **A** then: moving apart; now: moving apart
   - **B** then: moving apart; now: not moving
   - **C** then: moving closer; now: moving apart
   - **D** then: moving closer; now: not moving
3. The volcanic activity and mountain building along the west coast of South America is an example of which type of plate tectonic activity? *(1 mark)*
   - **A** subduction zone
   - **B** spreading zone
   - **C** collision zone
   - **D** transform fault zone
4. Which diagram in Figure 3.13 shows a collision zone at plate boundaries? *(1 mark)*

A

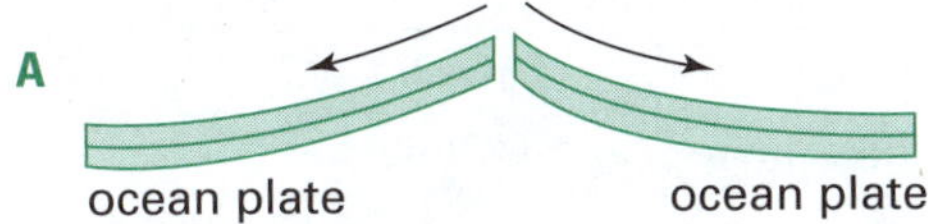

B

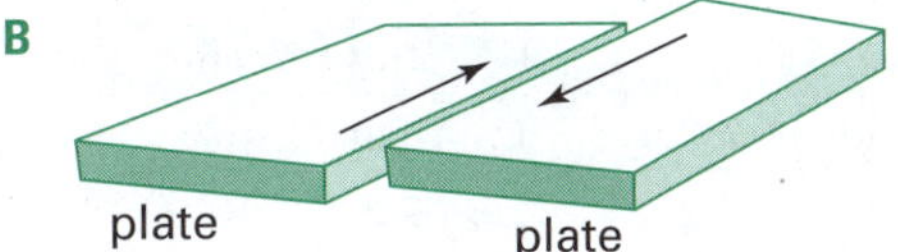

C

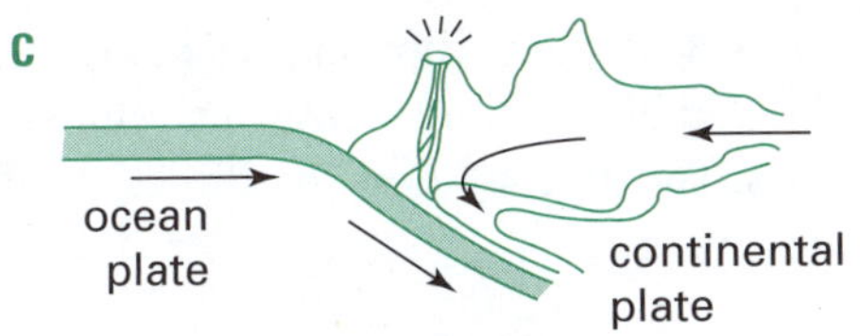

D

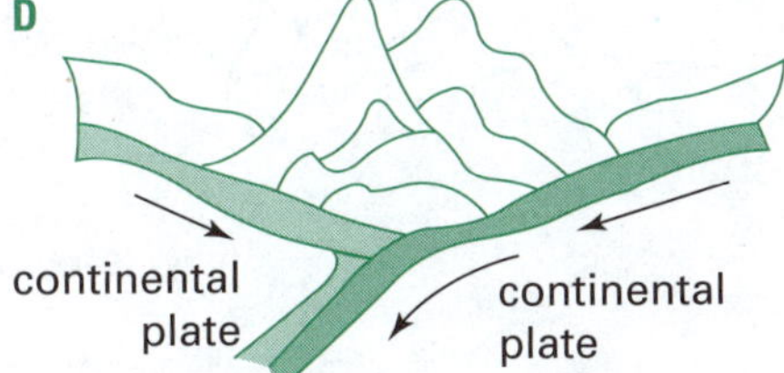

Figure 3.13 Plate interactions

5. Select the true statement about mid-ocean ridges. *(1 mark)*
   - **A** They are examples of divergent boundaries.
   - **B** They are examples of convergent boundaries.
   - **C** Mid-ocean ridges are examples of transform faults.
   - **D** Ocean plates become subducted at this location.

6. Complete the following restricted-response questions using the appropriate word. *(1 mark for each part)*
   a) Rocks in the Earth's .... have an average density 2.8 times that of water.
   b) In the upper mantle, below about 100 km, lies a zone of weak rocks called the ....
   c) Scientists believe that the core is mainly made of .... and nickel.
   d) Wegner argued that 225 million years ago a supercontinent called .... began to split apart.
   e) The word .... refers to the forces which lead to movement of the Earth's crust.

7. Use the code letters to match the terms or phrases in each column. *(1 mark for each part)*

Column 1	Column 2
A lithosphere	F liquid
B outer core	G convection
C subduction	H satellite
D LAGEOS	I crust recycling
E asthenosphere	J rigid

Part B: Skills

8. Figure 3.14 shows plate boundaries between Australia and New Guinea, and Australia and Antarctica. The arrows show the direction of plate movement.

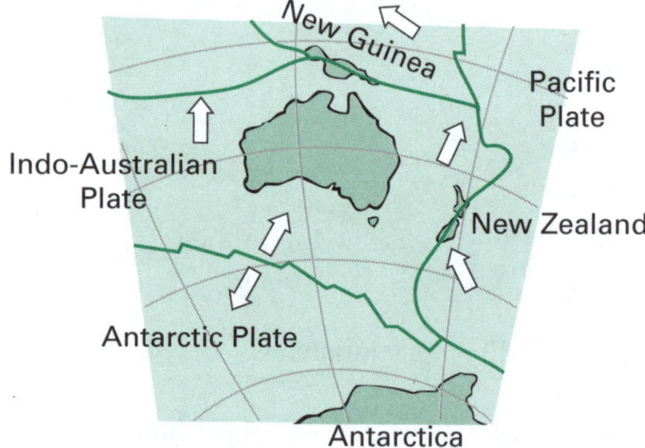

Figure 3.14 Indo-Australian Plate interactions with Pacific Plate

   a) What type of plate interaction between the Indo-Australian and Antarctic Plates is shown in this figure? *(1 mark)*
   b) The Indo-Australian Plate is moving north-east at an average speed of 67 mm per year. How far will it move in 1 million years? Convert your answer to kilometres. *(2 marks)*
   c) Explain why there are high mountain ranges in central New Guinea. *(1 mark)*

9. Use the theory of plate tectonics to account for the following observations.
   a) The rim of the Pacific Ocean, often called the Ring of Fire, has a large number of earthquakes and volcanoes located there. *(1 mark)*
   b) Flightless birds such as the ostrich, emu and rhea are found in Africa, Australia and South America respectively. *(1 mark)*
   c) The climate of the Sydney region has changed from cold and icy to warm and temperate over the last 135 million years. *(1 mark)*
   d) Earthquakes and volcanic activity are common in the Mediterranean area and the Middle East. *(1 mark)*

10. Figure 3.15 shows geological activity at a zone between two oceanic plates.

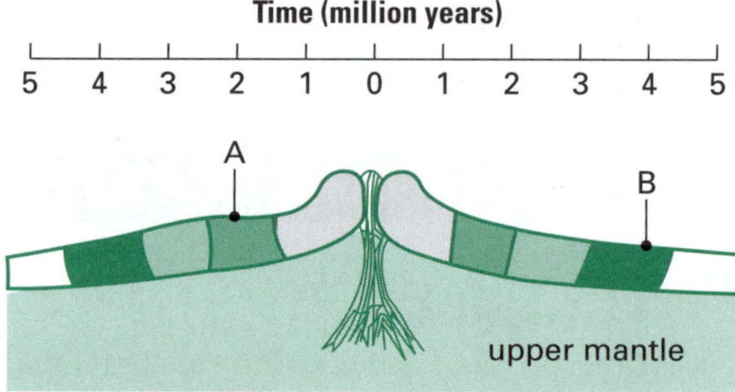

Figure 3.15 Geological activity at a plate boundary

   a) Name the type of plate tectonic activity shown. *(1 mark)*
   b) Use the scale to determine the age of the rocks at points A and B. *(2 marks)*
   c) Explain why rocks A and B have different ages. *(1 mark)*
   d) In which layer of Earth are the convection currents causing crustal movement found? *(1 mark)*
   e) If new crust is forming at these zones, why doesn't the crust of Earth expand? *(1 mark)*

11. Figure 3.16 shows a map of Gondwana before it split apart. Fossils of Mesosaurus, a freshwater reptile, were found at the locations marked on the map of Gondwana. Explain why the fossils

are not found in India, Australia and Antarctica. *(2 marks)*

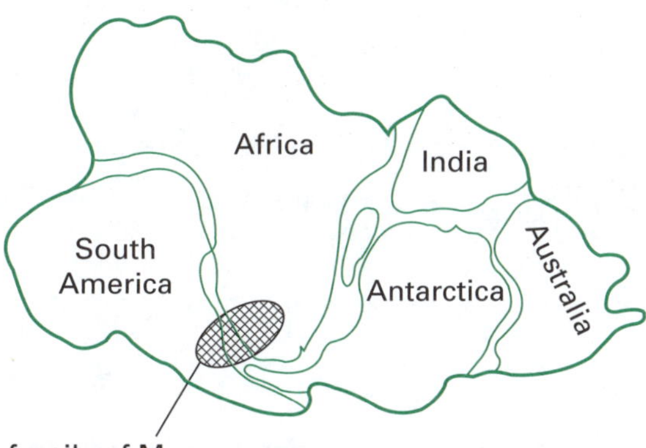

Figure 3.16 Map of Gondwana

12. a) In his book, *The Expanding Earth* (1977), Tasmanian Warren Carey stated that when he did his doctorate in 1937 he had to leave out many ideas of continental drift because they were too radical at the time and would have cost him his degree. What does this tell us about the view that scientists are open-minded? *(1 mark)*

b) Even as late as 1955, Carey couldn't get many of his papers published. So in 1956 he organised a continental drift symposium in Hobart. Why do you think he decided to organise this conference? *(1 mark)*

13. Use the theory of plate tectonics to predict which countries close to Australia could have major earthquakes, active volcanoes and geysers. Explain how you made this prediction. *(2 marks)*

14. Similar fossil plants and animals have been found on the east coast of South America and the west coast of Africa. Why is this important evidence for the theory of plate tectonics? *(1 mark)*

Go to pp. 238–239 to check your answers.

3.3 Earthquakes and volcanoes

A powerful earthquake has struck off the Indonesian island of Java this morning, killing hundreds of people, officials say. More than 3000 residents have been injured and it is feared the death toll will rise as many homes and villages have reportedly been buried by a landslide triggered by the quake.

You have possibly read newspaper reports like this. Earthquakes and volcanoes are regular features on this planet. As more and more people crowd together and cities grow, devastation and loss of life from such natural disasters will only increase.

In 1906 a disastrous earthquake struck San Francisco. While it lasted less than a minute, it destroyed most of the city and killed 700 people. Today San Francisco's population is much larger, so a similar earthquake could bring greater havoc. One of the worst earthquakes of the 20th century killed over 650 000 people in Tientsin Province, China in 1976.

Various types of movement occur along the different kinds of plate boundaries. These collisions create landforms such as coastal volcanoes, island arcs and mountain chains. When plates move apart, new ocean floor is formed as magma from the mantle rises up and deposits new rock along the plate boundaries. In some areas plates slide alongside each other, thus neither creating nor destroying land.

When the plates move, they cause vibrations that produce earthquakes. While thousands of earthquakes take place every year, only a few of them are destructive enough to be considered disasters.

Locations of volcanoes

Volcanoes form at subduction zones, at mid-ocean ridges and at hotspots in the Earth's crust.

Subduction zone volcanoes

Volcanoes often occur where the edges of two moving plates of the Earth meet. The Nazca Plate and the South American Plate form a subduction zone along the western edge of the South American continent as shown in Figure 3.17. The more massive plate, normally a continental plate (in this case, the South American Plate) will force the other plate, an oceanic plate (in this example, the Nazca Plate) down beneath it. This forms the subduction zone. The oceanic plate that is forced down during subduction enters into the magma and eventually is completely melted. The continent above builds up and layers of rock fold. The surface begins to look wrinkled.

The two plates grind together, creating enormous friction forces that melt the crust material. A pool of molten rock, called magma, forms. Because of the enormous pressures, the magma eventually rises through cracks in the crust to create a volcano on the surface. Magma is rich in silica and forms a sticky lava on the surface which is light in colour and slow moving. Such volcanoes tend to be explosive because the solidified lava often blocks the vent, leading to high internal gas pressure. Subduction zones can also occur offshore from a continental mass. The eastern end of the Asian continental crust, for example, is covered by shallow water. The subduction zone is located in a deep trench at the edge of this shallow zone. On the western side of the trench, magma formed by the frictional melting process occurring deep underground in the subduction zone moved up through the crust to create the islands of Japan.

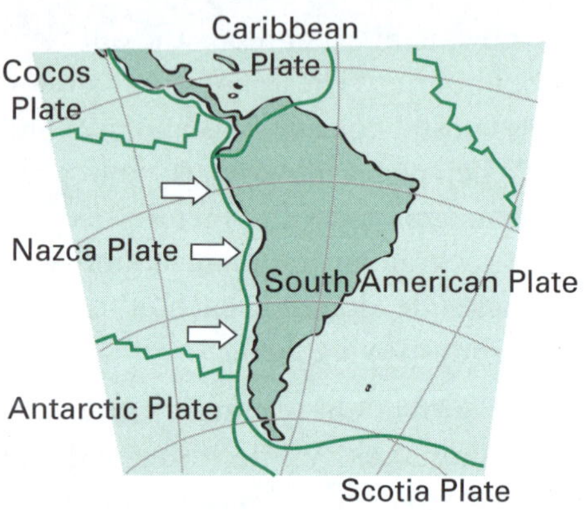

Figure 3.17 As the oceanic Nazca Plate pushes beneath the South American Plate, earthquakes result from the release of stresses. This subduction process (around 7.8 cm/year) is responsible for many earthquakes and volcanism, and is actively building the Andes mountains.

Figure 3.18 shows the active zone of volcanoes and earthquakes around the Pacific rim. This zone is often called the Ring of Fire. The Ring of Fire forms an arc stretching from New Zealand, along the eastern edge of Asia, north across to Alaska, then southwards along the coast of North and South America. Along this ring lies over 75% of the world's active and dormant volcanoes. This region of volcanic and earthquake activity was observed long before the theory of plate tectonics came about, and is found bordering the Pacific Plate and other major tectonic plates.

Mid-ocean ridge and hotspot volcanoes

At mid-ocean ridges the two plates are moving apart and new rocks form from the molten rock chamber deep in the mantle. The magma feeding these volcanoes is low in silica, dark and very runny.

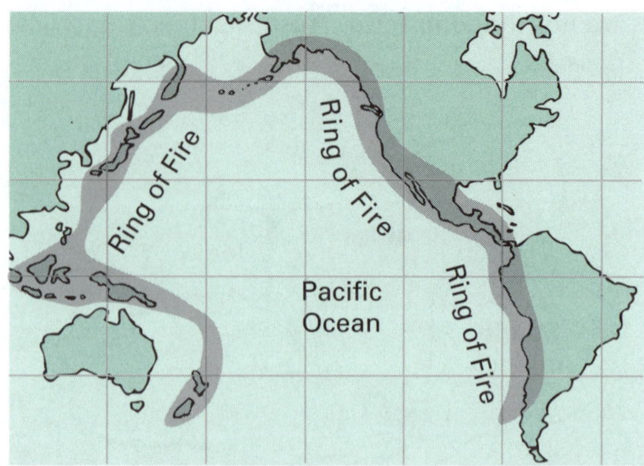

Figure 3.18 The most volcanically active belt on Earth is the Pacific Ring of Fire, a region of subduction zone volcanism. This ring occurs at plate boundaries and experiences many tremors and volcanic eruptions.

The red-hot magma is held in a chamber and it gradually pushes its way to the surface through cracks. Sometimes the molten rock hardens in the cracks before it reaches the surface to form a dyke.

If the molten rock reaches the surface, it produces a series of red-hot fire fountains along the crack. Just like water, hot runny lava forms rivers that run out of the ground. Successive eruptions form a wide basaltic volcano. As the magma chamber empties, the volcano may collapse into the chamber forming a caldera, or broad crater. The Hawaiian Islands were formed by hotspot volcanoes as shown in Figure 3.19; Iceland was formed by volcanic action at a mid-ocean ridge.

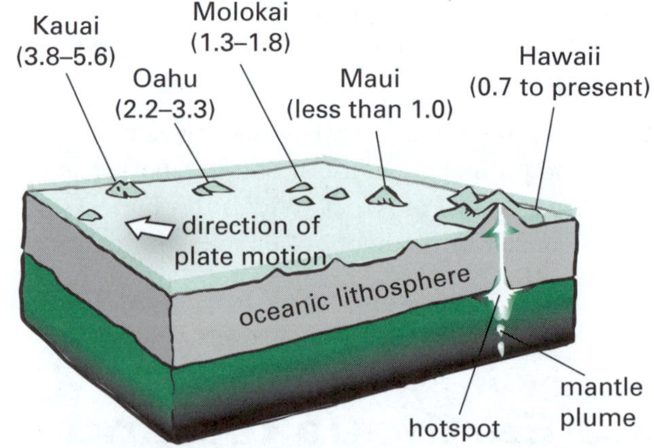

Figure 3.19 The Hawaii hotspot is a small, very hot area located under the Earth's surface. Over millions of years the Pacific Plate drifted across this stationary hotspot in a north-west direction forming a series of volcanoes that became the islands of Hawaii. (Ages of formation of islands are given in millions of years.)

Experiment 2

Volcanic activity

Aim

- To plot sites of volcanic activity on a world map

Method

1. Obtain a world map with latitude and longitude lines. These can be obtained on the internet, particularly from school geography sites.
2. Plot the location of the volcanoes from Table 3.1 onto your world map using the information on latitude and longitude.

Table 3.1 Volcanoes

Volcano	Location
Taal, Philippines	14°N, 121°E
Colima, Mexico	20°N, 104°W
Krakatau, Indonesia	6°S, 105°E
Etna, Sicily, Italy	38°N, 15°E
Montserrat, West Indies	17°N, 62°W
Guagua Pichincha, Ecuador	0°N, 79°W
Ruapehu, New Zealand	39°S, 175°E
Grimsvötn, Iceland	65°N, 17°W
Popocatepetl, Mexico	19°N, 99°W
Cerro Negro, Nicaragua	13°N, 87°W
Manam, Papua New Guinea	4°S, 145°E
Stromboli, Italy	39°N, 15°E
Iwate-san, Honshu, Japan	40°N, 141°E
Papandayan, Java, Indonesia	7°S, 108°E
Mount St Helens, United States	46°N, 122°W
Korovin, Alaska	52°N, 174°W
Arenal, Costa Rica	10°N, 85°W
Kilauea, Hawaii	19°N, 155°W
Rabaul, Papua New Guinea	4°S, 152°E
Eastern Gemini Seamount, Vanuatu	20°S, 170°E
Ruby Seamount, Mariana Island	33°N, 131°E

Analysis

Examine your map of the locations of volcanoes and list examples of volcanically active zones. Go to pp. 237–238 to check your answer.

Conclusion

Write a suitable conclusion.

Go to p. 237 to check your answer.

Types of volcanoes and historical accounts of volcanic activity

There are four main kinds of volcanoes.

- Cinder cones. These are the simplest type of volcano, growing from particles and congealed lava blobs ejected from a single vent as shown in Figure 3.20. Lava that is blown violently into the air breaks into small fragments, solidifying and falling as cinders around the vent to form a circular or oval cone. These ashes and small fragments are collectively called pyroclastic fragments. Most cinder cones have a bowl-shaped crater at the summit and rarely rise more than a few hundred metres above their surroundings.

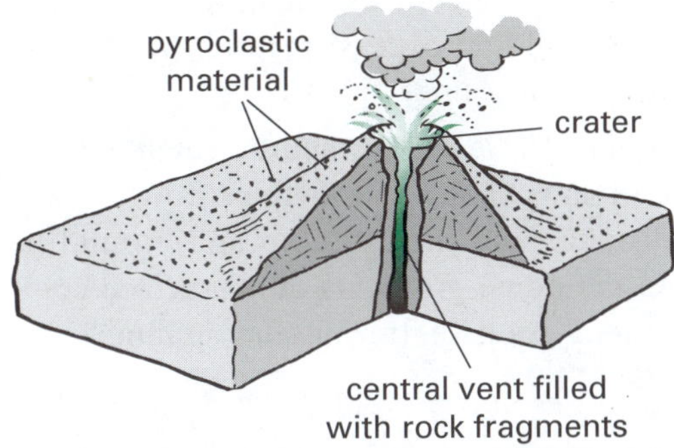

Figure 3.20 Cinder cones are made from deposited cinders and other pyroclastic material.

- Composite volcanoes. Sometimes called stratovolcanoes, they are typically large, steep-sided, with symmetrical cones consisting of alternating layers of lava flows, cinders, volcanic ash and blocks, rising as high as 3000 m. Many composite volcanoes have a crater at the peak containing a central vent or a group of vents as shown in Figure 3.21. Lavas either flow through breaks in the crater wall or squeeze through fissures on the sides of the cone. When lava

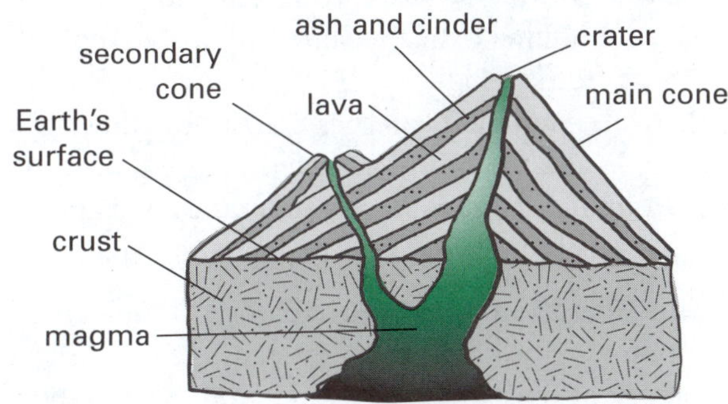

Figure 3.21 Composite volcanoes are formed with alternating layers of lava and rock fragments.

solidifies in the fissures, it can form dykes that greatly strengthen the cone. Many mountains are composite volcanoes, such as Mount Fuji in Japan and Mount St Helens in the United States (US).

- Shield volcanoes. These consist almost entirely of fluid lava flows. These repeating flows fan out in all directions from a central summit vent creating a broad, gently sloping cone as shown in Figure 3.22. They grow slowly by adding many multiple thin layers to their gently dipping sheets, spreading over great distances. Some of the largest volcanoes in the world are shield volcanoes. For example, the Hawaiian Islands are made of linear chains of these volcanoes. One of these is Mauna Loa, the largest of the shield volcanoes (and the world's largest active volcano), being 4170 m above sea level and over 8500 m above its base on the deep ocean floor.
- Lava domes. These are formed by small, blobby masses of lava that are too thick to flow any great distance and then pile up and around its vent. As the dome grows, its outer surface cools and hardens, soon shattering and spilling loose fragments down its sides. Some domes form craggy knobs while others form short, steep-sided lava flows (see Figure 3.24). Volcanic domes often occur inside craters or on the sides of large composite volcanoes. An example is Mount Pelée in Martinique (an island in the West Indies), which in 1902 demolished nearby St Pierre, a coastal town of 30000 people.

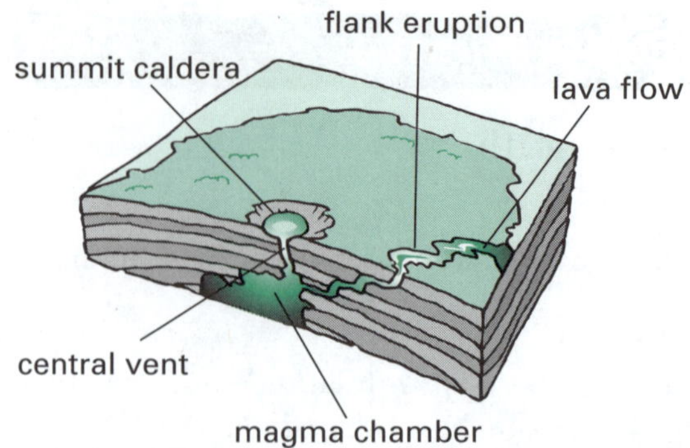

Figure 3.22 In shield volcanoes the fluid lava flows onto the surface, spreads out and cools forming a shape much like a warrior's shield.

Case study 1: Mount St Helens

After an earthquake occurred in May 1980, due to a sudden magma surge, the north face of Mount St Helens (a tall symmetrical mountain) collapsed in a mass of rock debris that slammed into the lake below. The rapidly released gas pressure within the volcano tore through the avalanche, creating a wind that swept over ridges and toppled trees. Hundreds of square kilometres of forest were destroyed. The mushroom-shaped ash column many tens of kilometres into the air drifted downwind blotting out the light with grey ash falling over the state of Washington and elsewhere (see Figure 3.23). Pumice poured from the crater and mud and rock slurries ran down the sides. While the eruption lasted only 9 hours, the surrounding landscape was dramatically changed fairly quickly, leaving a vast, grey landscape.

This eruption was the deadliest and most economically destructive volcanic event in the history of the US. Fifty-seven people were killed; and many buildings bridges and railway lines destroyed, costing billions of dollars. In 1982 the US government created the 45000-ha Mount St Helens National Volcanic Monument for research, recreation and education. Inside the national park, the environment is left to respond naturally to the disturbance.

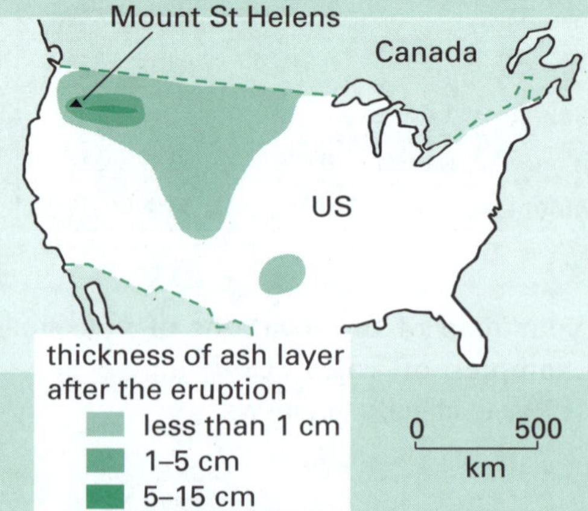

Figure 3.23 Location of Mount St Helens, and distribution of ash fallout in the US

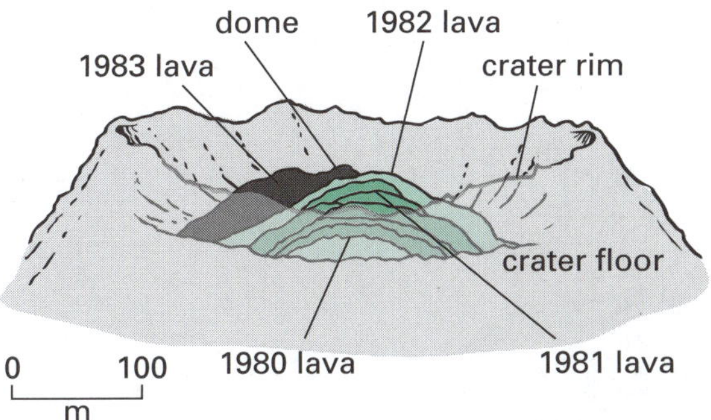

Figure 3.24 The development of the Mount St Helens lava dome between 1980 to 1983. Lava domes are one of the main structural features of many composite volcanoes.

Volcanic features

Even very ancient volcanic activity leaves traces in the form of lava flows, volcanic rocks and other features. Other evidence may be found in the shape of the land surface (landforms) that result from it. A volcano forms where erupted material builds up. Over time, weathering and erosion wear down and carry away surface materials, leaving behind remnants of volcanic rock that chill below the surface. While individual volcanoes vary, Figure 3.26 shows the basic features.

Some of the basic features of volcanoes are as follows.

- Dyke: sheet-like body of magma that cuts through and across layers of adjacent rocks and forms when magma rises into an existing fracture or forces its way through existing rock.
- Sill: a more or less horizontal layer of intrusive igneous rock forced between layers of sediments.
- Pluton: a body of igneous rock that solidified far below the Earth's surface.
- Batholith: a large mass of igneous rock that has melted and intruded surrounding rocks at great depths.
- Laccolith: igneous rock that did not find its way to the surface but spread laterally forcing overlying rock layers to bulge upwards.

Case study 2: Mount Vesuvius

Mount Vesuvius is a composite volcano on the Bay of Naples in Italy (see Figure 3.25). It is best known for its eruption in 79 AD that buried and destroyed the cities of Pompeii and Herculaneum. It was not lava that killed the thousands of people; it was hot ash and poisonous gases. Some researchers believe many died from an extreme heat surge produced by the volcano. Pompeii was buried 3 m deep, while Herculaneum was buried under 25 m of ash. The towns were eventually forgotten until they were re-discovered in the 18th century. Major changes happened to the mountain; vegetation was destroyed on its slopes and its summit altered due to the eruptive force.

In spite of the tragedies, the volcanic ash preserved and protected these cities against the elements until future archaeologists and historians unearthed this snapshot in time and are now studying them 2000 years later. Recent excavations in the slightly consolidated ash reveal the buildings, household objects and even the remains of some of the people and animals that lived in these flourishing Roman cities.

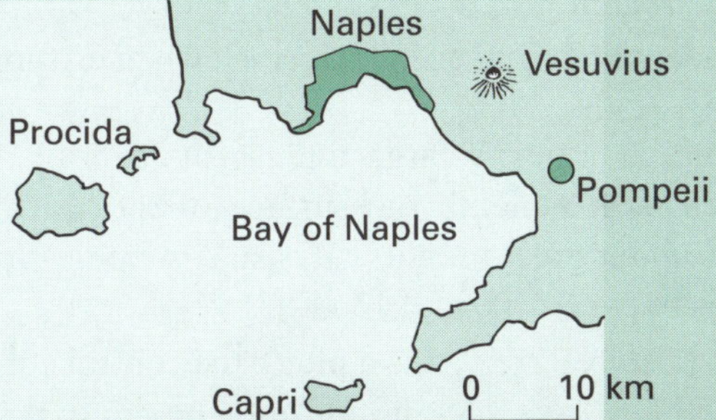

Figure 3.25 Location of Vesuvius in southern Italy, and the modern city of Naples nearby

Vesuvius has erupted many times since then, the last in 1944, and could erupt again making it dangerous to the 3 million people living near it. The Vesuvius crater now has a diameter of 700 m, and is 200 m deep, while the volcano itself is over 1000 m above sea level. While the volcano has been quiet for over half a century, some experts consider it the most dangerous in the world because at some time in the future it could roar to life with even more catastrophic results.

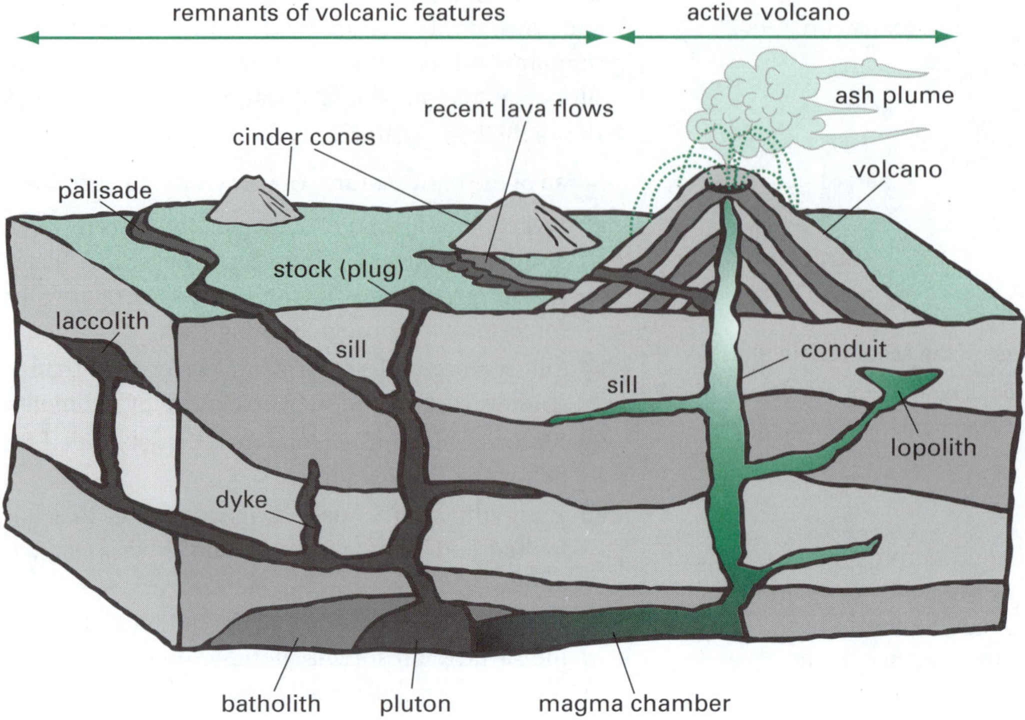

Figure 3.26 Volcanic features at different stages of their development

- Lopolith: a large igneous intrusion, similar to a laccolith, sagging in shape with a depressed central region.
- Palisade: a line of steep cliffs formed when erosion exposes a sill.
- Magma chamber: a large underground pool of molten rock beneath the surface of the Earth, that is under great pressure; its usual temperature ranges between 700 and 1300 °C.
- Stock (also referred to as a plug): the central core of solidified lava sticking up into the cone; it is usually harder than the surrounding rock.
- Lava: molten rock expelled by a volcano; it originates from magma but also includes surrounding rock that it has had to melt its way through, less volatiles (gases and water) that escaped when pressure was released.

Effect of volcanism on the atmosphere, biosphere and hydrosphere

Despite the obvious dangers of living near a volcano, there are many reasons why people continue to live and work near them.

- Volcanic ash is broken down and washed into the surrounding plains to form a rich, fertile soil.
- Valuable chemicals are brought to the surface in water, particularly sulfur and mercury. These can be mined.
- Geothermal power is available as hot water can be used to drive turbines to produce electrical power.

The effect of volcanism on the atmosphere, biosphere and hydrosphere are interrelated.

Effect on the atmosphere

When magma rises to the surface through cracks in the sedimentary strata (layers) it releases dissolved gases into the atmosphere, especially during eruptions. The most abundant gases are water vapour (H_2O), carbon dioxide (CO_2) and sulfur dioxide (SO_2). Ash and other solid particles are also thrown up into the atmosphere. This affects the weather near an erupting volcano as there is often much rain, lightning and thunder. Figure 3.27 shows some of the atmospheric effects caused by volcanic action.

However, volcanoes can affect more than just local weather patterns. As the atmosphere already contains a lot of water vapour and carbon dioxide, these gases usually don't cause global problems. But excessive increases in carbon dioxide levels may contribute to global warming.

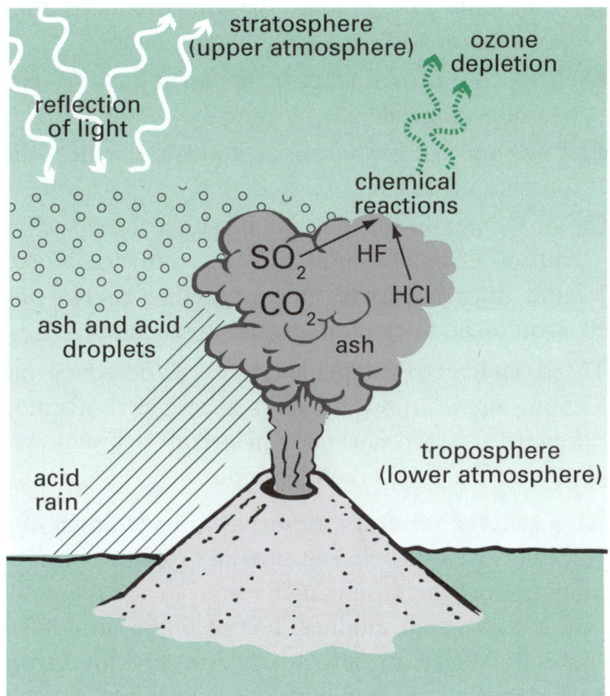

Figure 3.27 Some of the atmospheric effects due to an erupting volcano

The size of particles, and where they end up, can cause problems. Large particles let sunlight in but don't let heat radiated from the Earth's surface out, resulting in a warmer Earth (the greenhouse effect). But large particles can be quickly flushed out of the atmosphere by rain and gravity. Very fine particles can block some of the incoming energy from the Sun, causing cooling. Dust reaching the dry upper atmosphere can remain for weeks or months before finally settling, blocking sunlight and cooling large areas of the Earth.

Sulfur compounds can rise easily into the upper atmosphere where they combine with small amounts of water vapour, forming a haze of tiny sulfuric acid droplets. These droplets reflect large amounts of sunlight back into space for months, cooling the Earth. As the upper atmosphere is so dry, it takes time (up to several years) for the droplets to fall to the ground. If they only rise a few kilometres, in the lower atmosphere, they can be flushed out by rain. And acid rain can be harmful to plants and animals.

Effect on the biosphere

Exposure to acidic gases such as sulfur dioxide, hydrogen sulfide and hydrogen chloride produces a variety of health hazards to plant and animal life and, under extreme conditions, can lead to death. For example, in Hawaii with ongoing small eruptions since 1983, a volcanic fog of SO_2 gets carried in the trade winds around the island resulting in poor air quality, causing eye and mucous damage and respiratory problems.

Other common complaints include headaches, fatigue, respiratory difficulties and allergies. Under certain conditions volcanic CO_2 emissions, being

Case study 3: Mount Pinatubo

Mount Pinatubo, on a Philippine island, erupted in 1991 making it the second-largest volcanic eruption of the 20th century. It had a number of effects on the atmosphere, and on the people below.

- Up to 800 people were killed and 100 000 made homeless.
- About a half degree of cooling was noticed around the world for the next 2 years.
- Ash ejected from the volcano mixed with the water vapour from a nearby tropical storm, creating a rainfall of muddy rock and ash (up to 33 cm thick) covering an area of 2000 km^2. The weight of the ash collapsed roofs, killing hundreds of people.
- Between 15 and 30 million tonnes of SO_2 mixed with water and oxygen in the atmosphere to form sulfuric acid, which promotes complex chemical reactions leading to ozone depletion.
- Volcanic gases and ash reached an altitude of 34 km and over 400 km wide within 2 hours. The 400-km wide 'cloud' spread around the upper atmosphere of the Earth in 2 weeks and then upper atmospheric wind currents distributed the gases over the whole planet within a year.
- The eruption is believed to have had an effect on the 1993 Mississippi river floods and the drought in the Sahel region of Africa.

heavier than air, can collect in low and poorly ventilated places. Nearly 2000 people have died of CO_2 asphyxiation near volcanoes in the past two decades, mostly in Cameroon, Africa and Indonesia.

Volcanic gases can also severely damage vegetation. Direct exposure to concentrated volcanic gas or long-term exposure to dilute volcanic gas can kill most plant life. Areas of dead or dying forests abound near volcanoes.

Although most acidic droplets eventually fall to the ground as acid rain, they can corrode engines, farm machinery, communications equipment and vehicles while they remain in the atmosphere. Persistent acid rain causes galvanised nails or lead solder in water catchment systems to deteriorate, releasing toxic metals into drinking water.

Effect on the hydrosphere

Acid lakes can form near volcanoes as the emitted gases are dissolved. Some gases can dissolve by bubbling up through the lake or other body of water. Ash and other particles raining down from above pollute otherwise pristine water courses. Temperatures usually rise in waters that are close to volcanoes. And with the rise of temperature, the dissolved oxygen content in water decreases, affecting plant and animal life dependent on it. Increasing temperature also speeds up the metabolic activities of aquatic organisms, making them require more of the already depleted oxygen. Acid rain forms from dissolving SO_2 and other gases in atmospheric moisture. This was covered earlier.

In the Mount St Helens eruption, heat melted much of the snow on the mountain and, together with streams flowing down the side, created thick volcanic mudflows, which moved down the volcano's slopes at speeds up to over 140 km/h, carrying everything in their paths.

Effect on aircraft

A lot of volcanic ash consists of bits of pulverised rock and glass created by volcanic eruptions, which can find its way to areas of high winds where aeroplanes cruise. These ash clouds can spread many thousands of kilometres from the volcano and remain aloft for weeks and months. Downwind from an erupting volcano, pilots find it difficult to distinguish between volcanic ash clouds and weather-related clouds. They can damage aircraft by the following.

- Ash clouds can sandblast the windscreen, creating poor vision and damaging landing lights.
- Aircraft sensors, such as speed indicators can become clogged.
- Charged ash particles can interfere with radio communication.
- Very fine volcanic glass-rich ash particles sucked into a jet engine melt and fuse onto the blades and other parts of the turbine, fouling engines and eroding and destroying rotating machinery.

These can lead to aeroplane crashes. So detecting and locating these eruption clouds is crucial to aviation safety because ash can cause an aircraft to lose power in all engines in less than a minute.

Almost 100 aircraft are reported to have flown into volcanic ash cloud between 1980 and 2010, with results ranging from increased engine wear to simultaneous power loss in all engines. For example, in 1982 a commercial aircraft flew through an ash cloud from an Indonesian volcano and all four engines cut out. The plane then descended from 11 000 m to 3700 m, where the engines could be restarted. Luckily it managed to hobble to a safe landing, but the damage had been done and the repair bill ran into millions of dollars. In April 2010, airspace all over Europe was closed (for the first time) due to volcanic ash in the upper atmosphere from the eruption of the Icelandic volcano Eyjafjallajökull. This disrupted travel, causing great inconvenience to both airlines and the travelling public.

Recently a system based on infrared light was introduced that will allow pilots to detect volcanic ash plumes up to 100 km ahead and so safely fly around them. The system allows this information to be passed onto ground stations and so assist other aircraft. Satellites can also be used to detect, map and measure volcanic ash in volcanic clouds and this can be relayed to pilots.

Fault lines and earthquakes

A fault is a fracture or crack in the Earth's crust where one side has moved in relation to the other. While faults are classified according to how the blocks on either side of the fault have moved, it is usual to have some combination of fault movements occurring. Figure 3.28 shows how different types of forces acting within the Earth produce different types of faults.

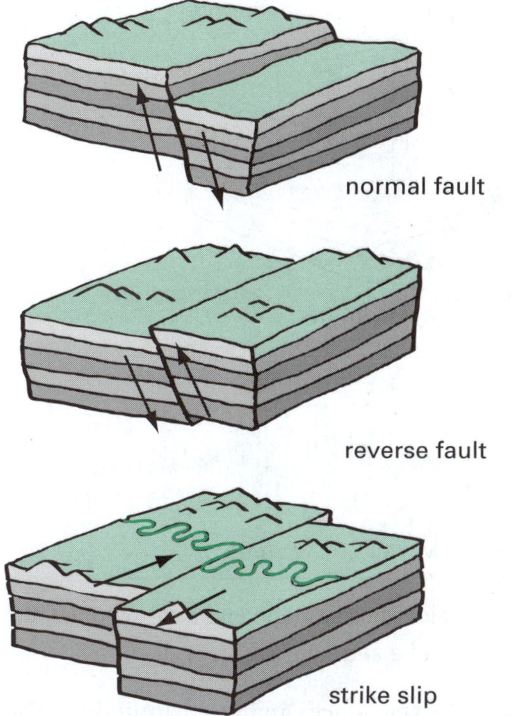

Figure 3.28 The type of movement reflects the kind of forces that are acting on the fault.

Strain in the crust builds up over a long period of time. Eventually a sudden movement occurs to lessen this strain (somewhat like a dry twig snapping if it's bent too far). An earthquake is the result of this sudden release of tremendous amounts of pent-up energy in the Earth's crust creating seismic waves. This is caused by underground movement along a fault plane or by volcanic activity.

Earthquakes occur at various depths underground. The site of the earthquake is called the focus. The epicentre is the point on the Earth's surface immediately above the focus. (See Figure 3.29.)

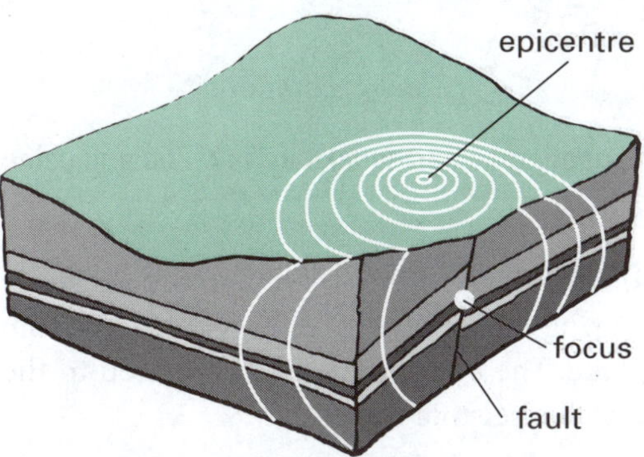

Figure 3.29 The focus is where the earthquake actually began; the epicentre is the point on the Earth's surface directly above the focus.

Measuring the magnitude of earthquakes

Seismology is the study of earthquakes. Most earthquakes occur on the edge of plates, which consist of many faults, especially where one plate is forced under another. Earthquakes also can cause a tsunami, or a series of waves that can cross an ocean and cause extensive damage to coastal regions.

Earthquake waves

Seismic (earthquake) waves can be classified into three groups (see also Figure 3.30).

- Primary (P) waves. These push waves are compression waves that can travel through solids, liquids or gases. These waves travel quite fast through Earth. The speed depends on the material they are passing through; for example, 330 m/s in air, 1450 m/s in water and about 5000 m/s in a hard rock like granite. As they pass through different materials, the waves bend (refract).
- Secondary (S) waves. These transverse shear waves travel through the solid Earth but not through liquids or gases. Consequently they do not penetrate the outer molten core of Earth. Shear waves cause rock particles to oscillate at right angles to the direction of the movement of the wave. These S waves travel more slowly and arrive at a given seismographic recording station later than P waves. Their speeds are around 60% of that of P waves in any given solid material.
- Land surface (L) waves. These waves, which originate at the epicentre of the earthquake on the Earth's surface, travel along the surface and cause the greatest damage. They are the slowest of all the seismic waves. They travel at about 90% of the S wave speed.

Measuring earthquake energy

A seismometer (or seismograph) is the scientific instrument used to detect and record earthquakes. These detect the ground motion and the results are recorded on a graph (seismogram). (See Figure 3.31.) The arrival times of the various waves are also recorded. Scientists use this information to calculate the size (amount of energy released) of an earthquake, which is then given a number: its magnitude. Data is usually received from several seismometers located at varying distances from the source of the earthquake and this helps to pinpoint its epicentre.

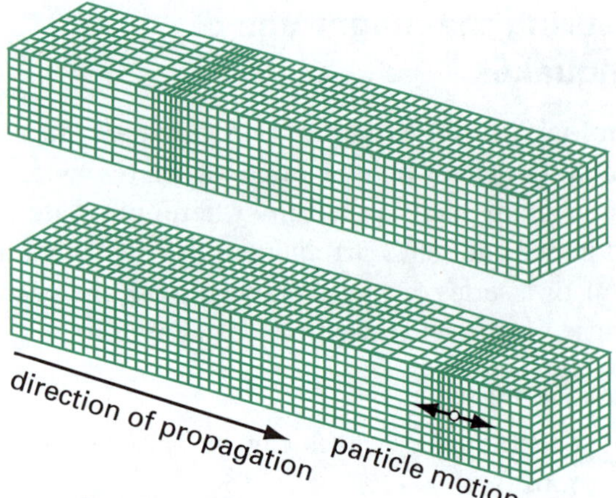

Movement of P waves
The motion of the particles is in the direction of propagation.

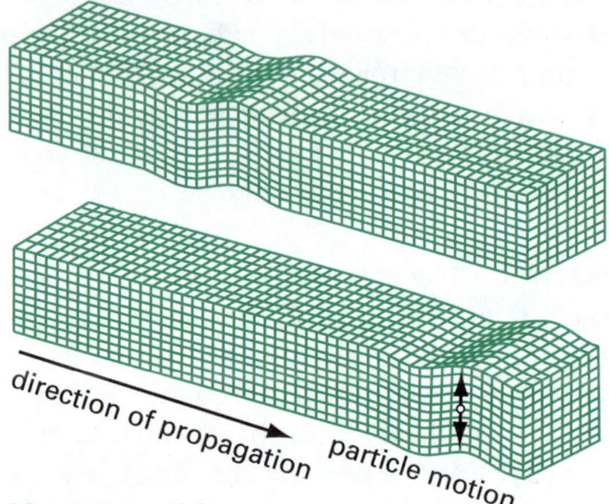

Movement of S waves
The particle motion is perpendicular to the direction of propagation.

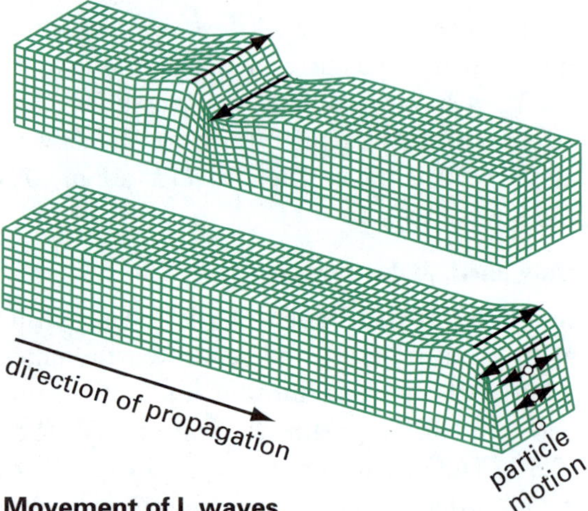

Movement of L waves
These surface waves move horizontally and perpendicularly to the direction of propagation.

Figure 3.30 Motion of P, S and L waves; the elastic material through which the wave travels returns to its original shape after the wave has passed.

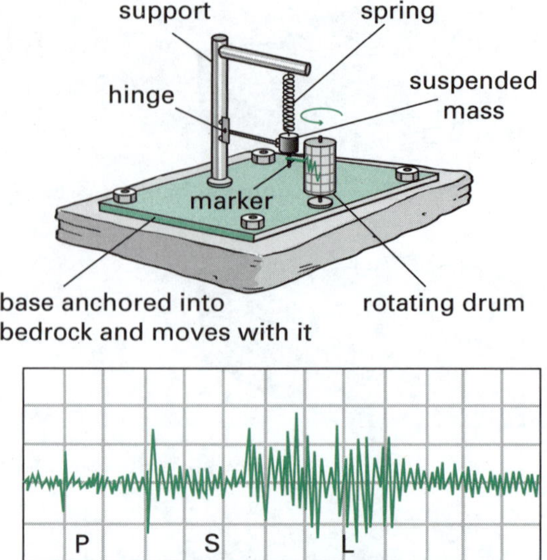

Figure 3.31 A typical seismograph and a seismogram, the graphical output of a seismograph

New Zealand experiences many earthquakes. The different arrival times of P and S waves for at least three different seismic stations can be used to locate the epicentre of an earthquake. An example is shown in Figure 3.32.

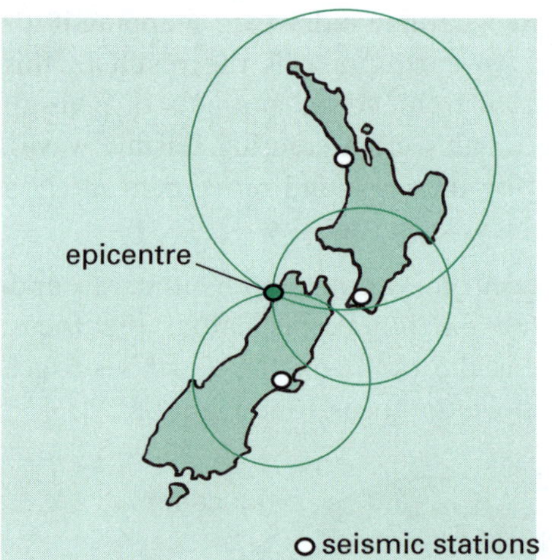

Figure 3.32 Locating an earthquake in New Zealand using seismic stations

Charles Richter devised the first magnitude scale in 1935, measuring the amount of energy released from an earthquake. The energy released is related to the amplitude of the seismic waves.

The Richter scale is not a linear scale. This is to accommodate the wide range of ground motions which earthquakes can cause. A change of one unit on the scale represents about a 31.6 times change in energy released by an earthquake. Thus an earthquake of magnitude 7 releases about 31.6 times more energy than one of magnitude 6. The largest known earthquake had a magnitude of over 9. The Richter scale does not measure damage because large earthquakes that happen in isolated areas cause little damage. Table 3.2 shows how earthquakes can be related to their Richter magnitude.

Table 3.2 The Richter scale

Earthquake	Richter magnitude
minor	0–3.9
light–moderate	4.0–5.9
strong	6.0–6.9
major	7.0–7.9
great	≥ 8

The worst earthquakes occur where there are large concentrations of people, such as near large cities. This causes great loss of life and huge devastation to properties and infrastructures. Sometimes more severe earthquakes occur in remote and sparsely populated areas, which pass mostly unnoticed by the world at large. Table 3.3 lists the top five worst earthquakes that have been recorded.

Experiment 3

Earthquake activity

Aim

- To plot sites of major earthquake activity on a world map

Method

1. Obtain a world map with latitude and longitude lines. This can be obtained on the internet, particularly on school geography sites.
2. Plot the locations of the major earthquakes from Table 3.4 (on the following page) onto your world map using the location information.

Analysis

Answer the following to analyse your results.

1. Compare your map of the major earthquakes with a map showing the tectonic plates. Is there any similarity?
2. Suggest a possible reason for earthquake activity there.

Go to pp. 237–238 to check your answers.

Conclusion

Write a suitable conclusion.

Go to p. 238 to check your answer.

Table 3.3 The top five worst earthquakes since records began

Where	Richter magnitude	When	Details
Valdivia, Chile	9.5	22 May 1960	Killed over 1650 people, injuring 3000 and making 2000000 homeless. Created a tsunami that spread and caused damage as far away as Hawaii, Japan and the Philippines. An earthquake rupture zone was more than 1000 km long. Nearby Puyehue volcano erupted two days later.
Prince William Sound, Alaska	9.2	28 March 1964	Killed 128. Most damage was the city of Anchorage, 120 km north-west of the epicentre. The tsunami created caused damage as far away as Hawaii.
Sumatra, Indonesia	9.1	26 December 2004	Around 227900 people killed or presumed dead, and around 1.7 million homeless over 14 countries in South Asia and East Africa. The epicentre was at a depth of 30 km. A few days later a mud volcano began erupting.
Sendai, Japan	9.0	11 March 2011	Estimated over 10000 dead from the combined effect of the powerful earthquake, aftershocks and the tsunami. Caused extensive problems with nuclear reactors.
Kamchatka, Russia	9.0	4 November 1952	No reported deaths. Earthquake generated a tsunami, causing damage in the Hawaiian Islands with waves over 9 m high.

Table 3.4 Major earthquakes

Place	Date	Intensity (Richter scale)	Location
Japan	11/3/2011	9.0	38°N, 142°E
New Zealand	4/9/2010	7.1	43°S, 172°E
Chile	27/2/2010	8.8	35°S, 72°W
Sumatra	6/4/2010	7.7	2°N, 97°E
China	12/5/2008	8.0	31°N, 103°E
Peru	15/8/2007	7.0	13°S, 77°W
Java	17/7/2007	7.7	9°S, 107°E
Greece	8/1/2006	7.0	36°N, 23°E
Kashmir	8/10/2005	7.6	34°N, 73°E
Sumatra	28/3/2005	8.6	2°N, 97°E
Sumatra	26/12/2004	9.1	2°N, 97°E
Iran	26/12/2003	6.6	29°N, 58°E
Japan	25/9/2003	8.3	42°N, 144°E
Alaska, US	3/11/2002	7.9	64°N, 147°W
India	26/1/2001	7.7	24°N, 70°E
Azerbaijan	25/11/2000	7.0	40°N, 50°E
Vanuatu	26/11/1999	7.3	16°S, 168°E
Ceram Sea, Indonesia	29/11/1998	8.1	2°S, 125°E
Mexico	9/10/1995	8.0	19°N, 104°W
Bolivia	9/6/1994	8.2	14°S, 68°W
Alaska, US	28/3/1964	9.2	61°N, 148°W
Chile	28/5/1960	9.5	38°S, 73°W
Russia	4/11/1952	9.0	53°N, 159°E
California, US	18/4/1906	7.9	38°N, 123°W
Ecuador	31/1/1906	8.8	1°N, 82°W

Living near plate boundaries: New Zealand, Japan and Indonesia

Countries located along tectonic plate boundaries, where plates grind together, will experience volcanic and earthquake activity (see Figure 3.33). Plates generally don't move smoothly past each other. They move in a series of small but rapid motions, each of which is accompanied by one or more earthquakes. Plates get locked and move very little for many years, even centuries, allowing pressure to build up. This is then suddenly and catastrophically released in the form of a major earthquake.

The Indian Plate and the Australian Plate fused together about 43 million years ago. Recent studies show that the Indo-Australian Plate may be in the process of splitting into two separate plates mainly due to the considerable stresses created by the collision of the Indo-Australian Plate along its northern border. So sometimes it is referred to as one plate (Indo-Australian) and at other times referred to as the Indian Plate and the Australian Plate.

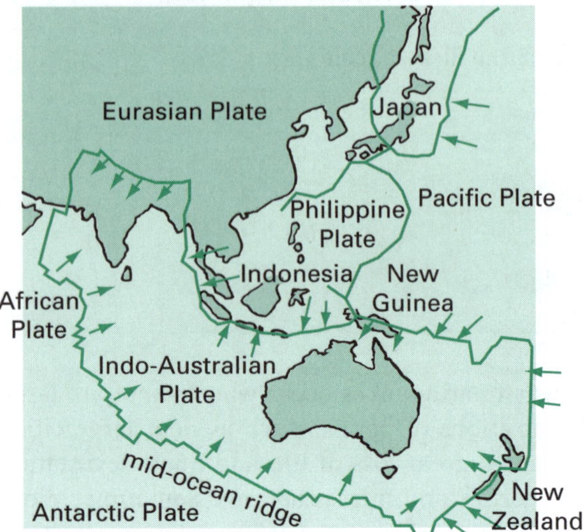

Figure 3.33 New Zealand, Indonesia and Japan are located on plate boundaries and will experience frequent earthquakes.

New Zealand

New Zealand is geologically active with high mountains, frequent earthquakes, geothermal activity and volcanoes. This is because New Zealand is located on the boundary of the Indo-Australian and the Pacific Plates (see Figure 3.34). As these two plates collide, the Pacific plate subducts (moves underneath) the Indo-Australian Plate that carries the North Island. The situation is reversed towards the south of the South Island; the Australian plate is being forced below the Pacific plate. On the South Island itself, the plates rub past each other horizontally.

Early white settlers to New Zealand did not know much about making buildings to resist earthquakes, so early brick buildings collapsed during minor tremors. Many older buildings in central Wellington (the capital) were built between 1880 and 1930, and were not designed to resist earthquakes. So in the 1970s the city council required such buildings to be strengthened. Today New Zealand's building codes specify how buildings must perform to withstand earthquake forces, giving directions to builders to create earthquake-resistant buildings.

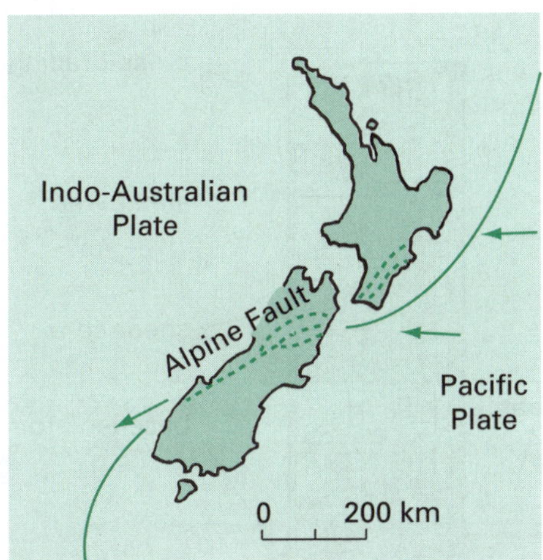

Figure 3.34 New Zealand lies at the edge of both the Indo-Australian and Pacific tectonic plates.

In September 2010 an earthquake (magnitude 7.1, at a depth of 10 km) hit the region around Christchurch, on New Zealand's South Island, damaging buildings but fortunately not causing any deaths. This was followed by about 10 aftershocks having magnitude greater than 5. In February 2011, another quake hit the region (magnitude 6.3) and is one of the most catastrophic quakes in New Zealand's history. The main difference is that in 2011 the ground broke almost directly under the country's second largest city, and at a shallow depth. Buildings that were weakened in 2010 would have crumbled this time. And New Zealand has earthquake building regulations that are among the strictest in the world.

Indonesia

Indonesia is a large archipelago consisting of some 17000 islands located on the edges of two tectonic plates, causing it to be very unstable (see Figure 3.35). There are some 400 volcanoes in Indonesia, of which 130 are active.

The intersection of these plates is responsible for the creation of a very complex geological setting. The Indo-Australian Plate is moving northward and becomes subducted (moves below) the Eurasian Plate at, on average, around 7 cm per year. Most of the earthquakes and volcanoes also concentrate in this subduction zone, a length of about 3000 km.

Like New Zealand and Japan, Indonesia is located on the Pacific Ring of Fire and experiences regular earthquakes. Indonesia is reputed to have more earthquakes than any other country in the world so it is imperative that steps be taken to ensure buildings and other structures are earthquake proof. A program has already begun to train some Indonesian scientists in identifying and analysing active faults to determine how big and how often earthquakes will shake the nation. Minimising earthquake risks and helping local governments and communities understand and prepare for future disasters will ensure lives can be saved and economic losses minimised when earthquakes occur. However, the large population (around 240 million) of mostly poor people often crowded together in substandard accommodation will make the task difficult.

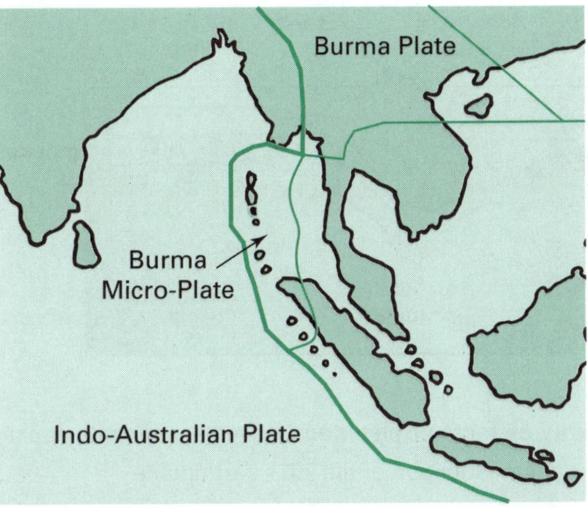

Figure 3.35 The plate boundaries around Indonesia

Japan

Japan sits atop the join of several huge tectonic plates on the Earth's crust. These plates mash and grind together, triggering regular earthquakes (see Figure 3.36). Around 1500 earthquakes strike this island nation every year, with minor tremors occurring almost daily. One important cause is the westward movement of the Pacific Plate, around 8.9 cm per year, which is responsible for major earthquakes in the past.

In March 2011 an earthquake of magnitude 8.9 occurred off the east coast of Japan's largest island under the Pacific Ocean. More than 150 aftershocks of magnitude 5 or greater followed. The quake was 160 times more powerful than the one that devastated Christchurch in 2011.

The undersea rupture (350 km by 100 km) at a depth of 24 km was shallow enough to trigger a tsunami. This offshore earthquake shook the ocean floor and produced a devastating tsunami, as high as 10 m, engulfing entire towns and reaching more than 5 km inland.

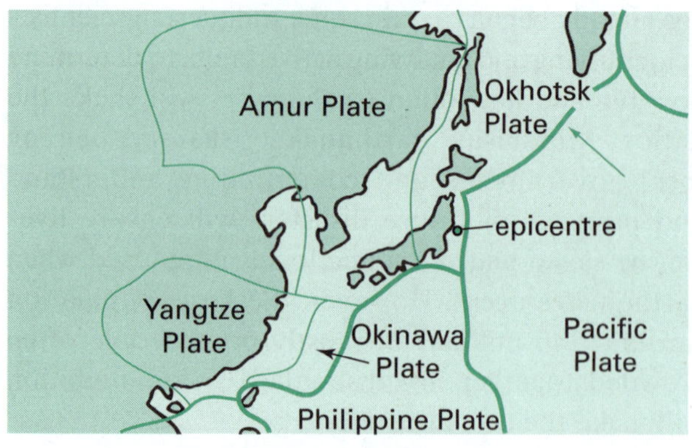

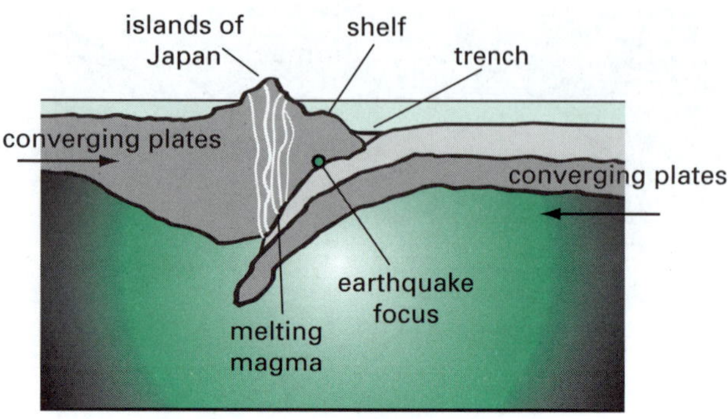

Figure 3.36 Tectonic plates and movement beneath Japan, showing the location of the 2011 earthquake

If an earthquake and tsunami killing tens of thousands of people, and obliterating cities and villages isn't enough, it also caused a nuclear emergency when it shut down the cooling systems at the country's largest electricity-generating nuclear plant. Radiation leaked into the air and waters around the plant, with people being evacuated to many kilometres away.

Japan has a tradition of innovation in earthquake-resistant buildings (see Figure 3.37) and trying out new disaster-proofing systems. But the double whammy of a giant earthquake and tsunami severely tested this, while ensuring the death toll wasn't much higher. Even the nuclear power plant was designed with earthquakes in mind, but no structure is ever 100% disaster proof.

3.4 Geological history of Australia

Stratigraphy and interpreting geological history from cross-sections

Stratigraphy is the study of rock layers and layering (stratification). It is mainly used to study sedimentary

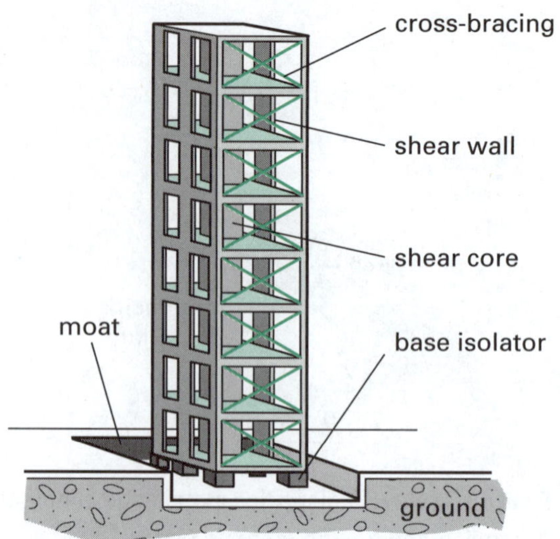

Figure 3.37 Japan is a world leader in creating earthquake-resistant buildings. A rubber core and deep foundations, to isolate the base, and cross steel bracing offer stability and support against seismic waves.

and layered volcanic rocks. The principle of superposition is a commonsense observation which states that, given normal conditions of deposition, sedimentary layers are deposited in a time sequence, with the oldest on the bottom and the youngest on the top (see Figure 3.38).

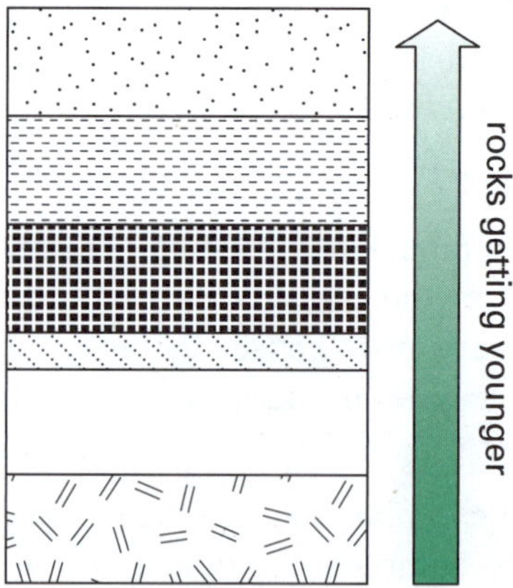

Figure 3.38 Younger rock layers are above older rock layers in this stratigraphic column.

A number of stratigraphic columns made from different areas in the same region can allow you to reconstruct its history. Columns can be made by considering outcrops (natural exposures, road and rail cuttings) of rocks or by digging down (bores, mine shafts) into the layers.

Figure 3.39 shows a common method of correlating by rock types.

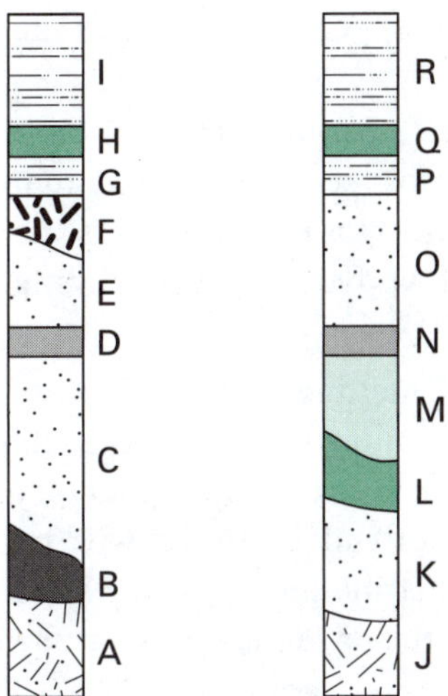

Figure 3.39 Correlation by rock types

Correlation allows you to compare the sequence of rock strata in a region and enables you to fix their relative ages. These two columns are taken from the same region, but several kilometres apart. While they do not match exactly, notice the following features.

- There are a number of matching rocks at the same levels (e.g. I and R, H and Q, G and P, D and N). These rocks are from the same layer and are the same age.
- Rock D is younger than rock C, C is younger than B, and so on.
- While rock C and rock E are in different layers, and therefore different ages, the same stippling indicates the same rock material. This is also true for rocks F and L.
- Rock layers are not the same thickness throughout. For instance, while rock C continues to rock K, it is thicker in the first columnar section than it is in the second.
- Rock B is not found in the second columnar section. This does not necessarily mean it was not deposited there; it could be that erosion before the overlying rock was put down removed any traces of it.
- Rock F is older than rock P, but younger than rock O.
- Any fossils found in rock E, for example, would be the same age as those found in rock O. If these fossil organisms existed for only a short time on Earth, they can serve as indicator fossils to date other rocks elsewhere they may be found in.

These columnar sections are just two 'snapshots' from the whole region. In working out the geology of large areas, the greater the number of columns made from different areas, the more accurate will be the reconstructed history. Figure 3.40 shows the cross-section from which these two columns were taken.

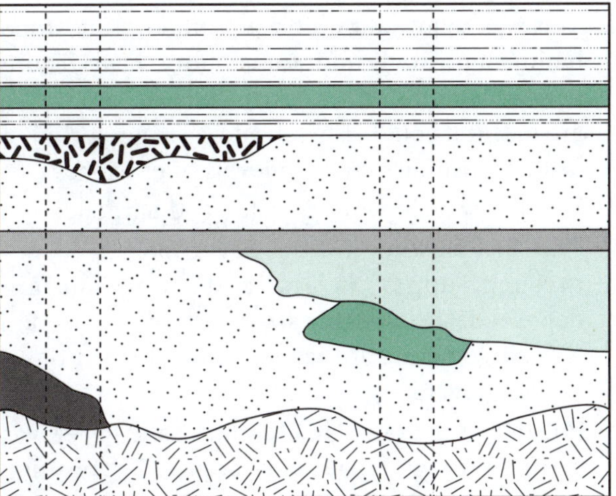

Figure 3.40 Cross-section of a region

Consider the cross-section shown in Figure 3.41.

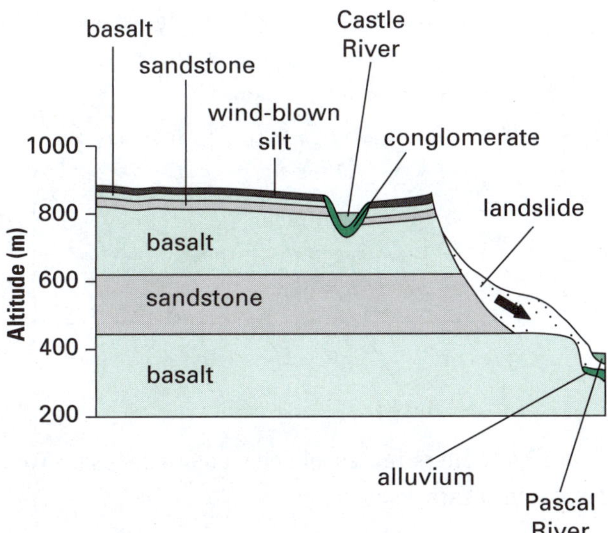

Figure 3.41 Cross-section near Pascal River

A brief geological history of the area could be as follows.

1. Basalt was deposited through volcanic activity in the region. (Basalt is a common extrusive volcanic rock. It is fine-grained due to rapid cooling of lava at the surface.)
2. Volcanic activity ceased.
3. Sandstone was deposited. (This sedimentary rock began as sand accumulated at the bottom of a stream, lake or sea by settling out from suspension,

compacted by the pressure of overlying deposits and then cemented together.)

4. Further volcanic activity deposited more basalt.
5. Volcanic activity ceased.
6. This was followed by a brief period of sandstone being deposited, and some more lava. These periods were brief, as rock thicknesses are not as great. But it could also indicate a longer period of little deposition with high levels of erosion.
7. Wind-blown silt was deposited and hardened.
8. The fast-flowing Castle River began to cut through upper layers and conglomerate deposited. (Reasonably sized rock pieces needed for conglomerate are carried as bed load at times of high flow rate.)
9. Possibly at the same time as the Castle River was cutting through, a landslide occurred depositing rock into the Pascal River.

Knowing the particular rock that was found at a location gives further information about the conditions that were necessary for it to be laid down.

Consider the cross-section shown in Figure 3.42.

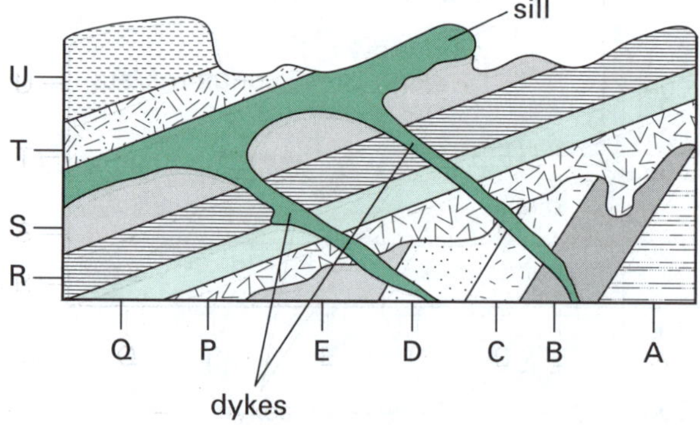

Figure 3.42 Letters represent sedimentary deposits. The dykes and sill are volcanic.

A brief geological history of the area could be as follows.

1. Sedimentary layers A, B, C, D and E were laid down horizontally in that order.
2. Tilting of the rocks occurred in an anti-clockwise direction. (Uplift and subsidence or both.)
3. Erosion wore down the landscape.
4. Sedimentary rocks P, Q, R, S, T and U were laid down horizontally in that order.
5. Two volcanic dykes intruded through these rock layers and a sill formed between S and T. (A sill is a sheet intrusion between older layers of sedimentary rock.)
6. Further tilting of all rocks occurred again in the same direction.
7. Erosion occurred to produce the current landscape. (From this cross-section it seems that sedimentary rocks R and U are harder, as is the sill, since these stand above the surrounding landscape.)

Intruded sills show partial melting and incorporate some of the surrounding country rock. On both the upper and lower contact surfaces of the sedimentary rock where the sill has intruded, there will be evidence of heating (contact metamorphism). Lava flows on the surface show this evidence only on the lower side of the flow.

Age of the Australian continent: fossil evidence and radiometric dating

Australia and nearby islands, including New Guinea, are all part of the same geological land mass being separated by seas overlying the continental shelf. Once these lands were joined with Antarctica forming part of the southern supercontinent Gondwana, until the tectonic plate began to drift north about 96 million years ago. For most of the time since then, Australia–New Guinea was a single, continuous land mass. At the end of the last ice age (about 10 000 BC) the sea level rose and created Bass Strait; Tasmania was separated from the mainland. Around 8000 to 6500 BC, the sea flooded the north lowlands, separating New Guinea from Australia.

As the continent drifted north, separated from the rest of the world, unique flora and fauna developed. While marsupials and monotremes existed on other continents, it was only in Australia and New Guinea that they got the upper hand on placental mammals and dominated. Bird life also flourished and diversified.

The main reasons for this enormous diversity include the following.

- Temperatures in Australia–New Guinea remained fairly constant for a very long time, as the continent drifted northwards towards the equator. (Most of the rest of the world cooled significantly during this time.) So a vast number of different plant and animal species were able to evolve to suit particular ecological niches.
- Very few outside species could colonise the continent as it was more isolated. This allowed our unique native organisms to develop unimpeded.

- Not having much glaciations or volcanic activity, Australian rocks and soils remained relatively untouched and infertile, except for a few pockets of high fertility. The overall climatic trend was towards greater aridity. This also slowed down the process of soil formation. Plants and animals had to compete on this basis for survival. For example, broad-leaf deciduous forest gave way to hard-leaved sclerophyll plants that are typical of the modern Australian landscape.

As Australia was isolated for millions of years, it has a rich and unique fossil record, going back as far as 3.2 billion years. This is an almost continuous record of its distant past. There are records of single-celled organisms ranging to large dinosaurs (see Figure 3.43). For example, through fossils we know that marsupials are relatively recent arrivals. Fossils at Lightning Ridge show that 110 million years ago Australia had a variety of different monotremes, but did not support any marsupials. The first evidence of marsupials in Australia comes from Murgon in southern Queensland with 55 million-year-old fossils.

Figure 3.43 Fossils of the plant-eating dinosaur Muttaburrasaurus that lived 120 million years ago were found in central Queensland (length 7 to 8 m, height at shoulder 2.4 m).

The age of a fossil, and its surrounding rocks, may be determined using radioisotope dating. As the evolution of life and landscapes occurs over vast time periods, scientists look for evidence in the rock layers. Events like volcanic eruptions, animal extinctions, periods when sediments were deposited and ice ages all leave distinctive traces in these layers. Radiometric dating methods compare the abundance of a naturally occurring radioactive isotope and its decay products. Among the best-known methods are radiocarbon dating (half-life 5730 years), potassium-argon dating (half-life 1.3 billion years) and uranium-lead dating (half-life 4.5 billion years). The longer the half-life, the older is the material that can be dated. This provides a significant source of information about the ages of fossils and rocks.

Stability of the Australian continent

Earthquakes are not common in Australia as the continent does not lie near the edge of a tectonic plate. Around 90% of all earthquakes in the world take place at plate boundaries. But occasional, and usually not so serious, earthquakes do occur. The 1989 Newcastle earthquake, Australia's most damaging, had a magnitude of 5.6. Although it was classified as a moderate earthquake, it caused much damage (over $1 billion) and death (13 people) because it happened in a populated area. An earthquake of Richter magnitude 5.5 or higher occurs every 15 months on average in Australia. An earthquake greater than magnitude 7 occurs somewhere in Australia every century or so. But since much of the Australian population hugs the east and south-east coast, even if a large earthquake occurred somewhere near the centre, it would not do much damage.

So why do we have earthquakes at all? No place in the world is completely immune to earthquakes. Just 5 months after European settlement began, the first recorded earthquake event occurred near Sydney Cove on 22 June 1788. Some new research is showing that stresses originating at points of collision between two plates are being dissipated back into our tectonic plate and generating enormous internal stresses. As much as 10% of the huge amounts of energy being created at plate connection points at Sumatra and Java are being transferred back into our plate and causing major stresses. Figure 3.44 details significant Australian earthquakes in the last hundred years.

As seismic waves radiate away from an earthquake, their amplitudes decrease. As well, some energy may be absorbed within the rocks, especially if they are soft or hot. Rocks in western and central Australia are old, relatively cold, and hard, so seismic waves do not lose much of their energy. Rocks in eastern Australia are younger and softer, and absorb energy. Hence earthquakes occurring in western and central Australia will be felt over greater distances than those in the eastern half of the country.

Many Australian buildings are not designed to withstand strong earthquakes, unlike those of New Zealand or Japan. This can explain why a moderate earthquake in Newcastle could cause such serious damage. In Australia the main earthquake hazard is the associated ground shaking. This shaking can

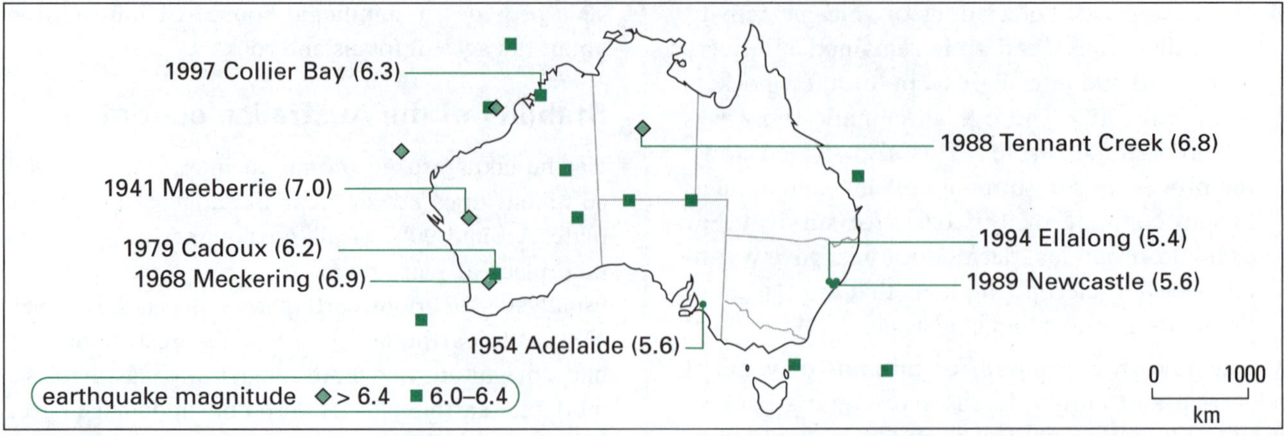

Figure 3.44 Epicentres of significant Australian earthquakes over the last century; those causing significant damage are labelled.

damage or destroy structures, leading to injuries or deaths.

Earthquakes have great potential to disrupt communities. They do this directly through damage to buildings and, less directly, through the damage they cause to the infrastructure on which communities rely. One of the main indirect effects from earthquakes is their ability to disrupt essential utility services, such as electricity, water and gas supply, along with transportation systems. For example, it is difficult for fire brigades to put out a fire caused by an earthquake if the road getting to the fire is destroyed, if the water mains are broken and if a broken gas main is feeding that fire.

There are no active volcanoes in Australia. No volcano has erupted for almost 5000 years and some are dormant, though possibly extinct. Mount Gambier in South Australia is the country's most recently active volcano, having last erupted around 4500 years ago. It is considered to be dormant, which means there is the potential for another eruption. Mount Gambier is thought to have formed by a mantle plume centre, an abnormally hot rock that originates at the core–mantle boundary and rises through the Earth's mantle.

Test yourself 2

Part A: Knowledge

1. The volcanic activity and mountain building along the west coast of South America is an example of which type of plate tectonic activity? *(1 mark)*
   - **A** a subduction zone
   - **B** a spreading zone
   - **C** a collision zone
   - **D** a transform fault zone
2. An earthquake that registers 6.2 on the Richter scale would be described as *(1 mark)*
   - **A** major.
   - **B** minor.
   - **C** light.
   - **D** strong.
3. An opening on the Earth's surface through which molten rock flows and builds up around that opening together form *(1 mark)*
   - **A** an earthquake.
   - **B** a volcano.
   - **C** a tectonic plate.
   - **D** a tsunami.
4. An earthquake's magnitude is measured with a machine called a *(1 mark)*
   - **A** Richter scale.
   - **B** telegraph.
   - **C** seismograph.
   - **D** volcanometer.
5. The seismic wave causing rock particles to move together and apart in the same direction as the wave is moving are *(1 mark)*
   - **A** P waves.
   - **B** S waves.
   - **C** L waves.
   - **D** transverse waves.

6. Complete the following restricted-response questions using the appropriate word. *(1 mark for each part)*
   a) In order to determine how far away from a seismograph station an earthquake occurred, scientists plot the difference in arrival times between P and .... waves.
   b) As seismic waves radiate away from an earthquake, their amplitudes ....
   c) The principle of superposition states that in a sequence of sedimentary rock layers, the bottom layers are .... than the top layers.
   d) Countries that are seismically active have .... about constructing earthquake-proof buildings.
   e) A .... is a crack in the Earth's crust resulting from the displacement of one side with respect to the other.

7. Use the code letters to match the terms or phrases in each column. *(1 mark for each part)*

Column 1		Column 2	
A	Mount St Helens	F	radioactive dating
B	secondary wave	G	hotspot
C	Hawaii	H	earthquake
D	carbon-14	I	volcano
E	epicentre	J	shear

Part B: Skills

8. Figure 3.45 shows a geological cross-section of an Australian landscape.

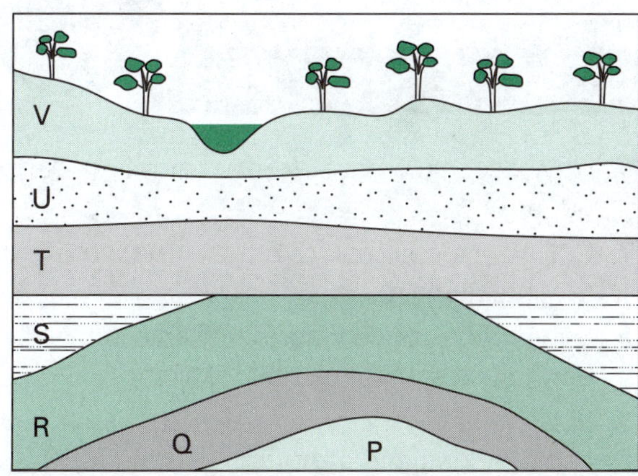

Figure 3.45 Cross-section of an Australian landscape

   a) Between which two layers has there been obvious erosion? *(1 mark)*
   b) Fossils were found in layers U and Q.
      i) What can be said about the relative ages of these fossils? *(1 mark)*
      ii) What can be said about the rock layers in which these fossils were found? *(1 mark)*

9. Figure 3.46 shows some volcanic features.

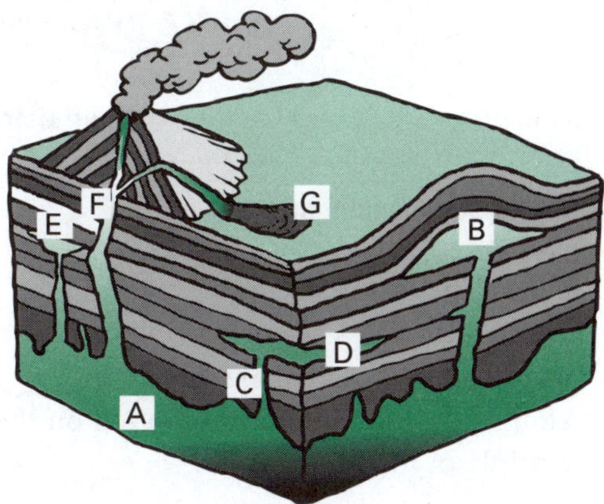

Figure 3.46 Volcanic features

Match the following words to their letters: lava flow; laccolith; sill; dyke; vent; magma chamber; lopolith. *(7 marks)*

10. Explain the main effects volcanic gases have on the atmosphere. *(7 marks)*

11. Figure 3.47 shows three seismograms (X, Y and Z) of an earthquake with the epicentre in New Zealand. The seismograms were recorded in Melbourne, Adelaide and Perth.

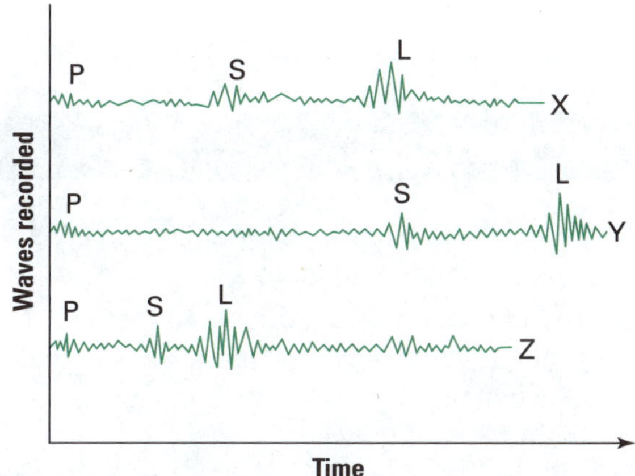

Figure 3.47 Seismograms X, Y and Z

Determine which seismogram was recorded in each city. *(2 marks)*

12. Figure 3.48 shows a section through sedimentary strata. List the strata from youngest to oldest. *(3 marks)*

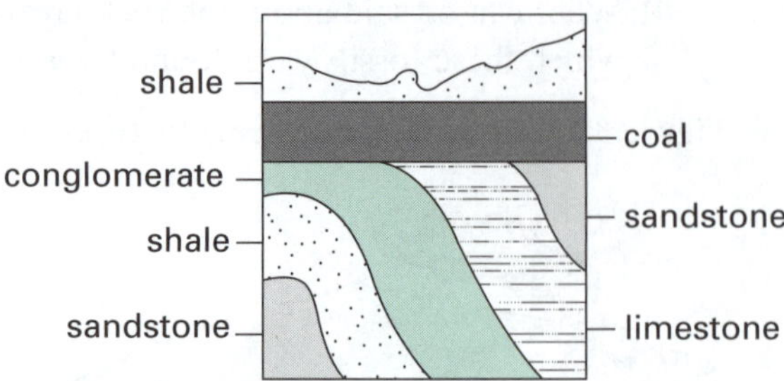

Figure 3.48 Sedimentary strata

13. a) What is the difference between a seismogram and a seismograph? *(2 marks)*
    b) Using Figure 3.49, explain the statement 'Almost all seismometers are based on the principle of inertia'. *(3 marks)*

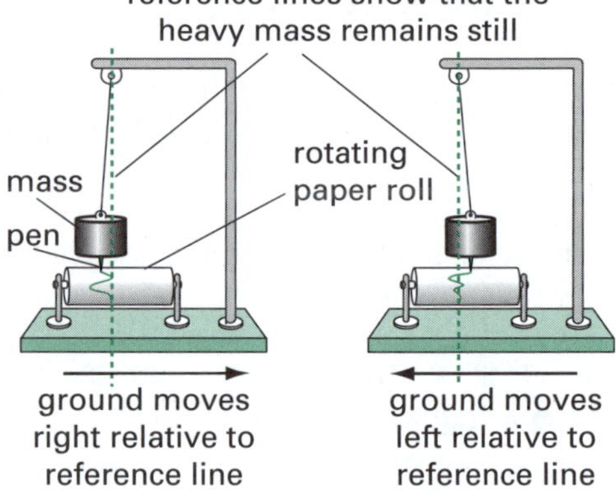

Figure 3.49 Measuring earthquakes

14. The geological cross-section shown in Figure 3.50 involves dolerite intruding into various sedimentary strata.

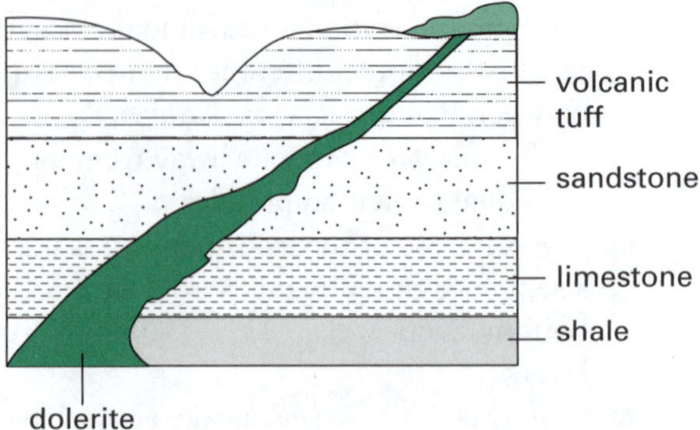

Figure 3.50 Dolerite intrusion in sedimentary strata

a) Name the type of igneous structure shown. *(1 mark)*
b) List the rocks in order from oldest to youngest. *(2 marks)*
c) Describe the effect of the intruding magma on the surrounding sedimentary rocks. *(1 mark)*
d) Compare the crystal size in the dolerite intrusion to the lava that flowed as an extrusion onto the Earth's surface. *(1 mark)*

Go to pp. 239–240 to check your answers.

Summary

1. The Earth is made up of four layers called the crust, mantle, outer core and inner core.
2. The Earth's crust is divided into tectonic plates that move due to the movements of convection currents in the asthenosphere and the effects of gravity.
3. Continental drift is the result of tectonic plate movement.
4. Satellite technology and radiometric dating has been used to measure the rate of continental drift.
5. Some plate boundaries are constructive and some are destructive.
6. New oceanic crust is formed at mid-ocean ridges.
7. Collision zones result in the formation of mountain chains.
8. Subduction zones lead to volcanic activity, earthquakes and mountain building.
9. Volcanoes form at subduction zones, at mid-ocean ridges and at hotspots in the Earth's crust.
10. There are various types of volcanoes: cinder cones, composite volcanoes, shield volcanoes and lava domes.

Summary cont.

11. Volcanic landforms produce a number of tell-tale features.
12. Volcanism has an effect on the atmosphere, biosphere and hydrosphere.
13. An earthquake occurs when a fault or fracture in the Earth's crust occurs.
14. There are different types of seismic waves produced by earthquakes.
15. Earthquake energy and location can be determined using a seismometer.
16. The Richter scale measures the amount of energy released from an earthquake.
17. Countries located along tectonic plate boundaries will experience volcanic and earthquake activity.
18. Geologists use stratigraphy to interpret geological history from cross-sections.
19. Fossil evidence and radiometric dating can be used to determine the age of the Australian continent.
20. Earthquakes are not common in Australia as the continent does not lie near the edge of a tectonic plate.
21. There are no active volcanoes in Australia.

Syllabus checklist

Are you able to answer every syllabus question in this chapter? Tick each question as you go through the list if you are able to answer it. If you cannot answer it, turn to the appropriate page in the guide as is listed in the column to find the answer.

	For a complete understanding of this topic	Page no.	✓
1	Can I state the four layers that make up the Earth?	89	
2	Can I recall what causes continental drift?	91	
3	Can I recall two examples of technology used to measure the rate of continental drift?	91–92	
4	Can I identify the two major reasons why tectonic plates are constantly moving ?	92	
5	Can I explain why some plate boundaries are constructive and some are destructive?	92–94	
6	Can I explain how new oceanic crust is formed at mid-ocean ridges?	93	
7	Can I identify the type of landform that is created at collision zones?	93	
8	Can I explain why subduction zones lead to volcanic activity, earthquakes and mountain building?	94	
9	Can I recall where volcanic features occur?	97–99	

	For a complete understanding of this topic	Page no.	✓
10	Can I recall the various types of volcanoes?	99–100	
11	Can I recall the various types of volcanic features?	101–102	
12	Can I explain the effects of volcanoes on the earth, air and water?	102–104	
13	Can I recall how an earthquake occurs?	104–105	
14	Can I recall the different types of seismic waves produced?	105–106	
15	Can I recall how the location and energy of an earthquake is measured?	105–107	
16	Can I recall what the Richter scale shows?	106–107	
17	Can I recall where volcanic and earthquake activity will be experienced?	108–110	
18	Can I explain what information can be extracted from a geological cross-section?	110–112	

	For a complete understanding of this topic	Page no.	✓
19	Can I recall how the age of the Australian continent is determined?	112–113	
20	Can I explain why Australia is reasonably stable?	113	

	For a complete understanding of this topic	Page no.	✓
21	Can I explain why there are relatively few earthquakes and no active volcanoes in Australia?	113–114	

Chapter test

Go to p. v for *Tips for tests and examinations*

80 MIN

Part A: Multiple-choice questions

(1 mark for each)

1. Select the statement that is true about the Earth's mantle.
   - **A** The upper mantle contains partly molten and easily deformed rocks.
   - **B** The lithosphere is a section of the lower mantle.
   - **C** The pressure in the mantle is about 4 million times that of air pressure.
   - **D** Rocks in the mantle are less dense that crustal rocks.

2. Who first proposed the existence of a supercontinent called Pangaea?
   - **A** Francis Bacon
   - **B** Antonio Snider-Pellegrini
   - **C** Alfred Wegner
   - **D** Alexander du Toit

3. The tectonic plate that is undergoing subduction at the western edge of the South American Plate is the
   - **A** Juan de Fuca Plate.
   - **B** Antarctica Plate.
   - **C** Pacific Plate.
   - **D** Nazca Plate.

4. Which of the following continents was not originally part of Gondwana?
   - **A** Asia
   - **B** South America
   - **C** Antarctica
   - **D** India

5. The Himalayan mountains have formed at what type of plate boundary?
   - **A** collision zone
   - **B** subduction zone
   - **C** constructive plate boundary
   - **D** spreading zone

6. In 1943 a cone started growing on a farm near the village of Parícutin in Mexico. Rapidly expanding and escaping gas from the molten lava produced fine particles that fell back around the vent, building up the cone to a height of 400 m. After the last explosive eruption, a funnel-shaped crater remained at the top of the cone. During its nine active years, Parícutin destroyed the town of San Juan and covered hundreds of square kilometres with ashes.

   This is an example of a
   - **A** cinder cone.
   - **B** composite volcano.
   - **C** shield volcano.
   - **D** lava dome.

7. Before a volcanic eruption, seismic (earthquake) activity seems to
   - **A** increase in frequency and increase in intensity.
   - **B** increase in frequency but decrease in intensity.
   - **C** decrease in frequency but increase in intensity.
   - **D** decrease in frequency and decrease in intensity.

8. Most earthquakes occur
   - **A** near volcanoes.
   - **B** where humans build cities.
   - **C** along plate boundaries.
   - **D** in the middle of the ocean.

9. Areas that have the highest concentration of earthquakes are located along
   - **A** Mediterranean Europe.
   - **B** the east coast of America.

C the west coast of Africa.
D the Pacific rim.

10. Figure 3.51 shows an example of a

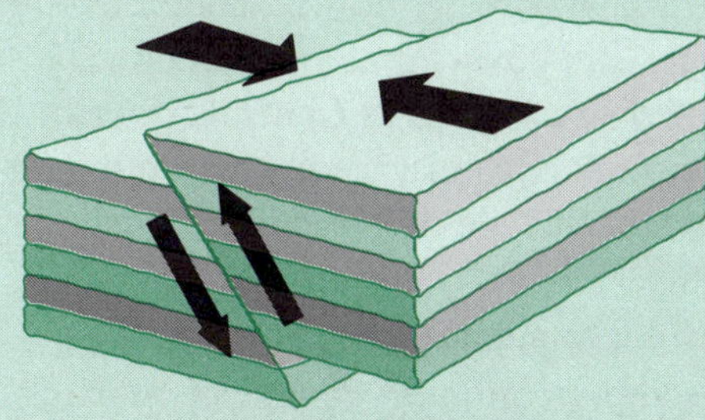

Figure 3.51 An example of a fault

A normal fault.
B reverse fault.
C strike-slip fault.
D hanging wall fault.

Part B: Short-answer questions

11. What does the history of ideas about continental drift tell you about the way scientific information is obtained and finally accepted? *(1 mark)*

12. Some of the views held by early scientists about continental drift were actually wrong. Does this make them bad scientists? Explain. *(2 marks)*

13. The theory of plate tectonics explains how the continents move.
   a) What is the major cause of the movement of the Earth's plates? *(1 mark)*
   b) How does this theory explain the cause of many earthquakes and volcanic eruptions? *(1 mark)*

14. According to the theory of plate tectonics, plates can interact in different ways. Explain each of the following. *(6 marks)*
   a) Japan has active volcanoes and a deep trench (10 000 m) off its east coast.
   b) Fossils, corals and shells can be found high on the Andes mountains along the west coast of South America.
   c) Italy, Turkey and the Middle East suffer many earthquakes.
   d) The island of Iceland in the north Atlantic is gradually growing in size.
   e) Mount Everest in the Himalayas is getting taller every year.
   f) A major earthquake devastated San Francisco early in the 20th century.

15. Various words have been replaced by code letters (a to i) in the following cloze passage. Use the list to identify which words match the code letters: found; movement; measurements; region; south; discovered; occurring; faster; speed. *(9 marks)*

Plate sets speed record

In 1996 American scientists a) .... evidence that moving tectonic plates on the Pacific Ocean floor had set b) .... records in a period between 11 and 18 million years ago. Douglas Wilson from the University of California used magnetic polarity c) .... to date the lava ridges on the sea floor. From this information he calculated the speed of d) .... of the spreading sea floor. Wilson e) .... that this sea floor spreading was f) .... at a rate of about 20 cm per year at that time. This is much g) .... than the movement of most plates today although it is similar to the rate of 15 cm per year in the h) .... of the ocean floor i) .... of Easter Island.

16. Consider Figure 3.52 about the distribution of a rock called tillite which is only found in regions that have experienced long periods of glaciation.

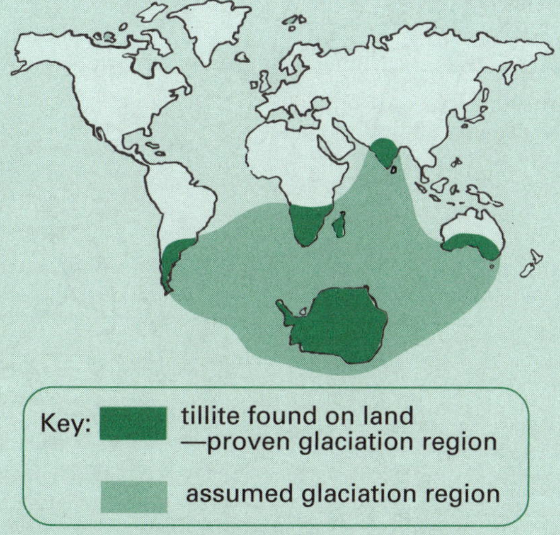

Figure 3.52 Tillite distribution

Tillite is a rock usually associated with glaciers. As icy glaciers melt, they release the boulders, sand and silt that they carry. This rock material is usually mixed without very much sorting or layering. Over 100 years ago, scientists discovered this tillite in many parts of Africa, South America, Australia, India and Antarctica. This tillite was all the same age, about 280 million years. How does this evidence support the theory of plate tectonics? *(2 marks)*

Use the information given to answer Questions 17 and 18.

Figure 3.53 shows the continents of the Earth at different times in the past as well as the present day.

17. Use the code letters for each drawing to match the following dates: 225 million years ago; 200 million years ago; 135 million years ago; 65 million years ago. *(4 marks)*

18. Match each map to one of the following jumbled captions. *(5 marks)*

I. The continents are now in their present locations, but motion is still occurring. Australia has now separated from Antarctica and is moving slightly east of north. North and South America are moving westwards.

II. The break-up of Pangaea is well under way. India is still at least 1000 km south of the equator on its trek northwards. North America and Eurasia had begun to separate as had South America and Africa.

III. The continents are starting to take on a more recognisable form. Australia is only just beginning to separate from Antarctica. India is still on a collision course with Eurasia, having by now reached the equator. Water still covers that area which will come to be known as Central America. The dinosaurs have just been wiped out, and the dawn of the age of mammals is occurring. Many varieties of flowering plants are starting to appear.

IV. The super-continent Pangaea is starting to break up. A rift is starting to develop between what is now known as Africa and Eurasia. The land is dominated by dinosaurs. The first mammals have appeared. Plant life is mainly palm-like cycads and ferns.

V. A great rift has opened between the north and south lands separated by the Tethys Sea. Gondwana is beginning to break itself up. Large plant-eating dinosaurs dominate, feeding off ferns and cycads while the smaller meat-eating dinosaurs feed off them. Trilobites are now extinct, but the sea is still full of fish. Early mammals roam the forest, and the first birds are seen flying.

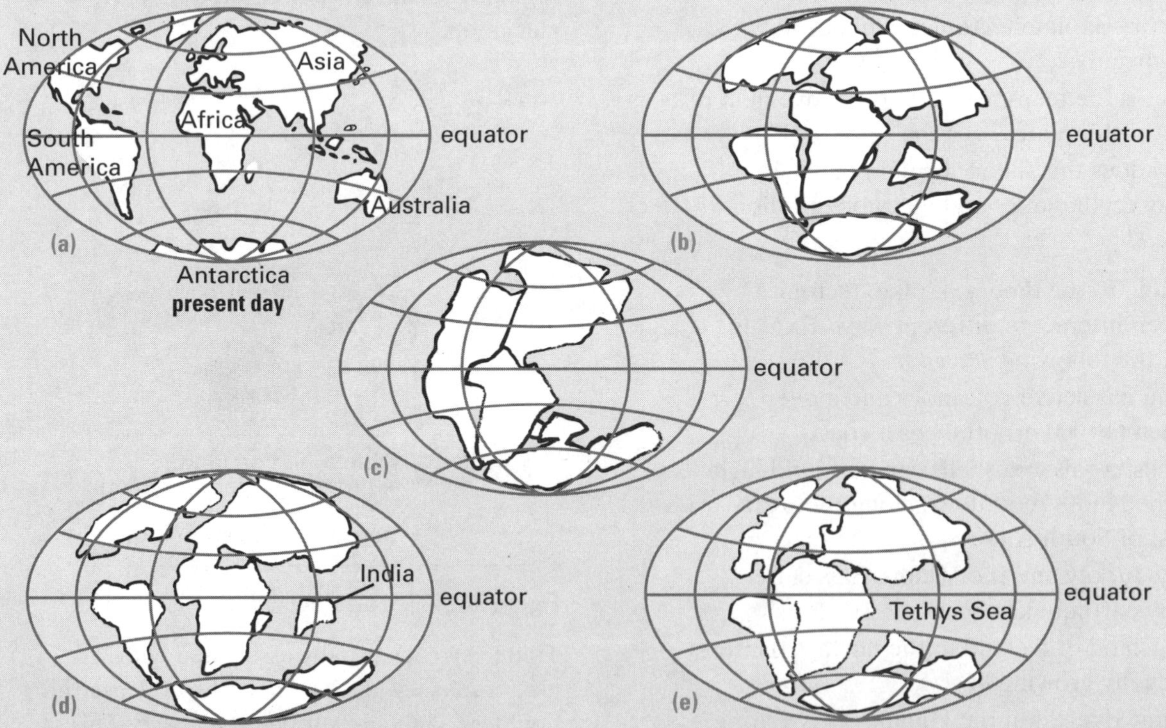

Figure 3.53 Continental drift

19. Scientists have found that the magnetism of the Earth has reversed itself many times over millions of years. Reversals happen at irregular times and last for tens of thousands to hundreds of thousands of years. A geomagnetic time scale has been established.

a) Figure 3.54 shows a portion of this scale developed from information collected on land.

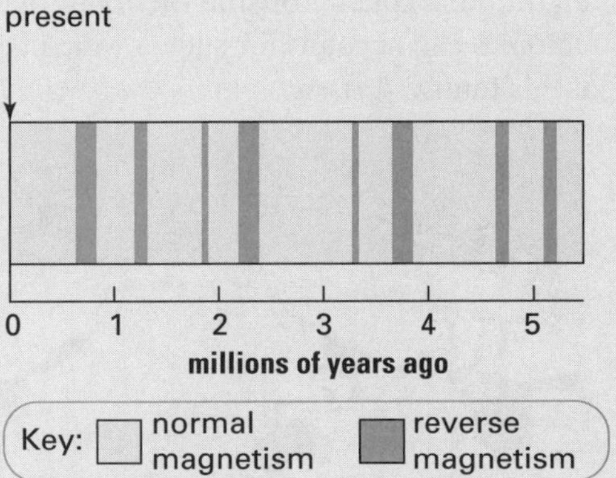

Figure 3.54 Geomagnetic time scale

Use the scale on the diagram and your millimetre ruler to answer some of these questions. Express your answers to the nearest 50 000 years.

i) The Earth is currently in a 'normal' magnetic state. How long has this state existed? *(1 mark)*

ii) Between what time periods was the most recent reversal of magnetism? *(1 mark)*

iii) How many magnetic reversals have there been in the last 4 million years? *(1 mark)*

b) Figure 3.55 shows the pattern of normal and reversed magnetism on either side of a mid-ocean ridge. This was measured by examining the lava flows produced from the trench in the centre of this ridge.

i) At what distance from either side of the central ridge is the most recent magnetic reversal? *(1 mark)*

ii) Use your answers to a) i) and b) i) to calculate the speed of sea floor spreading in recent times in this area. *(1 mark)*

iii) How old are the rocks at point X in the figure? *(1 mark)*

iv) How is this information about magnetic reversals used as evidence for the theory of plate tectonics? *(1 mark)*

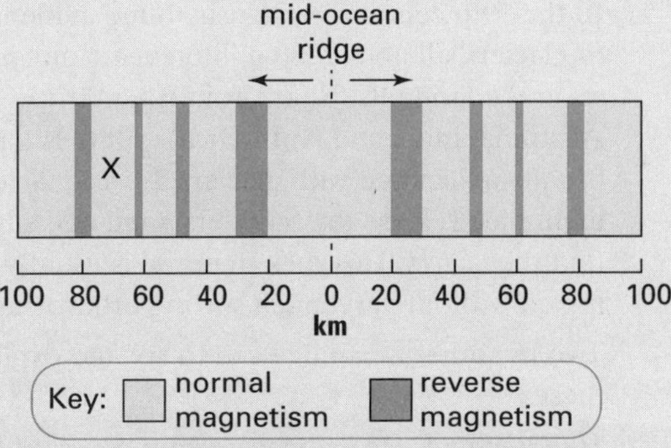

Figure 3.55 Patterns of geomagnetic change on each side of a mid-ocean ridge

20. Figure 3.56 shows the pattern of magnetic field reversals in the area south of the Great Australian Bight. Explain what features of the pattern show that the sea floor spreading has speeded up in more recent times. *(1 mark)*

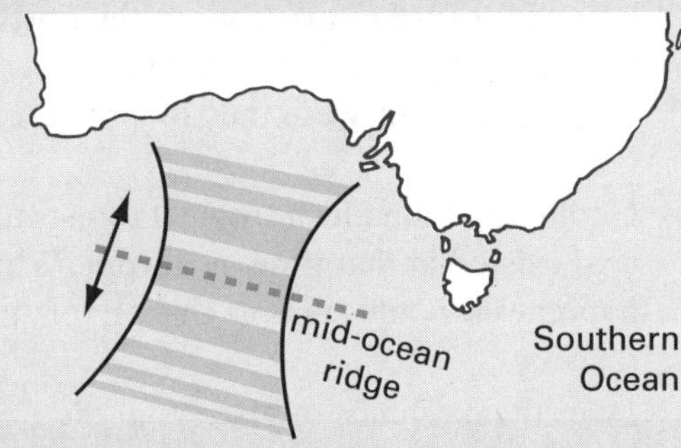

Figure 3.56 Magnetic field patterns in the mid-ocean ridge in the Southern Ocean

21. Australia is moving away from Antarctica at an average rate of 6.7 cm per year.

a) If it is now about 2900 km from Antarctica, calculate when the split occurred. *(1 mark)*

b) Other information tells us that Australia and Antarctica split about 96 million years ago (the oldest sea floor between the two continents is this age).

i) How well does your answer to a) compare to this figure? *(1 mark)*

ii) In doing your calculation, you made an assumption. What was this assumption? Was it a valid one? *(1 mark)*

c) Recalculate the average speed the Indo-Australian Plate has been moving north over this 96-million-year period. *(1 mark)*

22. In the 19th century, geologists found evidence of glaciers, all about 280 million years old, in several continents (South America, Africa, Australia, India and Antarctica). Tillite is a rock usually associated with glaciers. As icy glaciers melt, they release the boulders, sand and silt that they carry. This rock material is usually mixed without very much sorting or layering.

Two hypotheses can be used to try and explain these.

Hypothesis A: Ice had covered almost all of the southern hemisphere. The ice age was very cold.

Hypothesis B: Ice had covered only a small area around the South Pole. The ice age was not as severe.

a) Which hypothesis would be used by those geologists convinced of continental drift? Explain. *(2 marks)*

b) Geologists who supported hypothesis A looked for further evidence in the northern hemisphere to support their argument. What kind of evidence were they looking for? *(1 mark)*

c) Geologists found lush tropical forests and coal existed at that time in Europe. Which hypothesis would support this? Explain. *(2 marks)*

d) Does the existence of glaciers in the southern hemisphere prove the idea of continental drift? Explain. *(2 marks)*

Refer to Figure 3.57 to answer Questions 23 and 24.

Figure 3.57 Tectonic plates

23. Use the tectonic plate map to predict how the size of the Atlantic Ocean will change over the next few million years. Justify your answer. *(2 marks)*

24. A subduction zone exists along the western coast of South America. In which direction is the Nazca Plate moving? *(1 mark)*

25. The map in Figure 3.58 shows the worldwide distribution of a family of plants called the Proteaceae. This family includes plants such as banksias, waratahs, and grevilleas. The Proteaceae do not disperse their seeds well across water. They need land for continuous migration. Explain how the theory of plate tectonics can account for such a wide dispersal of this family. *(2 marks)*

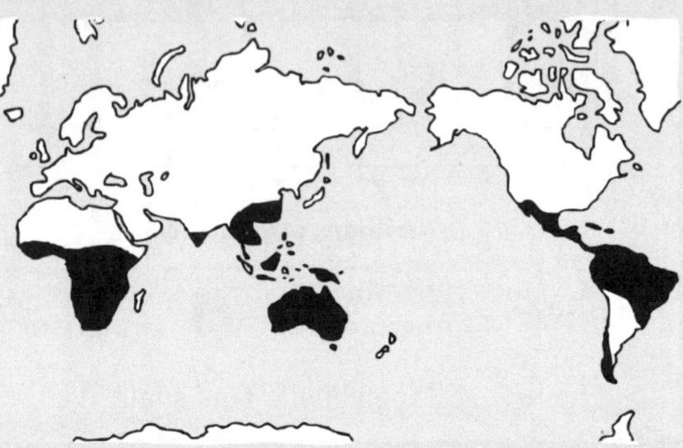

Figure 3.58 Proteaceae distribution

26. As a long-term average, volcanism produces about 500 million tonnes of CO_2 each year. Current fossil fuel and land use practices now introduce about 17 600 million tonnes of CO_2 each year into the atmosphere. What percentage of CO_2 released to the atmosphere comes from:

a) volcanoes? *(1 mark)*

b) human-caused sources? *(1 mark)*

27. Sometimes after a volcano erupts, it leaves behind a caldera—a crater lake (see Figure 3.59).

The following are steps in the formation of a caldera, but they are out of order. What is the correct sequence? *(3 marks)*

A. The weight of the fractured crust above now makes this empty magma chamber collapse, creating a large basin (a caldera).

B. Erosion starts to wear away at the caldera, forming rivers into and out of the lake.

C. The volcano explodes violently, releasing lava and firing an ash column high into the atmosphere.

D. The amount of magma accumulating beneath the Earth's crust increases, heating the rock material in the crust above it and building up pressure.

E. The erupted material partially empties the magma chamber.

F. The crust fractures, releasing this pressure with fissures many kilometres long appearing at the surface above the magma.

G. For months and years later, rain fills the deeper parts of the caldera.

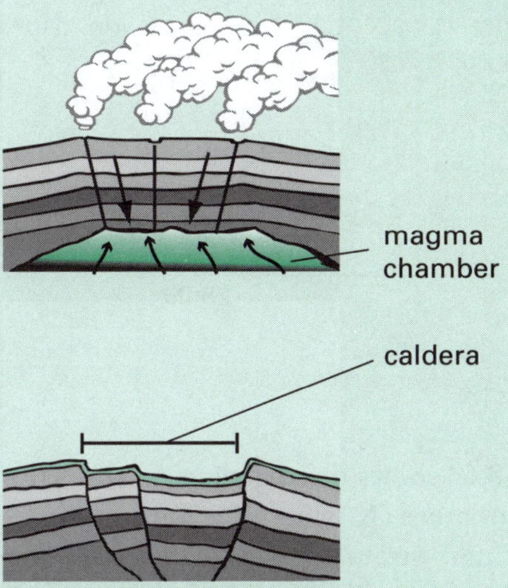

Figure 3.59 Forming a caldera

28. The timeline in Figure 3.60 shows the major eruptions of the volcano Mount Vesuvius since 1800.

a) What are the differences in the patterns of eruptions between the 19th and 20th centuries? *(2 marks)*

b) Vesuvius is not extinct. What can you say about the chance it might erupt again, and when? *(2 marks)*

c) A major eruption occurred in 1631, which was the worst for over a thousand years. Using the trends in the timeline, estimate how many major eruptions occurred between 1631 and 1800. *(1 mark)*

29. Kilauea is a volcano in the Hawaiian Islands, and one of five shield volcanoes that together form the island of Hawaii. It is currently very active, being virtually above the hotspot. The islands are slowly moving north-west along the Pacific Plate. Figure 3.61 shows the age of the Hawaiian Islands (being volcanic formations) and their distances from Kilauea.

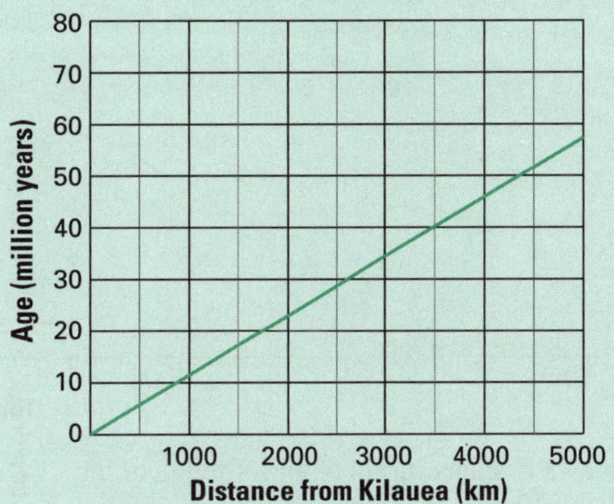

Figure 3.61 Age of Hawaiian volcanoes

a) Honolulu lies on the island of Oahu, some 400 km from Kilauea. How old is the rock at Honolulu? *(1 mark)*

b) Calculate how fast the Pacific Plate is moving, in cm/year. *(2 marks)*

30. Figure 3.62 shows the trace left by an earthquake.

a) What is the time difference between the arrival of the following?

i) P waves and S waves *(1 mark)*

ii) S waves and L waves *(1 mark)*

b) What does this tell you about the speed of each of these waves? *(2 marks)*

c) There is another seismic station closer to where the earthquake occurred. Would the time difference between the P and S waves be greater or less? Explain. *(2 marks)*

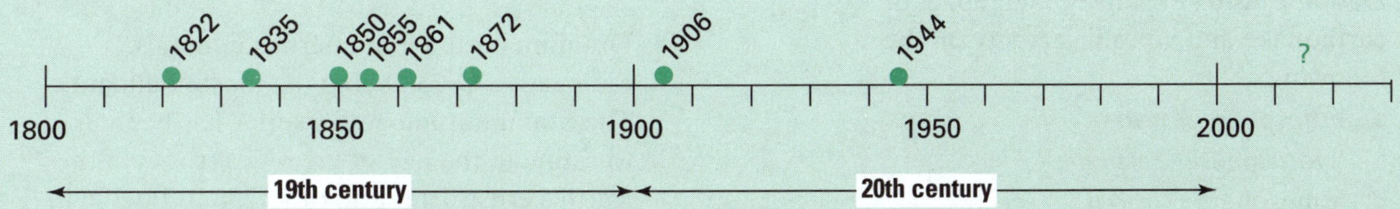

Figure 3.60 Mount Vesuvius eruptions time line

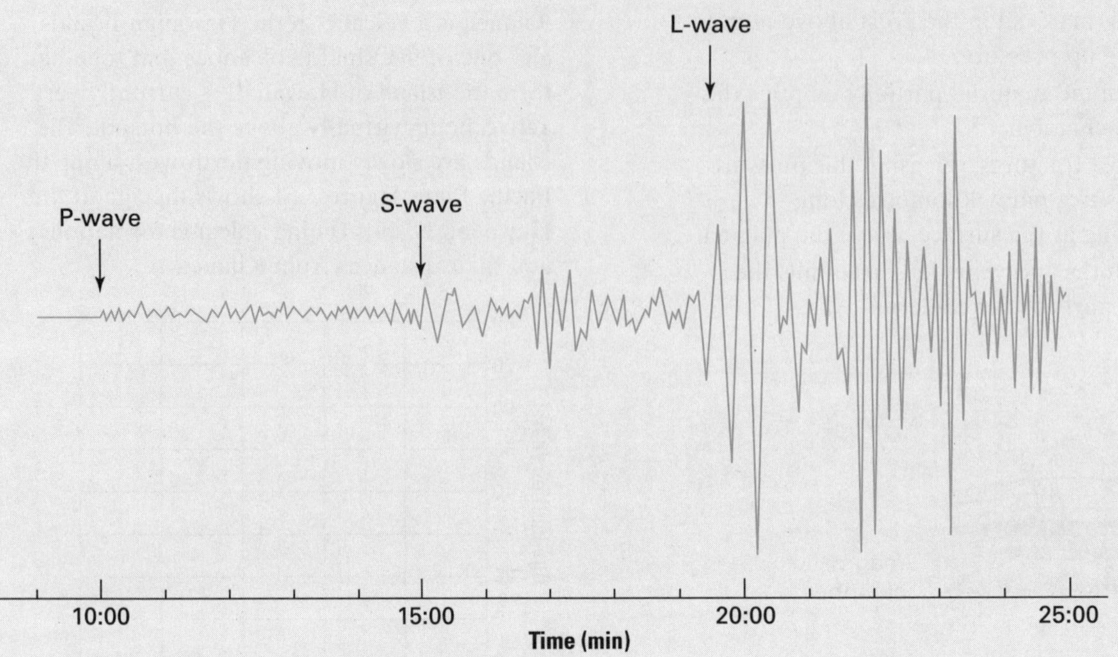

Figure 3.62 A seismogram of an earthquake

31. Table 3.5 shows the percentage of major gases released by the volcano at the summit of Kilauea (Hawaii), where there is a hotspot.

Table 3.5 Percentage of major gases released at the summit of Kilauea

Gas	Percentage
H_2O	37.1
CO_2	48.9
SO_2	11.8
other	

a) Complete the 'other' cell in the table. *(1 mark)*
b) Draw a pie (sector) graph of this information. *(2 marks)*

32. a) Explain how a volcano is formed at a subduction zone. *(2 marks)*
b) Name two locations in the Pacific rim where volcanic activity is due to plate movement at a subduction zone. *(2 marks)*

33. Describe, using examples, the impact of earthquake and volcanic activity on the following.
a) lithosphere *(2 marks)*
b) hydrosphere *(2 marks)*
c) atmosphere *(2 marks)*

34. Figure 3.63 shows three geological cores from different areas (X, Y and Z). Some of the sedimentary strata contain fossils. Assuming that the layers with the same fossils are of the same age, determine which core has the most recent fossil. *(2 marks)*

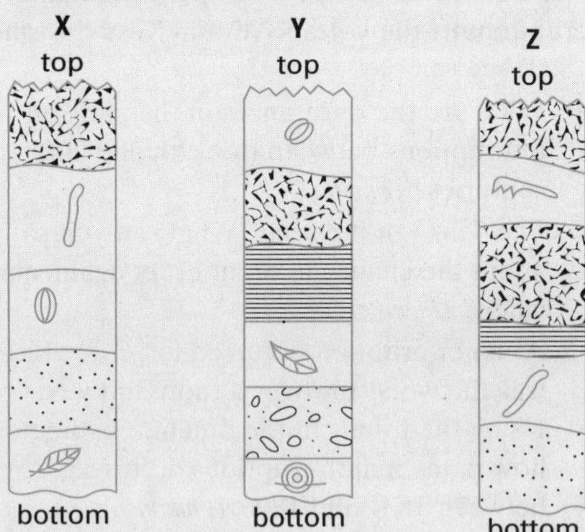

Figure 3.63 Core sections X, Y and Z

35. Uranium occurs in numerous minerals, radioactive U-238 being the most abundant. The uranium radioactive series has been used to estimate the age of the oldest rocks in the Earth's crust. The ratio of U-238 to Pb-206 in a rock changes slowly as the U-238 in the rock

decays, eventually becoming Pb-206. Figure 3.64 shows this uranium decay.

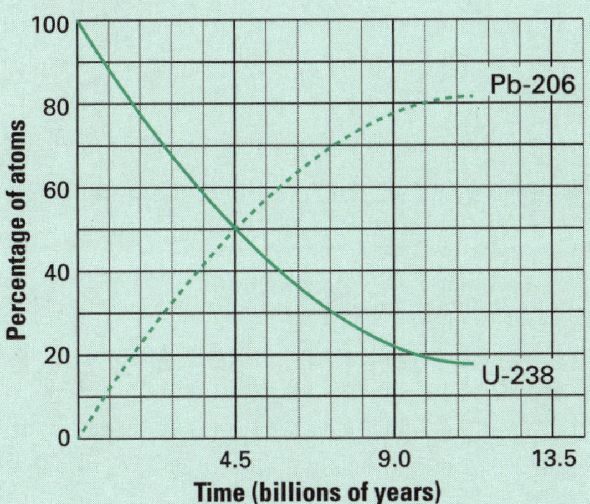

Figure 3.64 The decay rate of U-238

a) What is the size of each horizontal division on the axis? *(1 mark)*
b) What is indicated by the dashed curve? *(1 mark)*
c) What is indicated where the two curves meet? *(1 mark)*
d) In a sample, around 35% of U-238 remained. How old is this rock? *(1 mark)*
e) The Australian fossil record goes back some 3.2 billion years. Approximately what percentage of U-238 is there expected to be in the sample? *(1 mark)*

36. Figure 3.65 shows how a tsunami occurs and propagates.

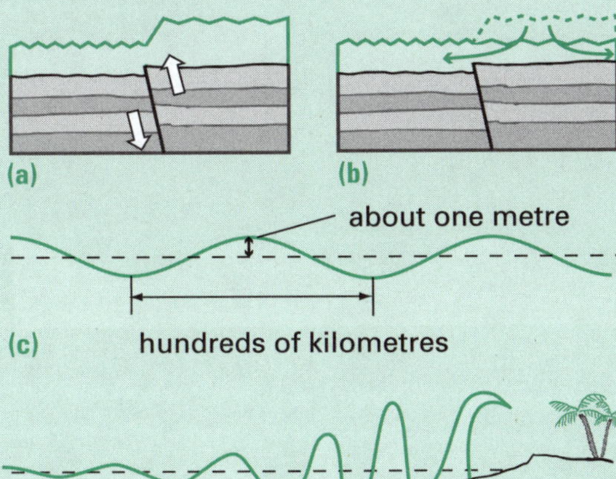

Figure 3.65 Features of a tsunami

Write captions describing what each part of this figure shows. You should have two or three sentences for each. *(8 marks)*

37. Use the earthquake travel-time graph in Figure 3.66 to determine the approximate distance that the seismic station is from the epicentre of the earthquake, given there is exactly 6 minutes between the arrival of the first P waves and first S waves. *(1 mark)*

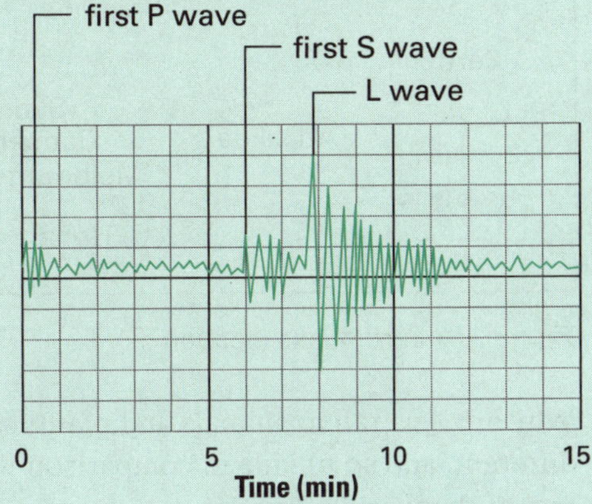

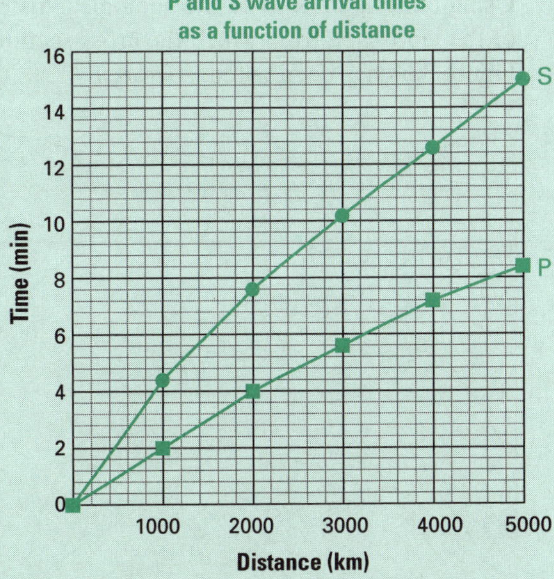

Figure 3.66 Earthquake seismogram and travel-time graph

38. An earthquake occurred somewhere in Australia. Three seismic stations calculated the distance of the epicentre from them. The results were as follows:

- Canberra, 1600 km
- Perth, 1800 km
- Darwin, 1400 km.

Use the map of Australia in Figure 3.67 to pinpoint where this earthquake occurred. *(3 marks)*

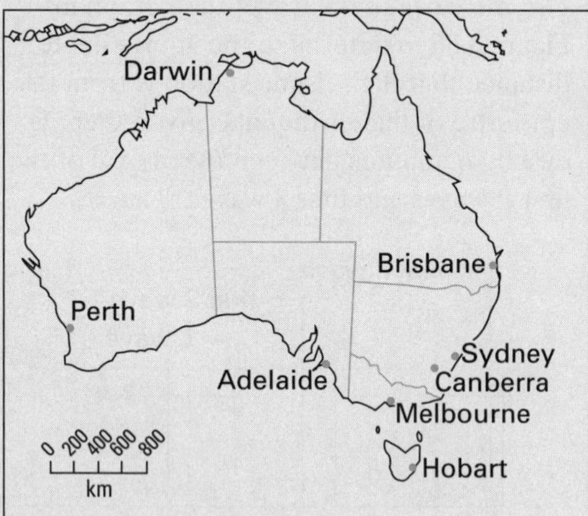

Figure 3.67 Earthquake location

39. Why are Australian animals and plants so different, and so unique in comparison with the rest of the world? *(3 marks)*

40. Using bullet points, write the geological history of the landscape from which the cross-section in Figure 3.68 has been taken. *(11 marks)*

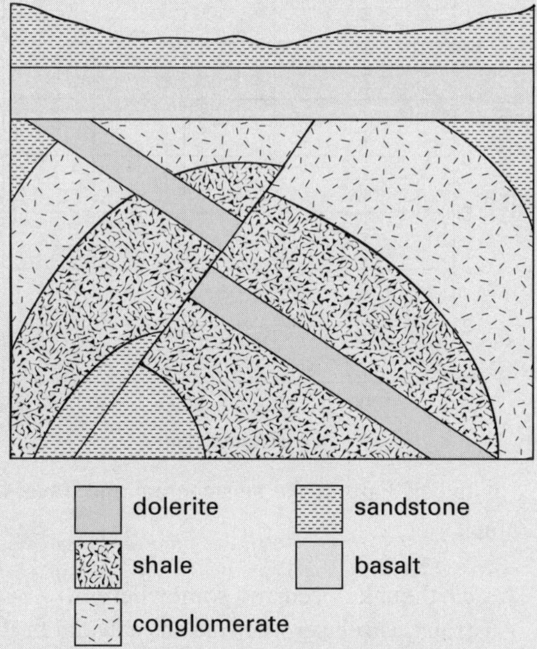

Figure 3.68 A cross-section of a landscape

Go to pp. 240–244 to check your answers.

CHAPTER 4

Energy on the move

Overview

In this chapter you will learn about:

- the nature of heat energy
- conduction, convection and radiation of heat
- mechanical waves and electromagnetic waves
- amplitude, wavelength and frequency of waves
- the speed of sound waves in different media
- reflection of sound waves
- loudness of sound
- hearing and cochlea implants
- electromagnetic spectrum
- absorption and transmission of light
- reflection and refraction of light
- the human eye and the bionic eye
- electrical circuits
- voltage, current and resistance
- electrical measuring devices
- modern communications
- mobile phones
- satellite and telecommunications
- fibre optic communications
- radar
- radio and television.

Glossary

Absorption—when light does not reflect but disappears into the material

Altimeter—an instrument used to measure the altitude (height) of an object above a fixed level

Battery—a portable electrical source consisting of a number of electrical cells in series

Cochlea—spiral cavity of the inner ear

Conductor—a substance that allows electricity to flow through it with low resistance

Converge—to come together

Current—the flow of electric charge (measured in amps, A)

Diverge—to spread apart

Electric cell—a device that produces electrical energy by chemical reactions

Frequency—number of cycles per unit of time

Incandescent—glowing due to being hot, such as a glowing wire filament in a light bulb

Incident ray—incoming ray of light that strikes a surface or boundary between two layers

Infra-red—invisible form of light which is transformed into heat when absorbed by a body

Insulator—a substance that prevents the flow of electric current or heat and has a high resistance to such flow

Longitudinal wave—a compression wave in which the particles of the medium oscillate to-and-fro along the axis of energy transfer

Mechanical wave—a wave form in which the particles of the medium oscillate in order to transmit the energy

Medium—the material space in which a wave travels

Normal—a line at right angles to a surface

Opaque—describes a material that does not allow light to pass through it

Oscillate—move to and fro

Ossicles—set of three bones in the ear

Prosthetic—an artificial device extension that replaces a missing or damaged body part

Rarefaction—a region of a sound wave in which the particles are widely separated

Refraction—the bending of light rays when they pass into different media at an angle

Resistance—property of a conductor that restricts current flow (measured in ohms, Ω)

Resistor—a device that exhibits resistance to the flow of current

Seismic wave—energy wave in the Earth caused by earthquakes

Sonar—acronym for sound navigation and ranging; a technique in which reflected sound waves are used to measure distance

Total internal reflection—light that is reflected back from the edge of the medium, such as an optical fibre, it is travelling through

Translucent—describes a material that partially transmits and partially scatters light rays

Transmission—passage of light rays into a medium and out the other side

Transparent—describes a material that allows light to pass through it without scattering

Vibration—an oscillation or shaking motion (a to-and-fro movement)

Voltage—the electrical pressure that causes currents to flow (measured in volts, V)

4.1 Heat energy

Heat energy is a very important form of energy. We use it to warm up in winter, to cook our food, to heat water for our showers and to make many reactions in industry occur. With just about every energy transformation, heat is released. Often this heat is lost to the environment. For example, we can write the energy transformation occurring in a light bulb as follows:

electrical energy → light energy + heat energy

Often we are only interested in the light energy from the bulb and forget that it gives out heat as well. We are quickly reminded if we touch a glowing light bulb.

The food we eat provides us with nutrients and chemical energy. But in the transformation, heat is also produced. This is important as it keeps us at a constant body temperature of 37 °C. We radiate extra heat away from our bodies. On cold days we want to retain this heat, so we dress with 'warm' clothes. Actually it is not the clothes that are warm; they just act as a barrier to prevent excessive heat loss from our bodies.

What is heat energy?

Figure 4.1 shows how heat energy flows from a hot to a cold object when in contact. The fast-moving atoms or molecules in the hot object hit the less-energetic atoms or molecules in the cold object. This speeds them up. In this way, heat energy passes from a hot object to a cold one. The degree of hotness can be measured with a thermometer.

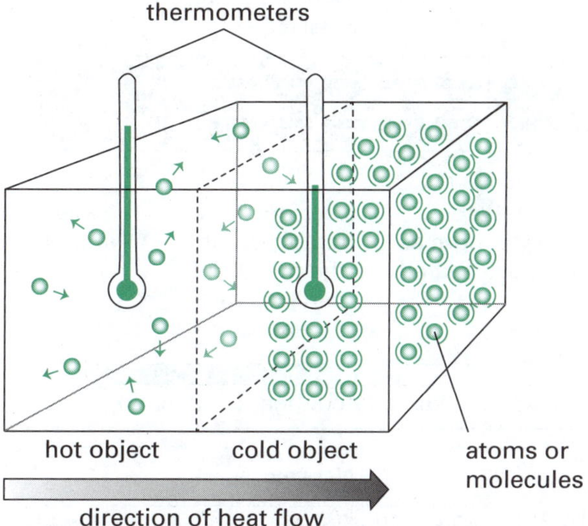

Figure 4.1 Direction of heat flow

Heat is a form of energy. All things are made up of atoms and molecules. These are always moving; often they just vibrate or jostle back and forth. This motion gives every object its internal energy. The more violently they move, the greater is this energy. How much internal energy they have is a measure of heat.

Temperature is a measure of how hot something is. We usually use an instrument called a thermometer to do this. It is also a measure of how much kinetic energy the particles in the substance have. Thermometers have a reservoir (bulb) where a liquid, such as mercury or alcohol, resides. The bulb is the temperature sensor. As the liquid in the bulb is heated, it expands along a fine tube. This is a physical change that occurs with temperature, so there needs to be some means of converting this change into a numerical value. This is where the scale on the thermometer comes in, as seen in Figure 4.2.

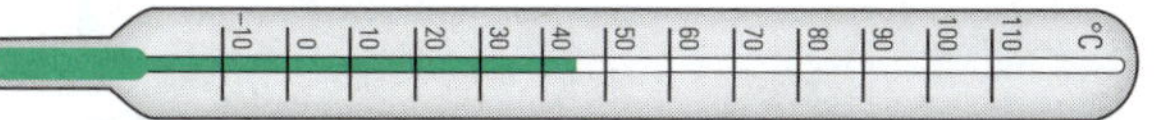

Figure 4.2 A thermometer consists of a liquid inside a narrow tube and a scale.

When energy is added to a substance, the added energy usually increases the kinetic energy of the motion of its particles. We say the substance gets hotter. While there is a relationship between heat and temperature, the two are not identical. For example, a beaker of boiling water is at a higher temperature than a bathtub full of warm water. Yet the bathtub full of water has more heat energy than the boiling water.

Consider this experiment:

Equal amounts of water are placed into two separate beakers. Their temperatures are the same. A red hot iron nail (mass 2 g) is added to one beaker and a warm cylinder of iron (mass 200 g) is added to the other. The iron cylinder is much heavier than the nail. Although the iron nail is at a higher temperature, the iron cylinder contains more heat. This can be seen when both are added to equal quantities of water. The temperature of the water with the cylinder placed in it increases by a greater amount than the temperature of the water containing the iron nail.

The important difference between heat and temperature is that heat is a form of energy and temperature is a measure of the amount of that energy which is present in a body.

Conduction

If you hold a steel rod in the flame of a Bunsen burner, the rod will become so hot that you will not be able to hold it without using a non-conducting material such as an oven cloth. What is happening here? Heat from the flame flows, or is conducted along, the cooler object (steel rod). If you leave the steel rod in the flame it will get hotter and hotter until it may eventually reach the temperature of the flame. We call this process heat exchange; the heat lost by the hot object equals the heat gained by the cold object. (Of course, we're assuming heat isn't transferred elsewhere.) A handle or oven cloth is made of a material that is a poor conductor of heat. We call this a heat insulator. The steel, on the other hand, is a good conductor of heat.

Figure 4.3 shows the insulated handle of a metal spatula. This allows the spatula to be held safely without burning the hand.

Figure 4.3 Why does this metal kitchen spatula have a plastic handle?

Heat moves from hotter objects to colder ones, never the other way around. When we say that dinner is getting cold, it is because heat from the (hotter) food moves to the (colder) surroundings. You could say that cold is the absence of heat. Cold things have less energy than hot things. A metal gauze is a good conductor of heat. In the laboratory you can place a gauze on a tripod just above a Bunsen burner. Now light the gas passing through it. You will see the gas burns above the gauze, but not below it. Why? The gauze quickly spreads the heat from the flame, preventing it from setting the gas below alight. For the gas to burn, it has to reach a certain temperature, but the gauze prevents this from being reached.

Heat conduction can be explained using a particle model. Our model must explain why heat conduction in metals is much faster than heat conduction in non-metals and other insulators such as glass and plastic. When heat is applied to a metal, it makes the nearby particles vibrate faster. These heated particles are the atoms of the metal as well as the free electrons within the metal crystal. While the metal atoms don't move along the solid, they vibrate and bump into their neighbouring atoms more violently. This makes the neighbouring particles vibrate faster, and so heat spreads throughout the solid. The free electrons are able to move (or diffuse) through the metal and so they transfer heat energy much faster than the metal atoms do. In the case of insulators, there are no free electrons and so conduction can only occur by the vibrating atoms.

Table 4.1 compares the relative heat conductivity of various materials. Silver is the best heat conductor and has been rated as 1000 units of conductivity. Metals have a high heat conductivity whereas non-metals, compounds and mixtures have low heat conductivity.

Figure 4.4 compares the conduction of heat along a metal and glass rod. The rate of heat conduction in a metal is much greater than in glass due to the mobility of free electrons in the metal.

Table 4.1 Heat conductivity of materials

Material	Relative heat conductivity
silver	1000
copper	947
aluminium	504
steel	123
lead	86
glass	2
water	1.5
brick	1.5
oxygen	0.05

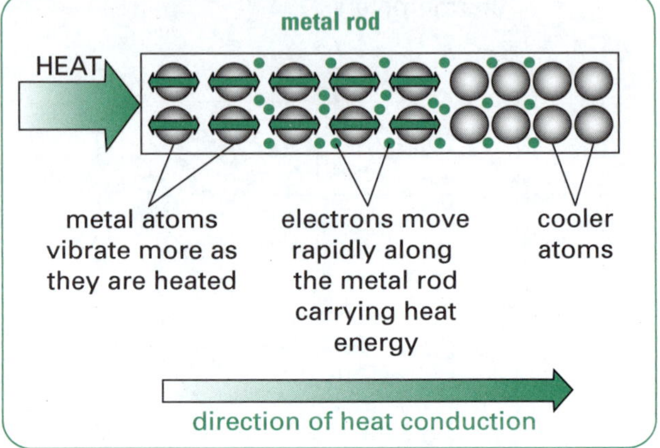

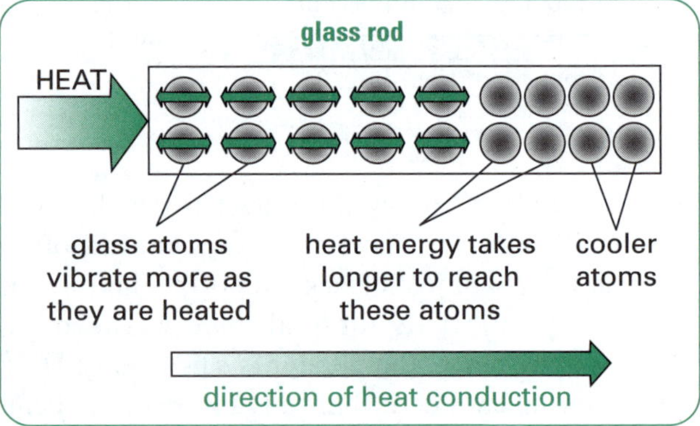

Figure 4.4 Conduction of heat along a metal rod and a glass rod

Convection

Heat moves in liquids and gases by convection. When air is heated, the molecules vibrate and move about more with greater speeds. The air starts to expand. This makes it less dense, and it rises. As warm air rises and becomes cooler, denser air moves in to take its place. In turn, this cooler air is warmed and it rises too. In the convection process, the moving material moves the heat. Convection is responsible for forcing smoke up a chimney. It can also explain sea breezes and cloud formation.

Figure 4.5 shows a simple experiment to demonstrate how heated air will cause smoke to move up a chimney by establishing a convection current. A piece of cardboard is cut into a T shape and placed in a gas jar containing a lit candle. There is a small air gap at the base of the jar. A smoking taper is placed at the lip of the jar and smoke particles are observed to move down into the jar and then up the side heated by the candle.

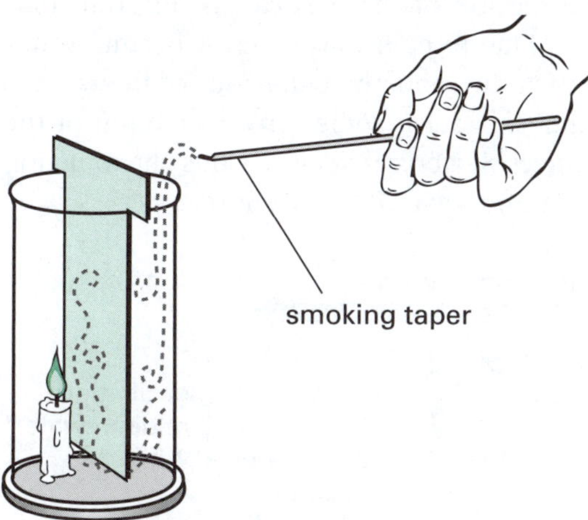

Figure 4.5 Convection current

Convection currents are set up in a heated liquid. If cold water is placed in a saucepan and heated on a hot plate, the water at the base of the saucepan becomes hot as it comes in contact with the hot metal. The heated water molecules vibrate faster and the water expands. This hot water is less dense and becomes buoyant. Cooler, more dense water at the surface descends and a convection current is set up. This convection current eventually leads to the complete heating of the water. See Figure 4.6.

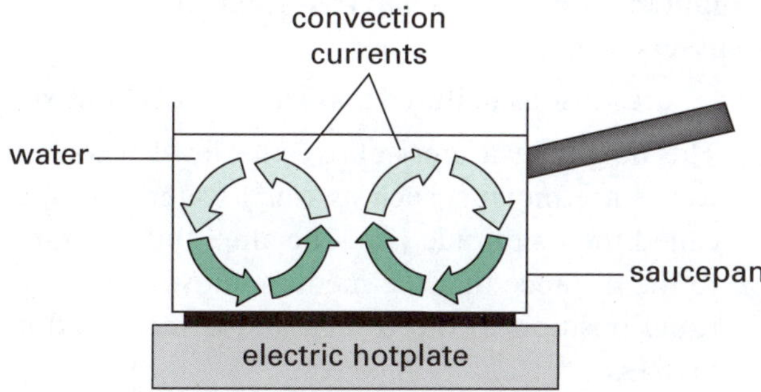

Figure 4.6 Convection currents in heated water

Radiation

Radiant energy is the energy carried by electromagnetic waves. Radiant energy can travel through the air or even through empty space where there are no particles. Infra-red waves are similar to visible light waves except they have a longer wavelength. Hot objects emit infra-red rays. The hotter the object, the more infra-red rays it gives off. When infra-red rays strike an object such as your skin, they speed up the particles in that object, thereby warming it. This is how the Earth gets its warmth from the radiant energy emitted from the Sun.

Figure 4.7 shows the amounts of radiant energy given off by two stars. The blue star is much hotter than the red star and the total radiation emitted is greater. Both stars emit ultraviolet, visible and infra-red radiation but the blue star gives out much more infra-red radiation than the red star. This radiant energy travels through space towards the Earth. Our atmosphere absorbs most of the ultraviolet radiation but visible and infra-red radiation mostly reaches the ground. Our Sun is much closer than these blue and red stars and so it is our major source of radiant energy.

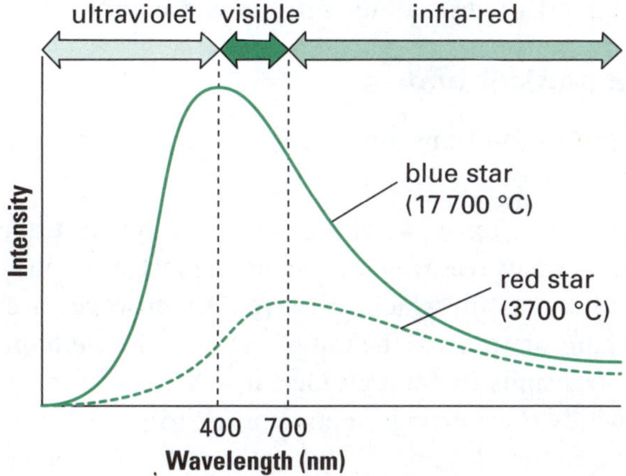

Figure 4.7 Stars emit radiant energy.

Microwave cooking

In conventional ovens, the food is cooked by infra-red radiation emitted from the electrical heating coils (or gas flame) as well as by convection currents. Many foods go brown when baked in this way. In a microwave oven, however, the food is cooked using microwaves. Microwaves are a form of radiant energy with longer wavelengths than infra-red radiation. No browning occurs in a microwave oven.

In a microwave oven a magnetron converts high voltage electrical energy into microwave radiation with a wavelength of about 12 cm. This microwave radiation is absorbed by water, fat and sugar molecules in the food and causes them to vibrate faster. The more vibrations that occur, the greater the increase in temperature of the food. The heat energy, therefore, cooks the food.

Microwaves can damage living cells and tissue, and so can be harmful to living things. So microwave ovens are surrounded by strong metal boxes preventing the waves from escaping. Never fool around with a microwave oven as microwaves can be very dangerous.

4.2 Sound energy

Sounds can be made in a variety of ways. Some examples include plucking a guitar string, bowing a violin string and thunder during a lightning storm. Each of these examples has one thing in common: they all involve vibrations. If you hold your voice box in your throat as you talk, you will feel these vibrations. Whenever something vibrates (such as a tuning fork), it disturbs the surrounding air molecules.

Mechanical waves

Waves are motions which carry energy from one place to another.

Sound is an example of mechanical wave motion. Mechanical waves require a medium (such as a solid, liquid or gas) in which to move. Ocean waves are mechanical waves, as the water surface is the medium that transmits the energy. Ocean waves are formed by winds transferring kinetic energy to the surface of the water. This energy makes the water molecules vibrate or oscillate. Some of this kinetic energy of vibration is transmitted to neighbouring water molecules, which also begin to move up and down.

- A wave crest begins to form where the particles move up relative to the normal undisturbed surface.
- When the particles move down, a wave trough is formed.

The crests and troughs move across the water surface because some of the energy of the vibrating water molecules is transferred away from the source of the original vibration. We see the waves travelling across the water surface. They carry energy with them due to their motion.

Motion of particles in water waves

The waves on the water surface carry energy but not matter away from the source. The water particles move to-and-fro (oscillate back and forth) but do not progress. If you watch a floating object in the ocean, it moves up and down as the wave passes by. The wave lifts it up and then it falls down as the wave passes. You can observe this using a cork floating in a bowl of water. At the beach where breaking waves occur, the motion of the particles is different (see Figure 4.8). Due to collisions of the vibrating water particles with the sand, the behaviour of the wave is altered and so there is some forward motion of the water particles. Surfboard riders use this phenomenon to ride breaking waves into the shore.

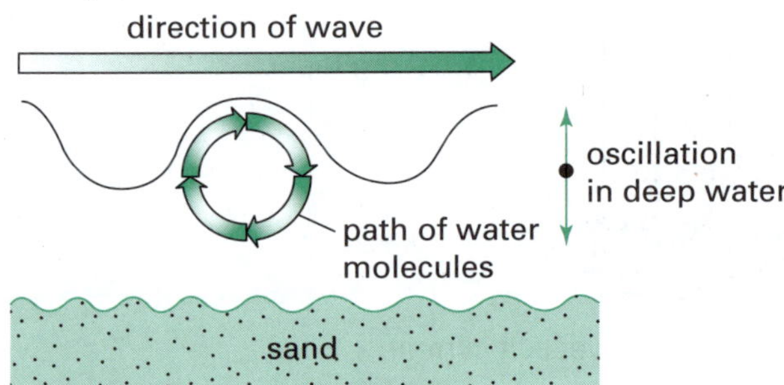

Figure 4.8 Motion of particles in a water wave near the shore

The motion of water particles in deep water is an example of a transverse wave. In mechanical transverse waves, the particles of the medium oscillate at right angles to the direction of energy flow. Other examples of mechanical transverse waves are surface earthquake waves and vibrating strings in musical instruments.

Amplitude, wavelength and frequency in a transverse wave

There are various features that characterise a wave.

- The height of a crest above the level when no waves are moving, such as when water is still, is called the amplitude (A). The amplitude is equal to the distance that the medium moves below its usual position to the trough and is measured in metres.
- The distance from one crest to the next (or from one trough to the next) is called the wavelength

(λ, the Greek letter lambda). Wavelength is measured in metres.

- The frequency (f) of a wave is the number of complete wave cycles that pass any point in one second. Frequency is usually measured in hertz (Hz). One hertz equals one cycle per second.

Figure 4.9 shows the main features of a transverse wave.

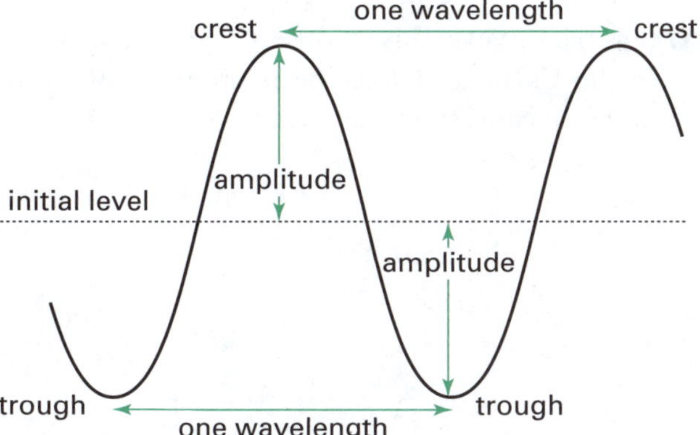

Figure 4.9 Features of a transverse wave

Motion of particles in a sound wave

Sound waves in air are caused by the transfer of energy from a vibrating source to the air molecules. For example, when a tuning fork is struck, the prongs vibrate and strike air molecules around them. The air particles also oscillate to and fro, and energy radiates out from the source as a sound wave.

Sound waves are different to water waves in that the particles of the air vibrate to and fro in the direction of energy transfer rather than at right angles to the energy transfer, as in water waves. As a result, the air particles alternately bunch up to form compressions and then spread out to form what are known as rarefactions, as shown in as shown in Figure 4.10. The sound wave thus consists of regions of compression and rarefaction. When the sound wave reaches our eardrum, it causes the drum to vibrate and energy is then transferred to the inner ear and finally the brain.

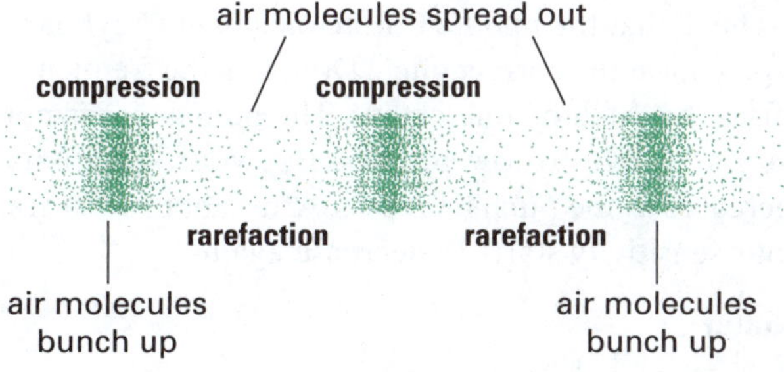

Figure 4.10 Compressions and rarefactions in air

Sound waves are classified as compression waves. Compression waves are sometimes called longitudinal waves.

The speed of sound waves

The speed of sound in air is quite simple to measure. In 1738, experiments were conducted in Paris to measure the speed of sound between two stations 30 km apart. Cannons were fired at constant intervals and the time between seeing the flash of gunpowder and the sound of the cannons was measured. These experiments and later ones showed that at 25 °C the speed of sound in still air is 350 m/s and that the speed decreased as the temperature of the air decreased.

Sounds travel at different speeds through different substances. In 1826, the speed of sound in the water of Lake Geneva in Switzerland was determined over a 13.5-km distance between two boats. The people in the first boat rang a bell underwater and at the same time exploded gunpowder on the surface so that the people in the second boat could start timing when they saw the explosion. A special ear trumpet covered with a membrane was used to detect the sound in the water under the second boat. These and other measurements have shown that at 25 °C sound travels at 1500 m/s in water. This is about five times faster than in air at the same temperature. The presence of more particles in the water helps to transfer the sound energy faster. Sound energy travels even faster through solids. Without any particles (in a vacuum) sound cannot travel at all. Not only does sound travel faster through solids and liquids than through air, less sound is absorbed (or transformed into other energy forms) by solids and liquids as well. This means that sound energy in these media can travel much greater distances. This is useful especially for sea mammals which can communicate through longer distances.

In the simple experiment shown in Figure 4.11, two students used two wooden poles about 300 m apart and a metal wire to show that sounds travel faster through solids than through air. One student hit one wooden pole with a baseball bat, the other student with his ear against the next pole heard two sounds: one through air and the other through the poles and wire. The sound that took the longer route (up the pole, across the wire space and down the next pole) reached the second student before the sound that travelled through air.

Table 4.2 shows the differences in speed of sound, depending on the temperature and the medium through which it is travelling.

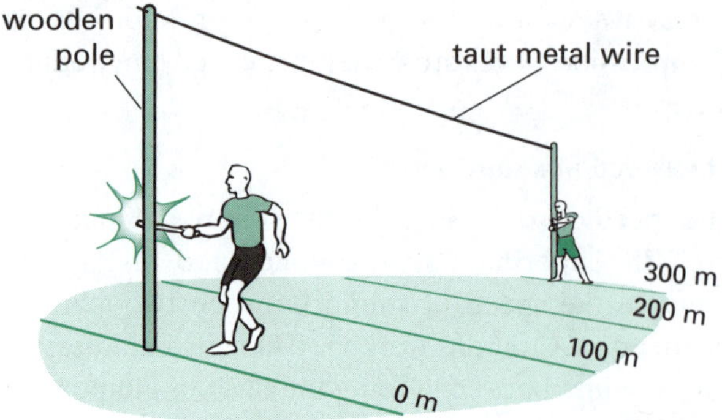

Figure 4.11 An experiment to show that sound travels faster through solids than through air

Table 4.2 Speed of sound in different media

Medium	Temperature (°C)	Speed of sound (m/s)
air	0	331
air	25	350
water	25	1500
iron	20	5100
glass	20	5500
granite rock	20	6000

Measuring the wavelength of a compression (longitudinal) wave

Figure 4.12 shows a compression wave travelling through a spring with a speed or velocity (v). The wavelength of these waves can be determined by measuring the distance between two compressions or two rarefactions.

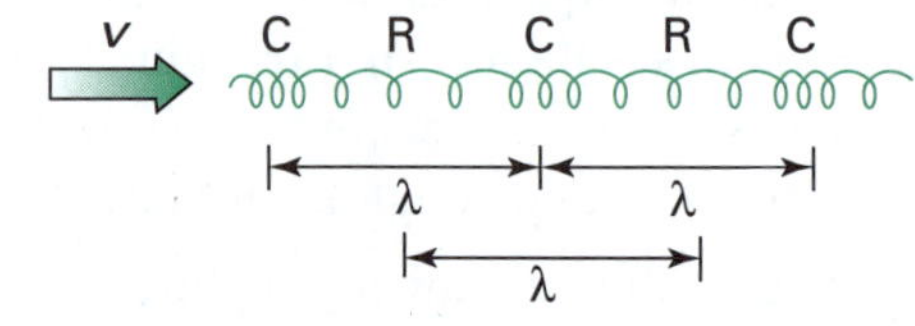

Key: C = compression R = rarefaction
λ = wavelength

Figure 4.12 Compression wave in a spring

Reflection of sound waves

Sound waves can be reflected from various surfaces. This property is used in a number of ways.

- Sonar. Boats use ultrasonic waves to test water depth and detect schools of fish.
- Echo location (or biosonar). Animals such as dolphins and bats produce ultrasonic waves (frequency ~200 000 Hz) for navigation. They listen for the return of the echo and thus determine how far away the reflecting surface is (see Figure 4.13). Smaller objects reflect fewer waves and thus the size of the reflecting surface can be determined.
- Obstetric ultrasonography. Sound waves with frequencies between 2 and 18 MHz can be used to image the foetus during pregnancy.
- Seismic surveys. This involves using seismic waves in the Earth to determine its internal structure and to locate structures such as oil deposits.

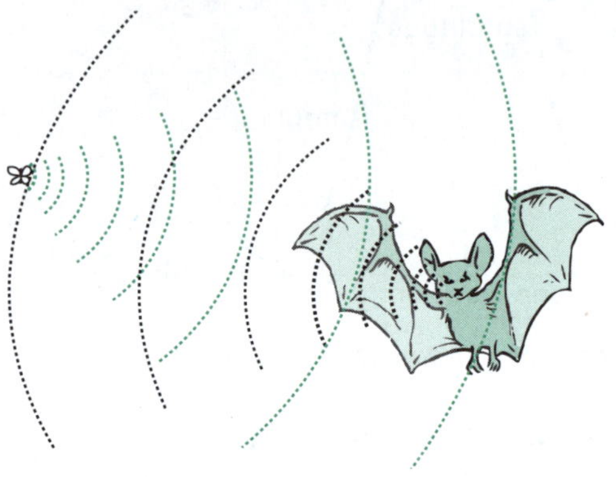

Figure 4.13 Biosonar in bats

Pitch of sound and musical instruments

The pitch (or frequency) of a sound measures how high or low the note sounds to your ears. A soprano sings high notes and a bass singer sings low notes. Pitch has nothing to do with loudness. A high-pitched sound can be either soft or loud, depending on the amount of energy involved. When a sound is high pitched, there are more compressions of air molecules striking your ear drum each second.

The pitch of most sounds you hear is usually due to a blend of various frequencies. But there are some instruments (such as tuning forks) that produce sounds of a single frequency. Middle C, for example, has a frequency of 256 Hz. A tuning fork can be used to check that the middle C note on a piano keyboard is playing at the correct pitch. Doubling the frequency raises the pitch by one octave. The human ear is not very sensitive to low pitches. But your sensitivity increases as the pitch increases. After about 1000 Hz, your sensitivity starts to decrease again.

Guitar

Consider a six-string guitar. Each string has a different length and mass. This will affect the pitch or frequency of the note when each string is plucked. String 1 is the thinnest string with the highest pitch and string 6 is the thickest string with the lowest pitch (see Table 4.3).

Table 4.3 Six-string guitar and frequencies

Number	Note	Frequency (Hz)
1	high E	330
2	B	247
3	G	196
4	D	147
5	A	110
6	low E	82

Woodwinds

Musical instruments in the woodwind section of an orchestra include flutes, oboes and clarinets. They are based on a long tube with a column of air inside that can vibrate. Along the length of the tube are various holes that can be opened or closed. The lowest pitched note is played when all the holes are closed. As the holes are progressively opened from the end of the tube, the pitch of the note rises. In a flute the air column is 66 cm long. The lowest pitch note is about 262 Hz and the highest pitch note is 2093 Hz.

Experiment 1

Musical tubes

Aim

- To make a musical instrument using test tubes

Background

An octave in music is a set of eight notes: doh-re-mi-fah-soh-la-ti-doh. Your test tube instrument will be able to play an octave.

Method

1. Set up the test tubes as shown in Figure 4.14. Various amounts of water are added to the tubes.
2. Purse your lips and blow across the top of an empty test tube or tap the top side of it with a pencil. It should make a sound. Practise this technique. This will be your 'doh'.
3. Take a second test tube and add a little water to it. Blow across the top of the tube to hear the note formed (or tap it with a pencil). Adjust the water level until you think you have the note 're'.
4. Continue on up the scale until you have built up your octave.
5. You may now be able to play a simple tune on your instrument.

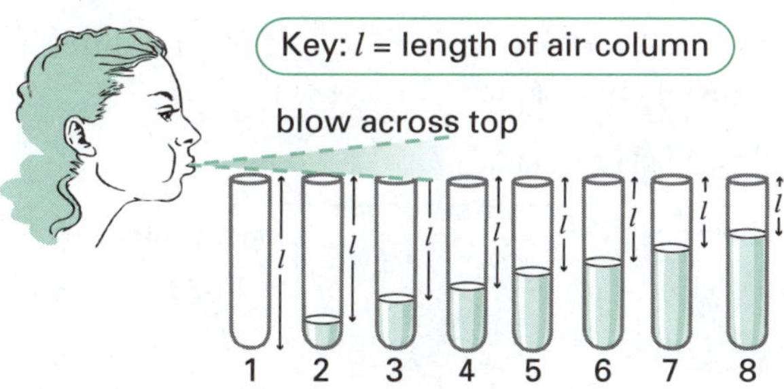

Figure 4.14 Musical test tubes

Analysis

1. What happens to the pitch of the note as the length of the air column in the tube decreases?
2. What happens to the wavelength of the sound wave as the length of the air column in the tube decreases?

Go to p. 244 to check your answers.

Conclusion

Write a suitable conclusion.

Go to p. 244 to check your answer.

Loudness of sound

The loudness of a sound depends on the amount of energy received each second by the ear. As the distance from the source of the sound increases, the loudness decreases rapidly. This is because sound waves spread out as they leave the source. The unit used to measure sound intensity is the decibel (dB). It is commonly used to describe noise levels relative to the threshold of hearing. An increase of 3 dB is a doubling of the sound intensity. An increase of only 1 dB (about a 25% increase in sound intensity) is just barely detectible.

The sound intensity of 0 dB is set at the threshold of human hearing. At the other end of the scale is the threshold of pain for humans, about 130 dB. Sound intensities over 90 dB (or even lower levels for longer, repeated periods of exposure) can cause permanent damage to the cochlea of the ear. Damage begins by people being less sensitive to certain frequency ranges and continues until deafness occurs. This damage is usually permanent. Unfortunately, there are very few pain receptors in the ear to warn you that damage is happening. This is why it is important to avoid being exposed to loud sounds for long periods of time.

Table 4.4 compares the loudness of several different sources of environmental noise.

Table 4.4 Environmental noise

Environmental source	Loudness (dB)
whisper (at 1 m)	20–30
conversation	60–70
city traffic (inside car)	80
power mower (at 1 m)	105
loud rock concert	115
jet engine (at 30 m)	140

In their regular working week, workers should not be exposed to noises any greater than 90 dB(A). The unit dB(A) is called 'A-weighted' and is used in workplaces as people are more sensitive to high frequency noises than low frequency noises. Thus, when two machines produce sounds of different frequencies, the higher pitched sound may appear louder. The A-weighting de-emphasises lower pitched sounds.

Table 4.5 compares the A-weighted loudness of various environmental noise sources.

Table 4.5 A-weighted noise levels

Environmental source	A-weighted loudness (dB(A))
conversation (at 1 m)	55
power mower (at 1 m)	92
newspaper press	95
textile machinery	102
circular saw (at 1 m)	115

Cochlear implants

Sound waves in the air are directed down the ear canal by the external ear lobe or pinna. The structure of the ear is shown in Figure 4.15.

The ear canal

At the end of the ear (or auditory) canal is the eardrum. This is a thin sheet of tissue about 7 mm in diameter. The sound waves cause the eardrum to vibrate. These vibrations pass across the middle ear via three small bones called the hammer, anvil and

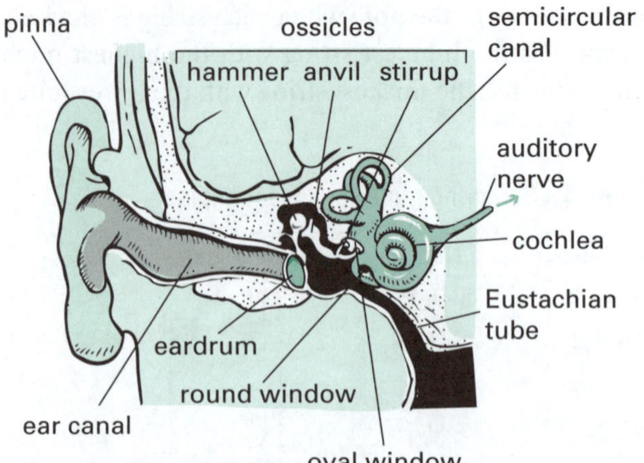

Figure 4.15 The ear

stirrup. Collectively these bones are called ossicles. The hammer is attached to the inside of the eardrum. The semicircular canals serve as the organ for our sense of balance.

The Eustachian tube

The middle ear is filled with air and is connected by a 4-cm tube, called the Eustachian tube, to the nasal cavity at the back of the throat. Its purpose is to equalise air pressure on either side of the eardrum, thus preventing it from rupturing. The 'popping' of our ears when we drive up the side of a mountain or travel in an aeroplane is caused by the sudden equalising of pressure when the Eustachian tube opens during yawning or swallowing.

Passing on the message

The vibrations set up in the ear bones are transferred to the oval window of the cochlea. The fluid in the cochlea is set in motion. The vibrations of the fluid move fibres or hair cells—about 24 000 of them! Sounds of different frequencies and intensities move these very sensitive hair cells in slightly different ways. The impulses stimulate nerve endings which send messages along the auditory nerve to the brain.

Deafness

Deafness can be caused by a birth defect, genetics, disease or injury. Here are some examples of causes of hearing loss.

- Conductive deafness is caused by the hammer, anvil and stirrup bones failing to conduct vibrations from the ear drum to the inner ear. Fluid in the ear canal can dampen the vibrations as well.
- Nerve deafness is caused by disease or injury of the auditory nerve to the brain.

- Genetic deafness may be caused by gene mutations that result in malformation of ear components.
- Loud noises can lead to hearing loss due to damage to the inner ear. Continued exposure to loud machinery sounds or loud rock music will cause hearing loss, even from a young age.

Cochlea implants were developed in Australia by Professor Graeme Clarke and his team at Melbourne University and the first implant was achieved in 1978. In the 1990s the implant was miniaturised so it could be worn like behind-the-ear hearing aids.

Cochlea implants do not cure deafness. They are prosthetic (artificial) devices that allow patients to hear. Deaf patients must have a working auditory nerve. A small microphone detects the sound waves in the external environment. Electrical signals are then sent from the microphone to a small speech microprocessor. The microprocessor encodes these signals and sends them to the transmitter. The external transmitter sends the encoded signals through the skin and into the implanted receiver. The receiver decodes the signals and the stimulator converts signals into impulses which are then relayed through electrodes implanted into the cochlea. This is shown in Figure 4.16.

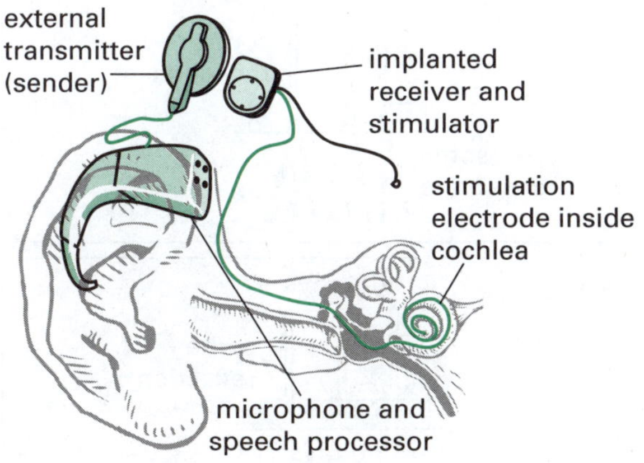

Figure 4.16 Cochlea implant

4.3 Light energy

Electromagnetic radiation

Unlike a mechanical wave in which the particles in a medium are disturbed, electromagnetic rays do not require a medium for their propagation.

- Electromagnetic waves involve the propagation of oscillating electric and magnetic fields.
- Electromagnetic waves are transverse waves.
- Electromagnetic waves move at their highest velocity ($v = 3 \times 10^8$ m/s, which is 300 million m/s, the speed of light) in a vacuum. They travel slightly more slowly through matter (e.g. air, glass).

Figure 4.17 shows the nature of an electromagnetic wave in terms of magnetic and electric waves

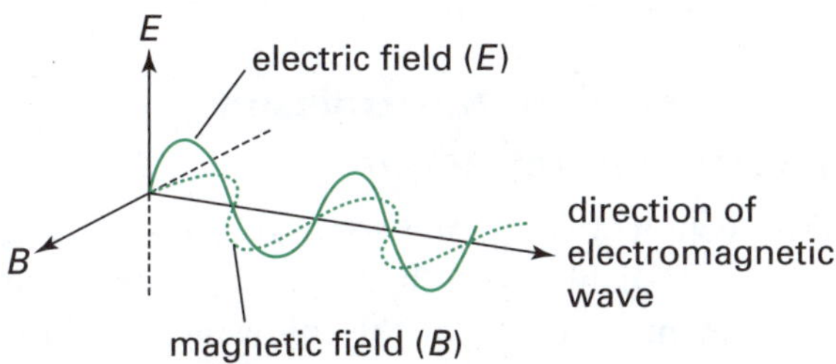

Figure 4.17 Electromagnetic wave

Electromagnetic waves vary in their frequencies and wavelength even though their velocities are the same in a vacuum. The collection of different frequency waves is called the electromagnetic spectrum. The various components of the electromagnetic spectrum are shown in Figure 4.18 and Table 4.6.

Our eyes are able to detect only the visible band which has wavelengths in the approximate range 400 nm (violet end) to 700 nm (red end). Photographic film or light meters are also able to detect light. Light is emitted from a variety of sources such as the Sun, incandescent and fluorescent light bulbs, burning materials and lasers.

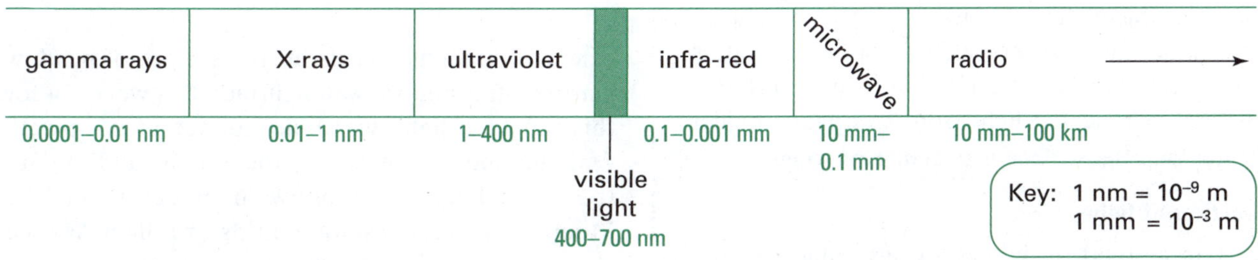

Figure 4.18 Electromagnetic spectrum—1 nanometre (nm) is one millionth of a millimetre

Table 4.6 Features of the electromagnetic spectrum

Band name	Wavelength band (approximate)	Sources of waves	Uses of waves
radio/television	10 mm–100 km	radio/television transmitters	radio/television communication, radio astronomy
microwave	0.1 mm–10 mm	radar transmitters, microwave ovens	satellite communication, cooking food
infrared	0.001 mm–0.1 mm	electric radiators	heating rooms, medical heat treatments, night vision systems
visible light	400 nm–700 nm	stars, electric lamps	human vision, photosynthesis, photography, astronomy
ultraviolet	1 nm–400 nm	UV lamps, stars	UV astronomy, sterilisation
X-rays	0.001 nm–1 nm	X-ray tubes, black holes	medical radiography (diagnosis and treatment), flaws in structural materials, X-ray astronomy
gamma rays	0.0001 nm–0.01 nm	radioactive minerals	sterilisation, killing cancer cells

(1 nm = 1 nanometre = 1 × 10–9 m)

Light absorption, transmission, reflection and refraction

In this section we examine some of the common properties of light rays. When light rays strike an object they may do one of the following (see also Figure 4.19).

- Light rays may pass straight through the material with some small loss of energy. This process is called transmission. Transparent materials such as window glass or the clear glass in spectacles are like this. The thicker the glass, the more light energy is absorbed.
- Light rays may be partially transmitted through the material and partially reflected or scattered at the surfaces. Translucent materials such as frosted glass in bathroom windows are like this.
- Light rays may be almost completely absorbed so that no light emerges on the other side. Absorption of light by a black body is almost complete.

Opaque objects such as brick walls and wood absorb most of the light rays that fall on them. Some of the light that is not absorbed is reflected off the surface of opaque objects. This allows us to see the object. Shiny or lustrous objects such as polished metals or mirrors reflect more light than rough surfaces. Scattering of reflected light from an opaque object indicates that the surface is irregular or rough.

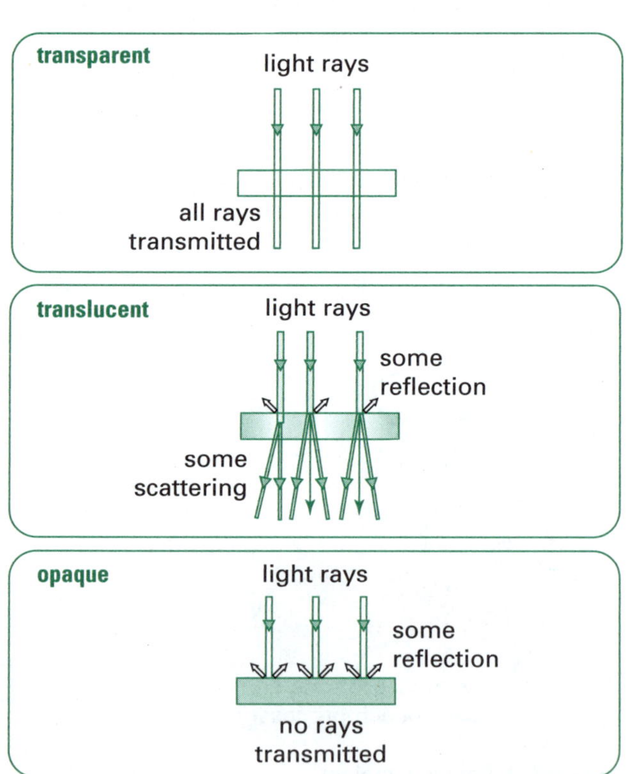

Figure 4.19 Properties of light rays

Reflection of light

A certain amount of light is always reflected from a surface that separates two different media. For example, sunlight reflects off the surface of window glass or off a smooth water surface. If it were not for this reflected light, we could not see most objects. Only luminous objects (e.g. the Sun, a candle) emit their own light, which allows us to see them. The Moon, however, does not emit its own light. We see it due to the reflection of the Sun's rays into our eyes.

- Smooth surfaces reflect light rays in one particular direction. This is called regular reflection.

If light from a torch is shone at an angle of 30° onto the surface of a mirror, the reflected light is brightest when viewed at 30° to the mirror's surface.

Rough surfaces cause the reflected light to be scattered in many directions. Light from a torch which reflects from a rough surface is bright when viewed from many directions. The surface of a white sheet of paper looks uniformly bright when illuminated by overhead lights, due to the scattering of light on reflection.

- When parallel rays of light reflect off a rough surface, the scattering of the rays is called diffuse reflection.

Regular reflection and diffuse reflection are shown in Figure 4.20.

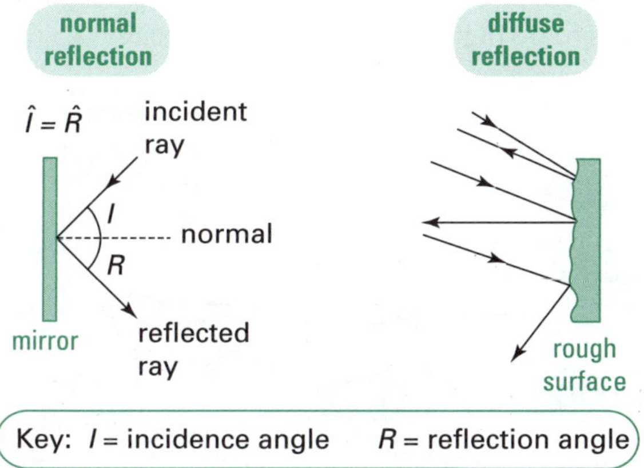

Figure 4.20 Regular and diffuse reflection

Curved surfaces reflect light rays in a different way to flat (or plane) surfaces. Figure 4.21 shows examples of plane, concave, convex and parabolic reflecting surfaces. Parallel light rays are made to converge (come together) when they reflect off concave mirrors, whereas with convex mirrors the rays diverge (spread apart).

Plane mirrors are used in our homes in the bathroom and the bedroom. Concave mirrors are used in many applications, including telescopic mirrors to reflect starlight, dental mirrors and shaving mirrors. In all cases the image formed by the concave mirror is magnified. Convex mirrors are commonly used as rear-vision mirrors on the passenger side of a car. The images are reduced in size and the objects appear closer than they really are but such mirrors give a wide field of view. Parabolic-shaped mirrors are useful in car headlights as they cause light rays to reflect off them to form parallel beams.

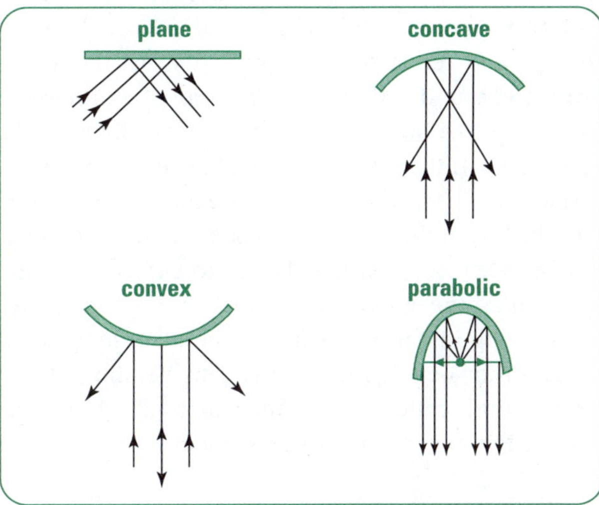

Figure 4.21 Reflection from plane and curved surfaces

When light rays reflect off a surface, they obey the law of reflection. This law states:

The angle of incidence is equal to the angle of reflection.

Figure 4.22 shows that the angle of incidence and the angle of reflection are both measured from an imaginary line at right angles to the surface. This imaginary line is called the normal. The incoming ray is called the incident ray. This law is obeyed when light rays reflect off any surface, including rough surfaces. The image formed in a plane mirror is also laterally inverted. This means that the left side of the object now appears on the right of the image. Because of this, some emergency vehicles such as ambulances and police cars have words written backwards on the front of the vehicle so that a motorist in front can read it correctly when looking in the rear vision mirror.

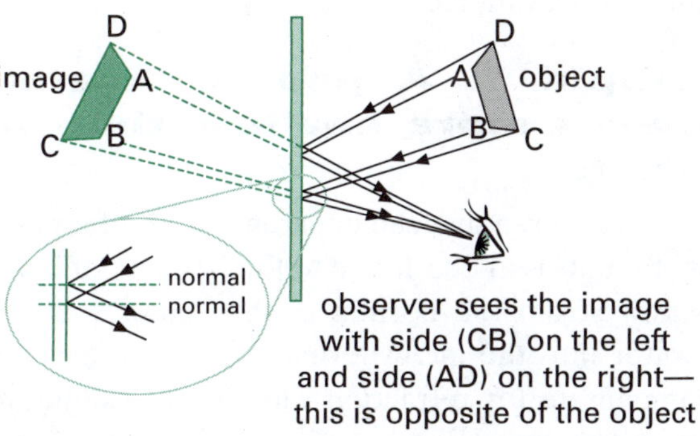

Figure 4.22 Law of reflection and lateral inversion

Refraction

Light rays only travel at 300 000 km/s in a vacuum. In gases the speed is slightly reduced, but in liquids and transparent solids the speed of the light is considerably reduced. For example, in glass the speed drops to around 200 000 km/s. If a ray of light passes from air into a block of glass, it will slow down while in the glass. If the ray is incident at right angles to the surface, it will continue on into the glass and emerge into the air on the other side. The emergent ray does not deviate from its original direction. If the incident ray strikes the glass surface at some other angle, the slowing of the ray while in the glass leads to a deviation or bending of the ray. This bending of the ray of light is called refraction. Figure 4.23 shows the path of two rays through a glass block.

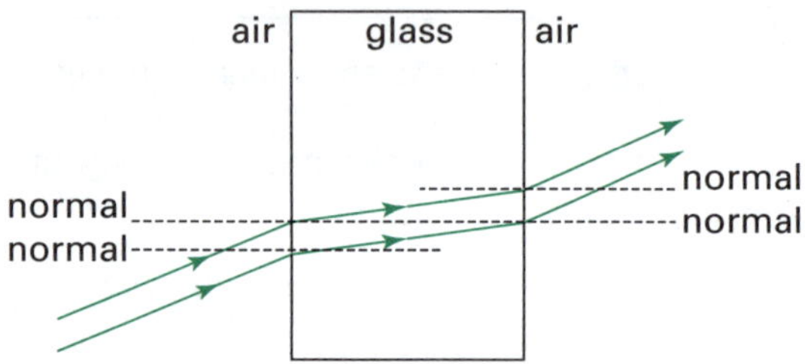

Figure 4.23 Refraction of light through a glass block

Refraction occurs with light coming in through a window but, as the two surfaces are parallel and close together, we don't normally notice it.

Generally when light rays are incident at an acute angle to a surface they:

- bend towards the normal when they pass from a less dense medium into a more dense medium (e.g. from air to water, from water to glass)
- bend away from the normal when they pass from a more dense medium into a less dense medium (e.g. from water to air, from glass to water).

These generalisations help us to explain some common observations.

Example 1: A pool of water appears more shallow than it really is

Figure 4.24 shows that the apparent depth of a pool is less than its real depth due to the bending of light rays away from the normal as they emerge from the water into the air. Aboriginal peoples learned to compensate for refraction when spear fishing in rivers and shallow lakes.

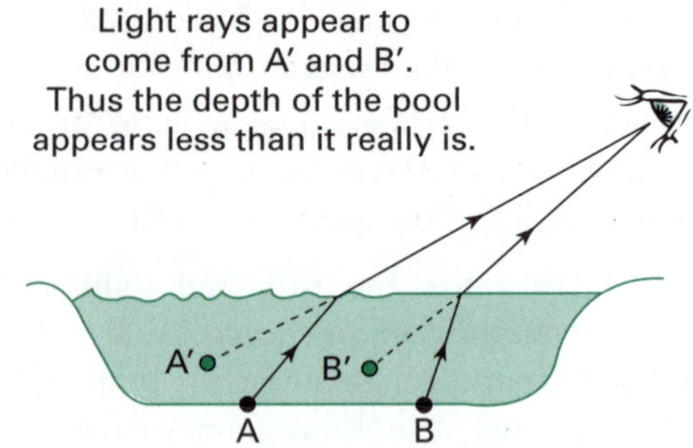

A and B are two points on the bottom of the pond

Figure 4.24 Apparent depth of a pool of water

Example 2: The apparent altitude of the Sun

Figure 4.25 shows the bending of rays of sunlight while they pass through different layers of atmosphere. Near the ground the density of the atmosphere increases and the refraction increases. The Sun appears to be higher in the sky than it really is.

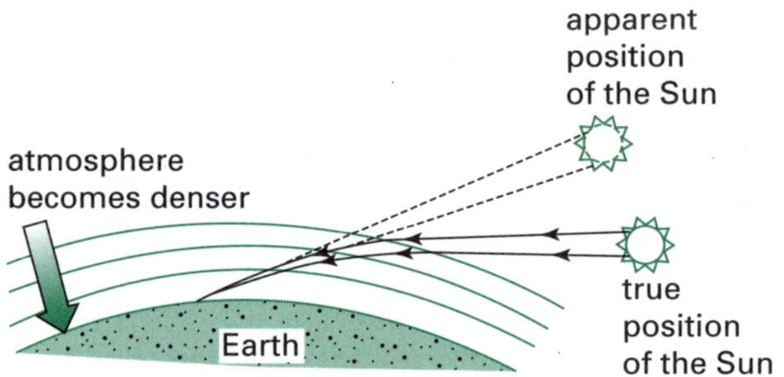

Figure 4.25 Apparent altitude of the Sun

Experiment 2

Refraction in water

Aim

- To investigate the bending of light rays as they travel from one medium to another

Method

1. Fill a 1-L beaker with water to within 1 cm of the top. Place a wooden stick or glass rod in the beaker so that it rests against the side as shown in Figure 4.26.
2. Observe the stick or rod from different angles. Look from above and at the side. Now hold it vertically. Observe again. Now put the stick or rod at a shallow angle to the water surface. Does

it always appear straight? Accurately draw what you see in each test.

3. Look at a picture or a newspaper through a glass or small beaker of water. What do you see? Why does the writing on the newspaper, or picture, look like this? Draw what you have observed.
4. Place a coin in a plastic or metal cup so that it is against one edge as shown in Figure 4.26. Now move your head back from the cup until the coin is just no longer visible. Hold your head still. Ask your partner to pour water slowly into the cup without disturbing the coin. Does the coin become visible? If not, repeat the experiment more carefully.

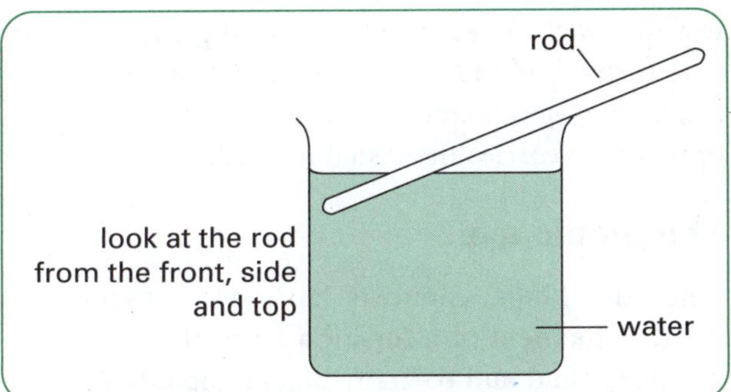

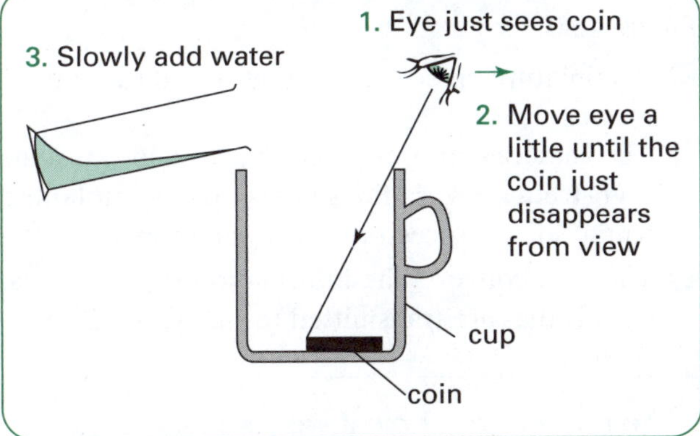

Figure 4.26 Refraction experiments

Analysis

Answer the following questions as part of your analysis.

1. Figure 4.27 shows how rays of light refract in the experiment involving a rod in water. Copy the diagram into your workbook.
   a) Use arrows to show on the diagram the direction in which the light rays are travelling.
   b) Use dotted lines to show the apparent position of the stick under the water.

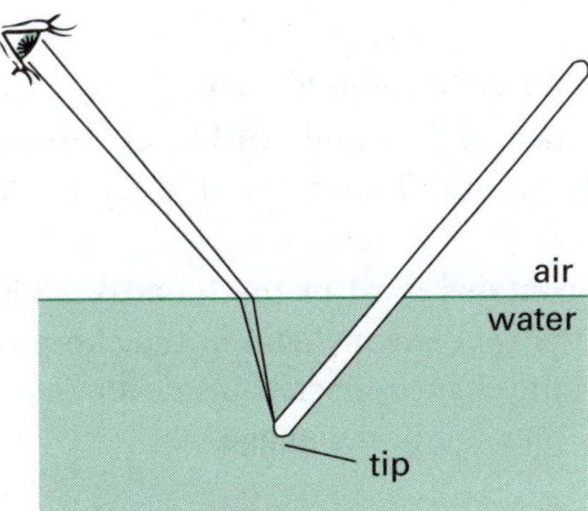

Figure 4.27 The bent stick

2. Figure 4.28 shows a ray diagram for the coin in the cup experiment. Copy the diagram into your workbook.
   a) Use arrows to show the direction in which the light rays are travelling.
   b) Use dotted lines to show the apparent position of the coin when the water fills the cup.

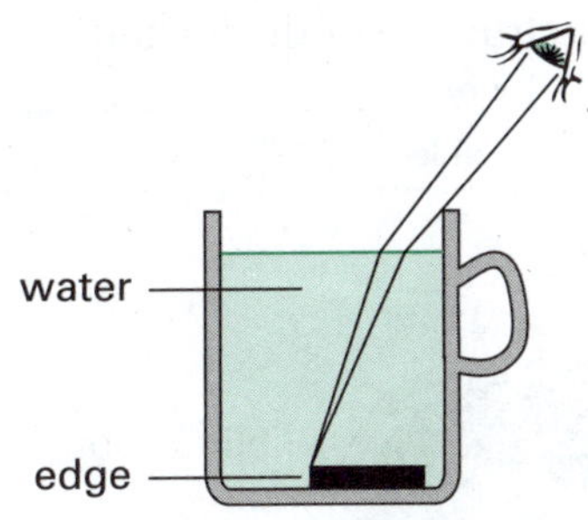

Figure 4.28 The appearing coin

Go to p. 244 to check your answers.

Conclusion

Write a suitable conclusion.

Go to p. 244 to check your answer.

The human eye

The human eye is roughly the shape of a sphere. There are three tissue layers: outer, middle and inner.

Outer layer

- The outer layer, the sclera, is extremely tough.
- The muscles of the eye are attached to the sclera and this allows eye movement.
- The sclera is white in colour, except at the front where it forms the transparent cornea.
- The cornea allows light to enter the eye.

Middle layer

- The middle layer is the choroid coat.
- The choroid has a rich supply of blood vessels and is deeply coloured with melanin (a black pigment).
- The choroid contains most of the blood vessels that nourish the eye. The melanin in the choroid absorbs any stray light reflected inside the eye. This prevents blurring of the image.

Inner layer

- The innermost layer is the retina.
- The retina contains two kinds of light sensitive cells: the rods and the cones.
- Rods are especially sensitive to dim light, but react to all wavelengths of the visible spectrum (the light you can see).They cannot detect colour.
- Cones respond best to narrower parts of the visible spectrum of light. They are sensitive to colours and details in bright light. Cones cannot detect very dim light.

Figure 4.29 shows the structure of the human eye.

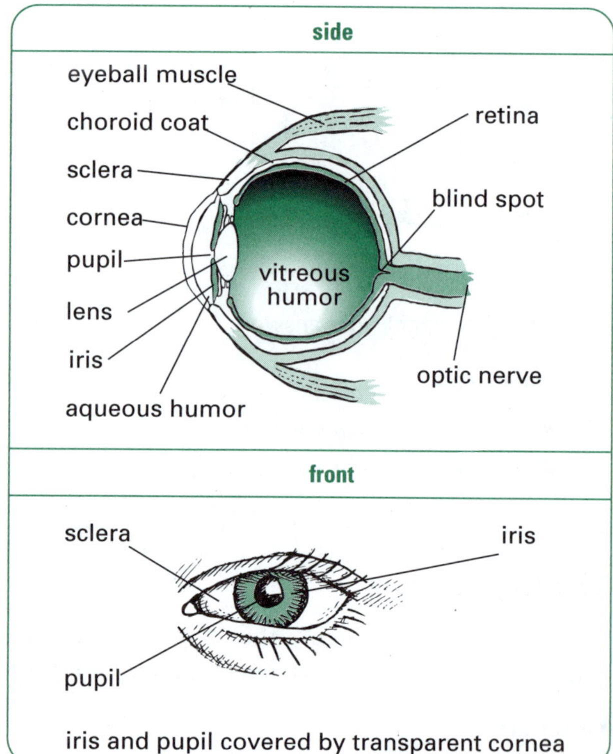

Figure 4.29 Structures in the eye

Controlling the amount of light

Light entering the eye first has to pass through the cornea. Refraction occurs here to produce some focusing. The amount of light entering your eye is controlled by the iris. Muscles in the iris control the size of the pupil (the opening within the iris), letting more light in when it is dim and constricting to cut out light when it is too bright. This occurs so that just the right amount of light falls on the retina.

Focusing with the lens

Just behind the iris is the lens. Muscles attached to it can alter its thickness and curvature so that the light is refracted more or less to produce a sharp focus on the retina. In a camera, light is focused by moving the lens back and forth. Humans, other mammals and birds can't do this.

Retina

Light rays passing through the cornea and pupil bend towards each other. This bending increases as they pass through the lens. When the light rays strike the retina, they should meet. This is called the focus. The optic nerve carries the visual impulses to the brain.

The bionic eye

Since the 1980s, scientists have been involved in the development of a functional prosthetic eye that will help blind and partially blind people to see. The bionic eye would work in the following way (see also Figure 4.30).

- A miniature video camera is attached to a pair of glasses.
- The camera captures video images and transmits high-frequency radio signals to an implanted 5 mm^2 microprocessor chip in the retina.
- The chip converts the information into electrical signals that are transmitted to the optic nerve.

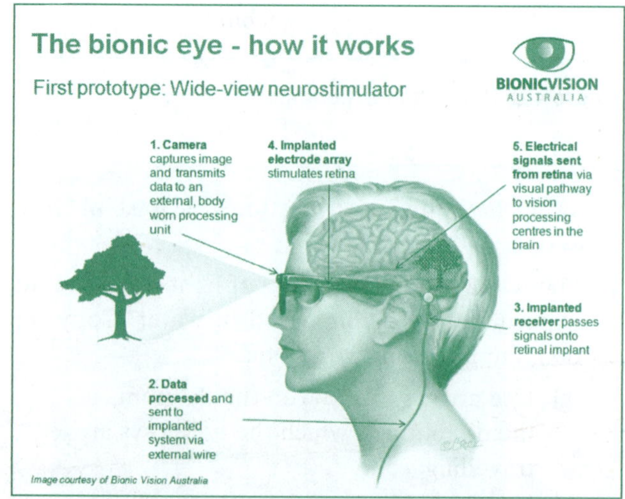

Figure 4.30 Bionic Vision Australia's bionic eye

This technology will only be suitable for patients with some functioning retinal cells and a functioning optic nerve. An Australian company called Bionic Vision Australia plans to have a prototype ready for testing in 2013. The wide-view design will have 98 electrodes in the microchip which will be used to stimulate the retina.

Test yourself 1

Part A: Knowledge

1. Two identical metal cans are filled with hot water and thermometers inserted to measure the water temperature over time. One can is painted silver and the other painted black. Select the correct response. *(1 mark)*
   A The water in the black can will cool down faster as radiant heat is lost at a faster rate.
   B The water in each can will cool at the same rate.
   C Heat is lost from both cans by conduction.
   D The water in the silver can will cool down faster as radiant heat is lost at a faster rate.

2. Which of the following waves is *not* an example of a mechanical wave? *(1 mark)*
   A seismic wave
   B gamma wave
   C ultrasonic wave emitted by a dolphin
   D vibrating violin string

3. Select the statement that is true of infra-red waves. *(1 mark)*
   A Infra-red rays have much higher frequencies than radio waves.
   B Infra-red rays are used in radiography for treating cancer.
   C Visible light rays travel at a much greater velocity than infra-red rays in space.
   D Green plants use infra-red rays for photosynthesis.

4. When light passes through a window pane it is *(1 mark)*
   A transmitted completely without loss of energy.
   B scattered and the glass becomes translucent.
   C slowed down as glass is more dense than air.
   D totally absorbed.

5. Refraction is the process in which light rays *(1 mark)*
   A are scattered.
   B are reflected and absorbed.
   C converge after reflection from a concave mirror.
   D are bent as they travel at an angle through media of different density.

6. Complete the following restricted-response questions using the appropriate word. *(1 mark for each part)*
   a) A .... wave is one in which the particles vibrate at right angles to the direction of wave propagation.
   b) The height of a wave above the equilibrium level is called the ....
   c) Gamma waves have the highest frequency in the .... spectrum.
   d) Rays of light bend .... the normal when they pass at an acute angle from air into water.
   e) In metals, free electrons are able to move (or diffuse) between the atoms and so they transfer heat energy much .... than the metal atoms do.

7. Use the code letters to match the terms or phrases in each column. *(1 mark for each part)*

Column 1	Column 2
A crest	F mechanical wave
B sound	G bending
C X-rays	H trough
D refraction	I oscillation
E vibration	J transverse waves

Part B: Skills

8. About 400 years ago, Galileo discovered that the density of liquids changes as the temperature varies. He used this discovery to develop a thermometer as shown in Figure 4.31. When the temperature rises, the liquid in the glass tube becomes less dense, and one by one the glass spheres sink. The sealed glass spheres are carefully weighted and a medallion that indicates the temperature is attached.
   a) Explain how Galileo would work out the air temperature on a particular day. *(1 mark)*

b) Would such a thermometer be responsive to rapid temperature changes? Explain. *(2 marks)*

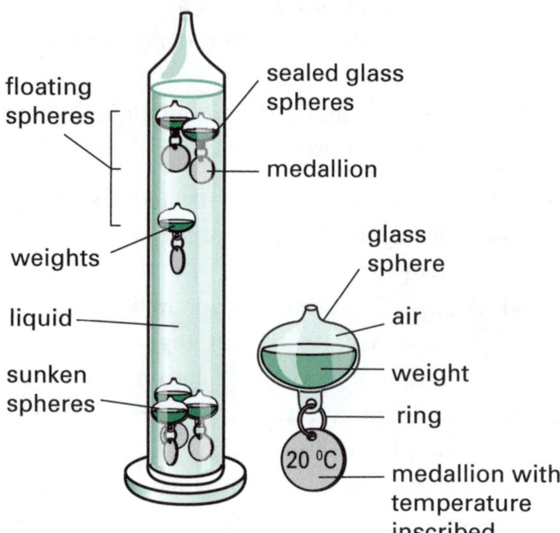

Figure 4.31 Galileo's thermometer

9. A test tube of water contains an ice block that has been weighed down so it sinks, as shown in Figure 4.32. The water at the top of the tube is heated with a Bunsen flame until it boils. During this process the ice block does not melt. Explain why the ice block does not melt in this short time interval. *(1 mark)*

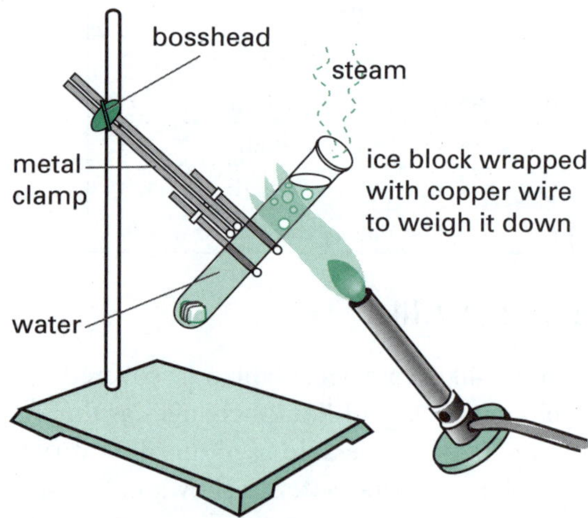

Figure 4.32 Boiling water in a tube with ice

10. Two radio stations (A and B) transmit programs on 702 kHz and 954 kHz respectively in the AM band.
    a) Which station transmits waves of greater wavelength? *(1 mark)*
    b) A third station (C) transmits at 104.9 MHz in the FM band. How do the waves from this station compare with those transmitted from station A in terms of:
        i) wavelength? *(1 mark)*
        ii) velocity? *(1 mark)*

11. Determine the amplitude and wavelength of the transverse wave in Figure 4.33. *(2 marks)*

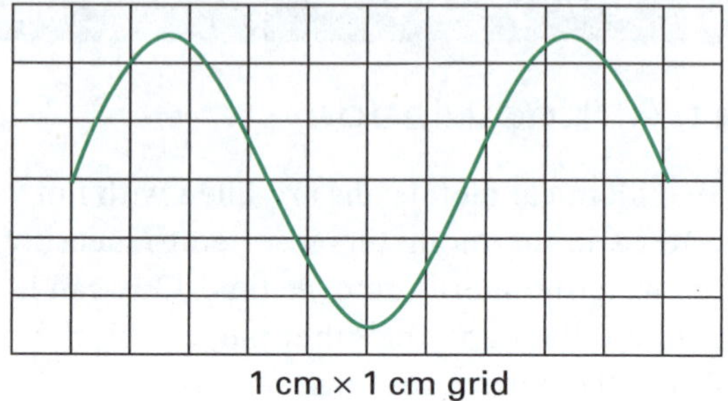

Figure 4.33 Transverse wave

12. A student places a ruler on the edge of a desk as shown in Figure 4.34 and causes it to vibrate by bending the ruler and letting it go.

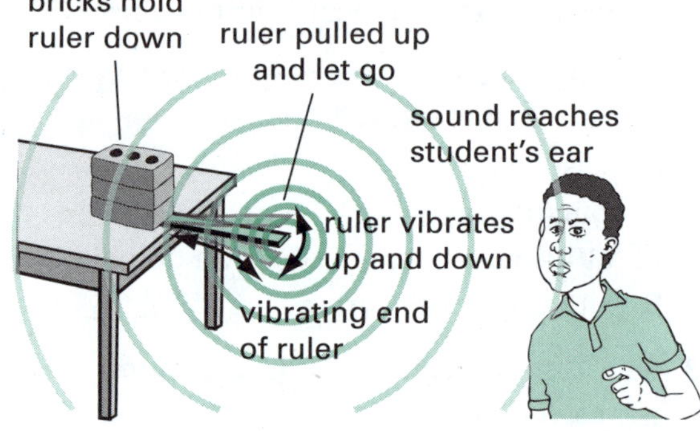

Figure 4.34 Transmitting energy to the student's ear

    a) Explain how the energy from the vibrating ruler is transmitted to the student's ear. *(1 mark)*
    b) What do you think would happen to the loudness and pitch of the sound in the following circumstances?
        i) if the ruler were to vibrate faster *(1 mark)*
        ii) if the length of the vibrating section of the ruler were to be increased *(1 mark)*

13. The greater the frequency of light, the greater the refraction as it passes from one medium to another. In the visible spectrum, red light has the longest wavelength and violet light has the shortest wavelength. White light is a mixture

of all the colours of the visible spectrum. When white light is passed through a triangular glass prism, it is separated or dispersed into the visible spectral colours due to differing extents of refraction. Only the red and violet components are shown. Explain which one of the diagrams in Figure 4.35 correctly shows that white light is dispersed when passing through a prism. *(4 marks)*

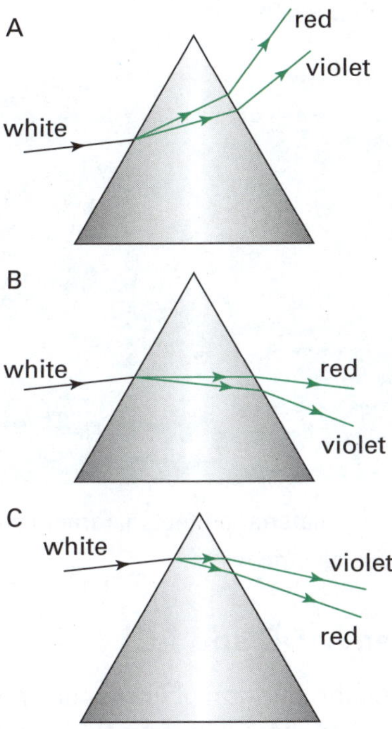

Figure 4.35 Refraction of white light through a prism

14. Figure 4.36 shows a student measuring the delay time between a sound reflecting off a distant cliff and returning as an echo. Use the information in the diagram to estimate the distance between the cliff and the student. *(2 marks)*

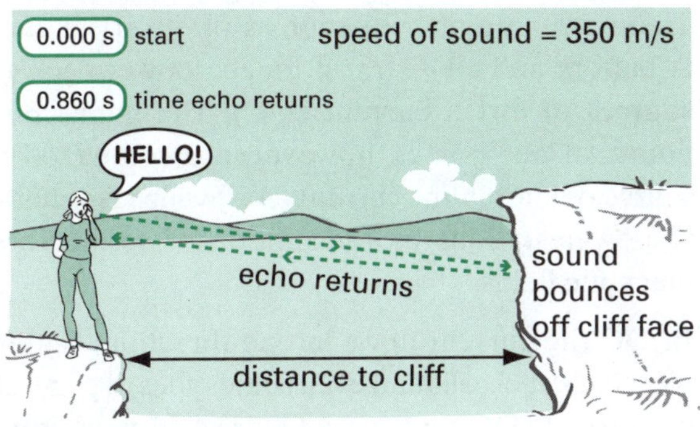

Figure 4.36 Echo

Go to p. 245 to check your answers.

4.4 Electrical energy

Electrical energy refers to the flow of power or the flow of charges along a conductor creating energy, and is often simply called electricity. We obtain electrical energy by converting other forms of energy, such as coal, natural gas, oil, solar, wind or nuclear energy. The energy of electricity is found in a variety of phenomena such as static electricity, electromagnetic fields and lightning. Humans have been able to harness these and store the electrical charge for later use.

Long before humans learned to generate and harness electricity, electric fish evolved a specialised structure called an electric organ consisting of modified muscle or nerve cells, usually located in the tail (see Figure 4.37). Some fish have a discharge that is powerful enough to stun their prey. This has a current of up to 1 amp and a voltage from 10 to 500 volts. Examples include the electric eel (*Electrophorus electricus*), which is really a knifefish, the electric catfishes (family *Malapteruridae*) and electric rays (order *Torpediniformes*).

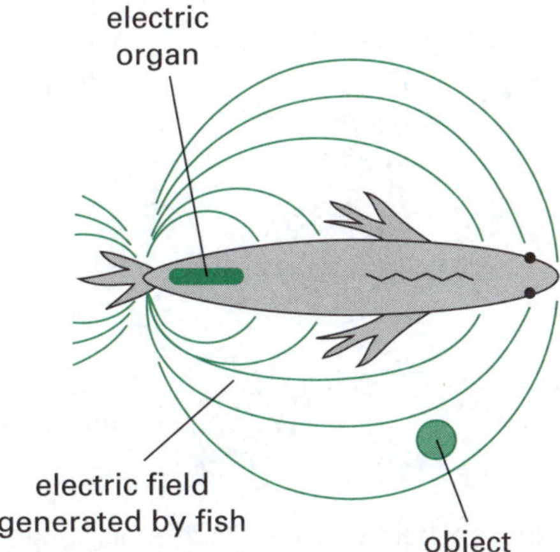

Figure 4.37 Electric fish can detect nearby objects by how they affect their electric field.

A few other fish can generate a discharge, often less than 1 volt. While this is too weak to stun prey, it can be used for navigation, to detect objects and to communicate with other electric fish. For example, electroreceptors embedded in the skin of the fish can be used to detect slight changes in the electric field due to nearby objects.

Conductors and insulators

Just as with heat conductors and insulators, there are electrical conductors and insulators. A substance is a conductor if it allows an electric current or flow of electrons through it fairly easily. Most metals are good electrical conductors, copper being among the most suitable for electrical wires. The electrical properties of copper are exploited in copper wires and other devices such as electromagnets, integrated circuits and printed circuit boards because of its superior electrical conductivity. (Only silver beats copper for conductivity among the materials commonly used in electrical applications.)

A copper atom has only one electron in its outermost (valence) shell. If this electron gains enough energy, it can break away from the parent atom to become a free electron. This is shown in Figure 4.38.

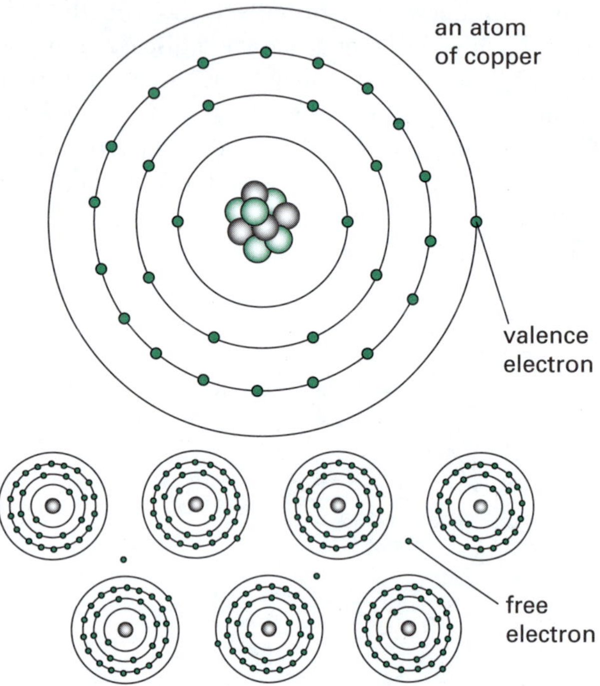

Figure 4.38 The loosely held valence electron in copper can drift randomly to neighbouring atoms.

Atoms with few electrons in the valence shell often have them loosely bound to the nucleus. This allows them to drift randomly from one atom to a neighbouring atom in the substance. In copper, electrons are so loosely held and so close to other atoms that it is difficult to determine which electrons belong to which atom. The electrons also drift randomly in all directions.

If an outside force (a voltage) is applied to the material, the flow of electrons is directed through materials in a specific direction. This is called an electrical current.

Many other materials do not have free electrons and so do not share their electrons around. These are called electrical insulators or non-conductors. Some common insulators include air, glass, plastic, rubber, cloth and wood. These materials have their electrons tightly bound to the nucleus and so are not free to roam around to be shared among neighbouring atoms. Insulators may be used to protect from the dangerous effects of electricity flowing through conductors (see Figure 4.39).

Figure 4.39 Insulating material protects us from coming into contact with live electric wires.

Electric currents: DC and AC

Electricity is an important form of energy in modern society. It is used to power machines, including large and small household appliances. It can be transported quickly to these devices via conducting wires in electrical circuits. The free electrons in the wire move relatively slowly (a few millimetres each second) but the electrical energy is transmitted very rapidly along the conductor. In this way the energy reaches an appliance almost instantaneously when you turn on the switch.

There are a number of sources of electrical energy. A battery and a DC transformer ('power pack') are sources of direct current (DC). The mains power points in our homes, however, are connected to a source of alternating current (AC) which is generated in power stations by the motion of conductors in magnetic fields.

In DC the current flows in one direction. Batteries, solar cells, fuel cells and even some types of generators can provide DC current. In Figure 4.40, work is being done in this circuit because the light bulb glows. If the terminals of the battery are switched around, the current would flow in the opposite direction but the globe would still light.

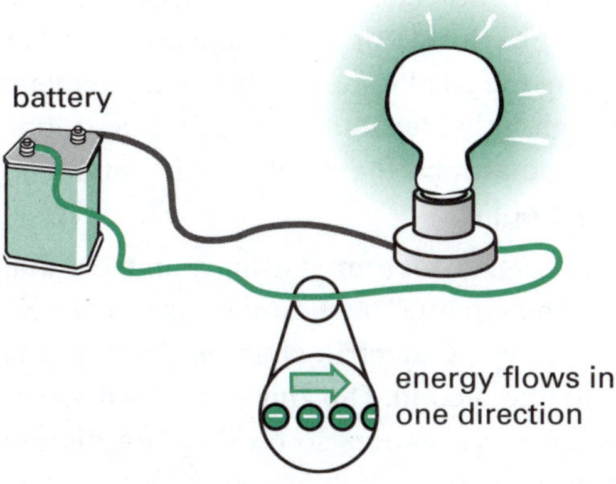

Figure 4.40 In DC current electrons, and energy, flow in one direction.

In AC the movement of electric charge periodically reverses direction, as shown in Figure 4.41. In Australia, the mains power changes direction 50 times each second. That is, its frequency is 50 Hz. Australian household power is supplied at 240 volts, although in some industrial sites and hospitals this might be higher at 415 volts.

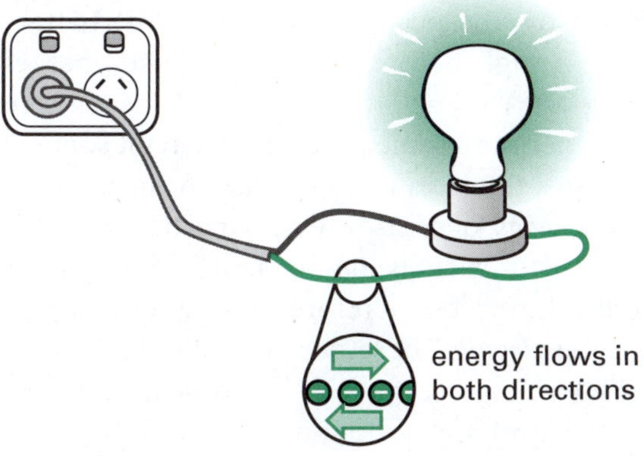

Figure 4.41 In AC current, electric charge flow changes direction periodically.

In this topic we will consider mainly DC circuits.

Electrical circuits

An electrical circuit is a complete conducting pathway for the electric current to flow from one terminal of an electric cell or battery (energy source) back to the other terminal. A battery consists of a number of simple electric cells joined in series with one another. The electrical energy is generated inside the electric cell by chemical reactions in each half of the cell.

As the electric current flows around the circuit, its energy is transformed into heat and sometimes light. If a motor is connected into the circuit, the electrical energy is partly converted into mechanical energy (a form of kinetic energy).

A simple electrical circuit such as that shown in Figure 4.42 consists of the following.

- A DC energy source: for example, a battery that stores separate electric charges and gives energy to the charges as they leave it; it also sets up an electric field in the connecting wires.
- Switches: used to open and close the circuit, allowing current to flow.
- Connecting wires: conductors (such as insulated copper wires) to transfer the current between each components of the circuit.
- Resistor(s): materials that resist current flow (e.g. filaments in light bulbs).
- Measuring devices: ammeters (meters which measure the current flow in the circuit) and voltmeters (meters which measure the voltage or potential difference across resistors or other components).

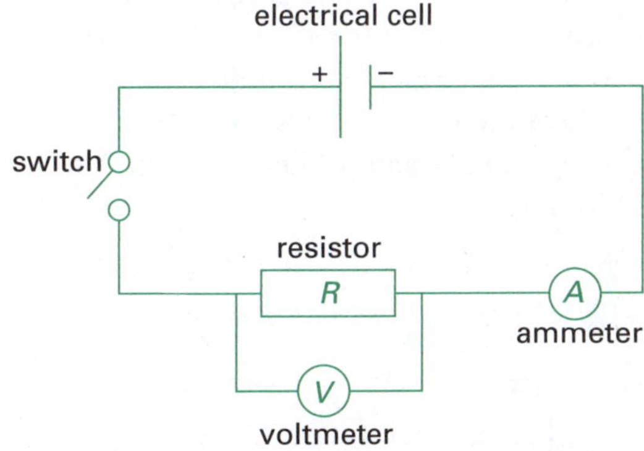

Figure 4.42 Simple electric circuit

A household circuit also includes fuses (or circuit breakers). This is a safety feature in household circuits to prevent overheating and fires from currents that are too high.

The following terminology is used for electrical circuits.

- Open circuit: the switch is OFF and no current flows.
- Closed circuit: the switch is ON and current flows.

Electrical circuits use circuit symbols which are a type of shorthand. Table 4.7 shows examples of the symbols used in electrical circuit diagrams.

Table 4.7 Electrical circuit symbols

Component	Symbol	Component	Symbol
conducting wire		electric cell	
battery		resistor	
lamp		switch	
ammeter		voltmeter	
fuse		variable resistor	

Some books still use the older resistor symbol: .

Current, voltage and resistance

Current

Before electrons were discovered at the end of the 19th century, physicists believed that electrical currents consisted of moving positive charges. Eventually it was discovered that electrical currents are due to the motion of (negatively charged) electrons. Although electrons flow out of the negative terminal of an electric cell and through the circuit to the positive terminal, physicists and electricians continue to use the original convention that electric current is the flow of positive charges out of the cell's positive terminal and around the circuit to its negative terminal.

An electrical (conventional) current is the flow of positive charge and is shown in Figure 4.43.

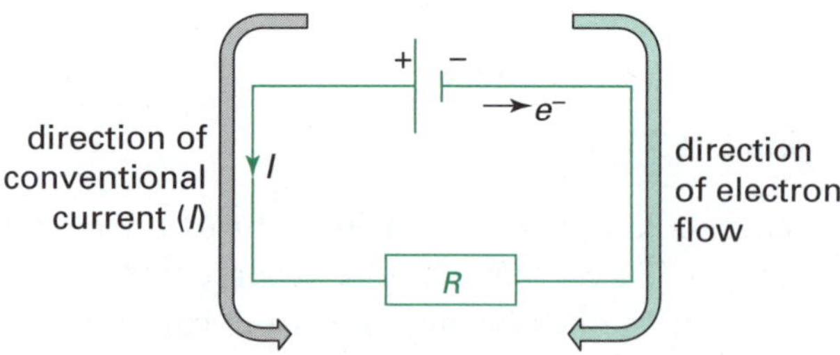

Figure 4.43 Conventional and electron currents

As energised electrical charges flow out of the positive terminal of the battery and into one end of the circuit, energy-depleted charges move back into the battery via the negative terminal.

It is often useful to visualise electrical currents in terms of an analogy.

- Observation: Current flow is almost instantaneous once the switch is turned on.
- Analogy: A long line of marching students is ordered to begin marching along a circular road, starting at the long bridge which links the two ends of the road. They all start together and observers at each side of the bridge and at other points along the road note immediate movement, even though individual students will take a long time to move around the whole road system and pass a fixed point.

Electrical currents are measured using ammeters that are placed in the circuit. These ammeters effectively count the number of electric charges passing any given point in one second. The ammeters themselves have low internal resistances so they do not disrupt the current flow.

Electrical current is measured in units called amperes or amps (unit symbol = A). This unit of current is named in honour of the French physicist André Ampère.

Voltage

Electrical charges flow around a circuit in response to an electrical force. The force is caused by the existence of an electric field that is established in the conducting wires when they are connected to the battery terminals. We can think of the charges emerging from the positive terminal of the battery as having a high potential energy. Those charges returning to the negative terminal have zero potential energy. In some parts of the circuit some of this energy is transformed into other forms and the potential energy decreases. This is particularly true across a resistor such as the filament in a light bulb. Potential differences therefore exist across various parts of the circuit. These potential differences are also called voltage drops (see Figure 4.44). Voltage is a measure of the potential energy differences between any two points in a circuit.

Some people like to think of the voltage of a battery as the electrical pressure it puts on the charges that flow around the circuit. It is often useful to develop an understanding of the term voltage in terms of an analogy.

- Observation: High-voltage batteries make model electric motors spin faster than do batteries with low voltages.
- Analogy: Water flows rapidly out of a tap if the water pressure behind the tap is high. If the water is stored in a tank and the tap is opened, the pressure of water flow decreases as the level of

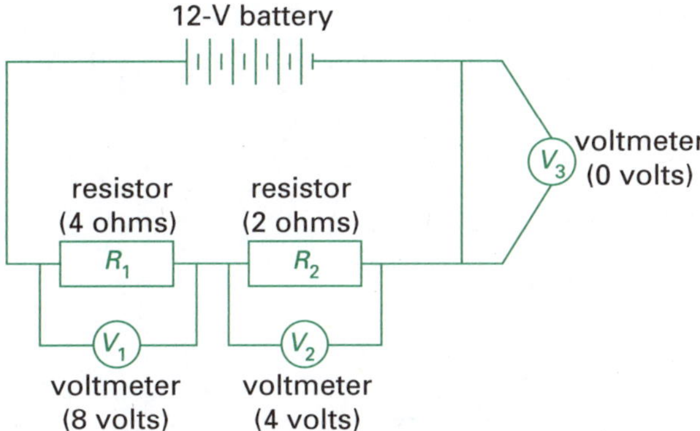

Figure 4.44 Voltage drop around a simple circuit

water drops in the tank (see Figure 4.45). In this analogy, the height of the water column behind the tap creates a pressure (due to the water's gravitational potential energy) which drives the water out when the tap is open. Similarly, the high-voltage battery gives the charges a high electric potential energy. When the switch is closed, the charges begin to flow as a response to the electrical pressure.

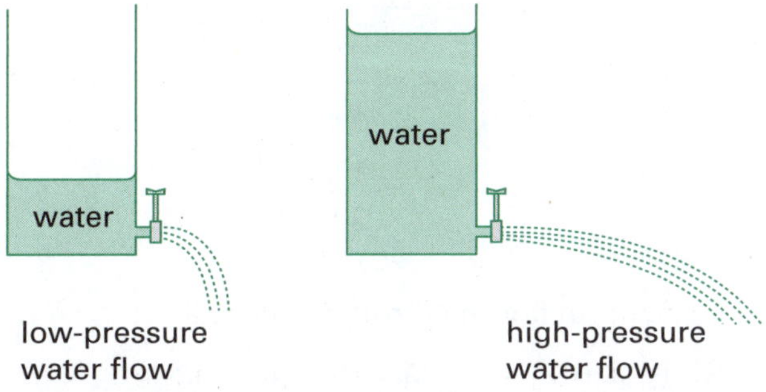

Figure 4.45 Tank of water analogy for voltage

Voltmeters are used to measure the voltage drop across any two points in a circuit. Because voltmeters have high internal resistance to current flow, they are placed in parallel to the circuit and so they only sample part of the current flow at the points being measured.

Voltage is measured in units called volts (unit symbol = V). The unit of voltage is named in honour of the Italian physicist Alessandro Volta.

Resistance

Not all materials are excellent electrical conductors like silver and copper. Some materials reduce the flow of electric currents when they are present in a circuit. Nichrome alloy wires, for example, reduce the flow of a current, compared with the flow in similar copper wires. Such materials are said to offer resistance to current flow. Materials that completely block current flow, such as plastics and glass, are called insulators.

Resistance is a measure of the electrical conductivity of the conductor. A wire that is a good conductor has a low resistance. A wire that is a poor conductor has a higher resistance. The resistance of any wire also depends on its length and diameter as well as its temperature.

Devices that are manufactured to provide resistance to current flow are called resistors. The element in an electric radiator or the tungsten filament in a car light bulb are resistors. As current flows through them, considerable energy is transformed into heat energy. In an electric radiator or electric kettle, the element is designed to generate heat. In an incandescent light bulb, such as those used in car headlights and torches, the tungsten filament glows white hot and emits light. In some other appliances, resistors are used to limit current levels in certain parts of a circuit.

Resistance is measured in units called ohms (unit symbol = Ω, the Greek letter omega). The unit of electrical resistance is named in honour of the German physicist Georg Ohm. Ohm also showed that the voltage, current and resistance are related mathematically. He showed that the resistance (R) of a conductor was equal to the voltage drop (V) across the conductor divided by the current (I) flowing through it.

Additional content: mathematical extension

Ohm's Law, as it is known, is expressed mathematically as:

$$R = \frac{V}{I}$$

That is, resistance = voltage ÷ current.

Example: Calculating resistance

A piece of nichrome wire is included in a simple electrical circuit. The ammeter registers a current of 2.5 A flowing through the wire and a voltmeter registers a voltage drop of 7.5 V across the wire. Calculate the resistance of the nichrome wire.

Answer:

$$I = 2.5\text{ A}$$
$$V = 7.5\text{ V}$$
$$R = \frac{V}{I}$$
$$= \frac{7.5}{2.5} = 3.0\ \Omega$$

It is often useful to develop an understanding of the term resistance in terms of an analogy.

- Observation: Narrow-diameter wires have greater electrical resistance than wide-diameter wires.
- Analogy: Narrow water pipes slow down the flow of water. In a garden hose, the nozzle can be narrowed to restrict the water flow. (See Figure 4.46.) Water is transported from reservoirs to cities in huge pipes with large diameters to reduce the resistance to flow.

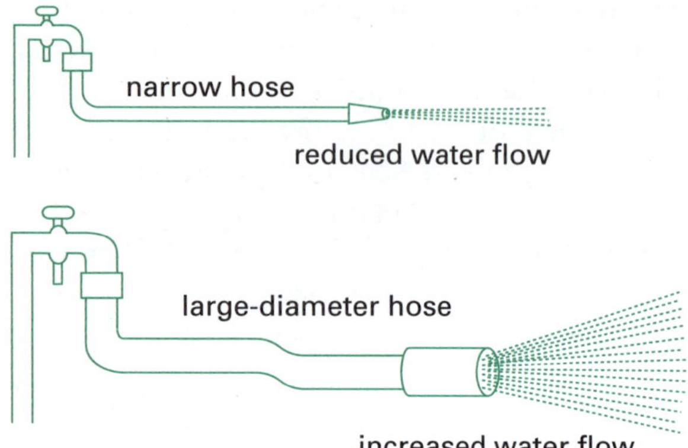

Figure 4.46 Water pipes and resistance analogy

Series circuits

Resistors can be connected in various arrangements in electrical circuits. They can be connected in series, in parallel or in combinations of the two. In a series circuit all the resistors are placed one after the other so that the electrical current passes in turn through each resistor as it flows around the circuit. The following points should be noted.

- The current is the same at all points in the series circuit.
- The voltage drop across the two battery terminals is equal to the sum of the voltage drops across each resistor.
- The greater the number of resistors in series, the greater is the total resistance.

Additional content: mathematical extension

Figure 4.47 shows a typical series circuit for two resistors R_1 and R_2.

The total resistance (R_T) of this circuit is:

$$R_T = R_1 + R_2$$

As the current (I) is the same at all points in the series circuit, Ohm's law can be used to calculate the voltage drops across each resistor.

Resistor 1: $V_1 = I \times R_1$

Resistor 2: $V_2 = I \times R_2$

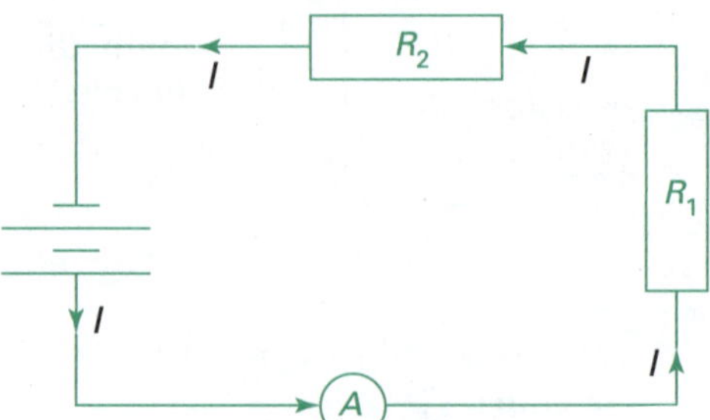

Figure 4.47 Two resistors in series

Example: Resistors in series

An 8-Ω and a 12-Ω resistor are placed in series with a 12-V battery. Calculate the:

1. current through each resistor
2. voltage drop across each resistor.

Answer:

1. Total resistance $= R_T = R_1 + R_2$

$$= 8 + 12$$
$$= 20\ \Omega.$$

So total current $= I_T = \dfrac{V}{R_T}$

$$= \frac{12}{20}$$
$$= 0.6\text{ A}$$

The current flowing through each resistor is 0.6 A.

2. Resistor 1: $V_1 = I \times R_1 = (0.6)(8) = 4.8\text{ V}$
   Resistor 2: $V_2 = I \times R_2 = (0.6)(12) = 7.2\text{ V}$
   (Note: $V_T = V_1 + V_2 = 4.8 + 7.2 = 12.0\text{ V}$)

Parallel circuits

In simple parallel circuits, the resistors are arranged so that:

- the battery supplies current to each resistor at the same time
- the voltage drop across each resistor is the same as the voltage drop across the battery terminals.

Thus for two resistors (R_1 and R_2) in parallel:

$$V_T = V_1 = V_2.$$

The total current which divides into each parallel resistor is equal to the sum of the currents in each resistor.

For two resistors (R_1 and R_2) in parallel:

$$I_T = I_1 + I_2.$$

Two lamps in parallel are very bright (as bright as a single lamp), but two lamps in series are dimmer. They are, however, as bright as each other. Figure 4.48 shows a simple parallel circuit containing two resistors.

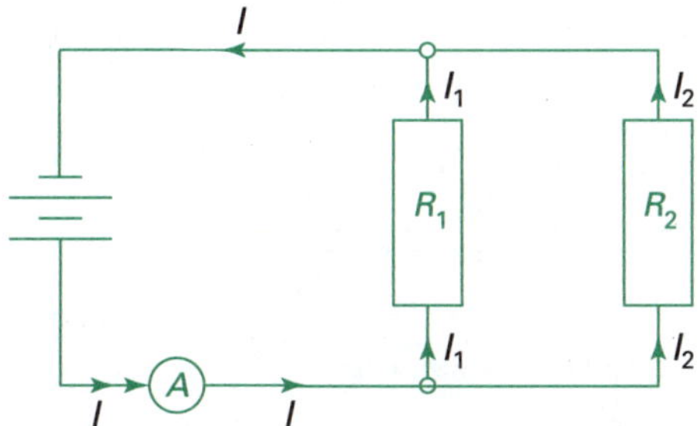

Figure 4.48 Simple parallel circuit with two resistors

Additional content: mathematical extension

Example: Resistors in parallel

A 3-Ω and a 4-Ω resistor are placed in parallel with a 12-V battery. Calculate the:

1. voltage drop across each resistor
2. current flowing through each resistor
3. total current in the circuit
4. total resistance of the circuit.

Answer:

1. The voltage drop across each resistor is the same as the voltage drop across the battery.
   Thus: $V_T = V_1 = V_2 = 12\text{ V}$
2. Resistor 1: $I_1 = \dfrac{V_1}{R_1} = \dfrac{12}{3} = 4\text{ A}$
   Resistor 2: $I_2 = \dfrac{V_2}{R_2} = \dfrac{12}{4} = 3\text{ A}$
3. The total current is the sum of the currents in each branch.
   $I_T = I_1 + I_2 = 4 + 3 = 7\text{ A}$
4. The total resistance is the total voltage divided by the total current.
   $$R_T = \frac{V_T}{I_T} = \frac{12}{7} = 1.71\ \Omega$$

This last calculation shows that the total resistance in a parallel circuit is less than the resistance of each individual resistor.

Modelling the resistance of series and parallel circuits

From the information and calculations discussed so far, we can see that the total resistance of a circuit increases when the resistors are placed in series, but it decreases when the resistors are placed in parallel. We can use a simple model or analogy to explain why this is so.

Modelling a series circuit

Figure 4.49 shows a crowd of people waiting to go into a football match at a poorly designed stadium. Initially, queuing for a ticket hinders the flow of people. Having purchased a ticket, they are made to pass through one turnstile before reaching a holding area. The turnstile slows down the flow of people. Once in the holding area, the fans have to move through a narrow tunnel under the stands to get to a point where they can go to their seats. The narrow tunnel also reduces the flow of people. Thus the total resistance to people flow is the sum of all the individual points of restricted flow.

R(total) = R(ticket queue) + R(turnstile) + R(tunnel)

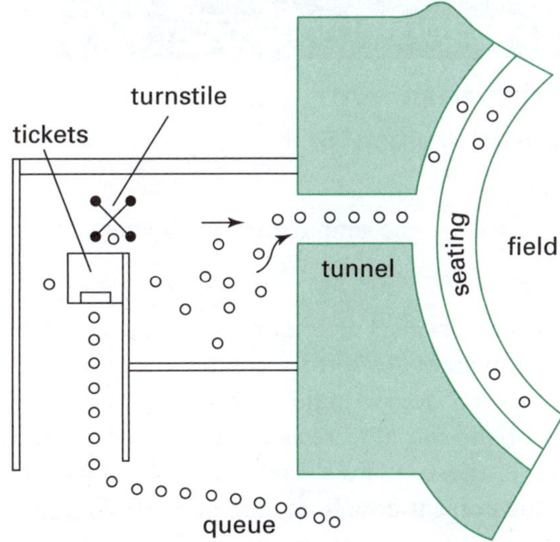

Figure 4.49 Series resistance: modelling resistance to a crowd flow at a football match

Modelling a parallel circuit

Figure 4.50 shows a remodelling of the football stadium access to speed up crowd movements. In the new design there are more ticket offices, to reduce waiting time for a ticket. The ticket offices are in parallel. There are also multiple parallel turnstiles so that the fans can move into the holding area faster. There are more parallel access tunnels that are wider than the old narrow tunnel to allow more people to reach their seats faster. The old narrow tunnel is still there and carries some people but the flow through it is not as great as in the wider tunnels. The wider tunnels therefore have lower resistance. The net result of these changes is a larger flow of people to their seats due to decreased resistance.

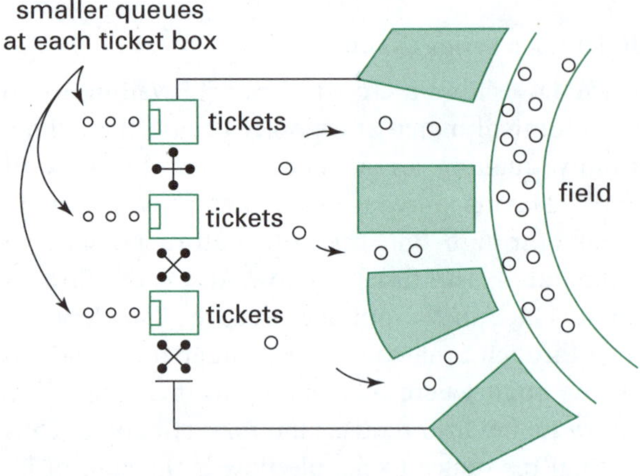

Figure 4.50 Parallel resistance: modelling resistance to crowd flow at a redesigned stadium

Experiment 3

Experiments with simple electrical circuits

Aim

- To construct some simple electric circuits

Method

1. Examine Circuit 1 in Figure 4.51 and connect the components to form a complete circuit. The arrows on a device indicate variable. Move the slide of the variable resistor to the centre and leave it there. (The variable resistor ensures that the current flow is not too large. Too large a current will blow the lamps.)
2. Before turning on the power, have your teacher check the circuit.
3. Increase the voltage slowly, one step at a time. Observe the brightness of the lamp as the voltage is increased. Place your fingertips near (but not touching) the lamp. Is any heat produced? Turn the power off.
4. Set up Circuit 2 with two identical lamps. Don't change the variable resistor's settings. Turn the power back on.
5. Are both lamps as bright as each other? Is the brightness of each of the two lamps the same as the brightness of the single lamp at the same voltage?

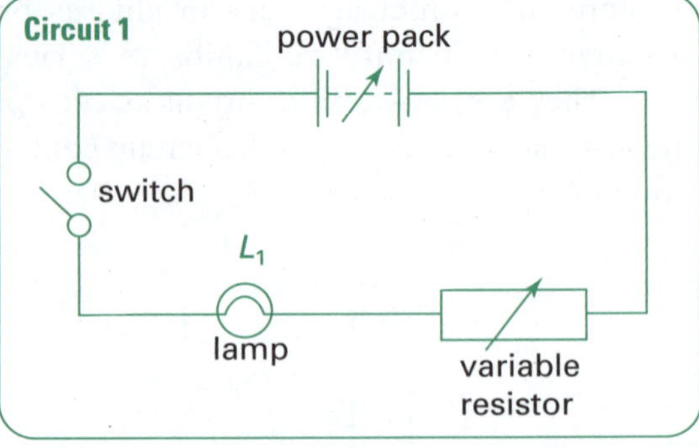

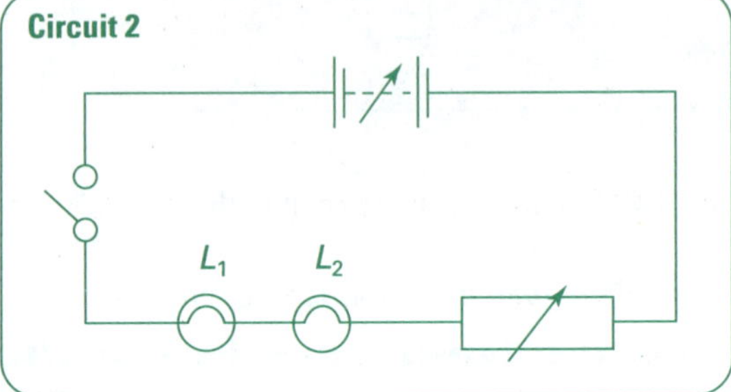

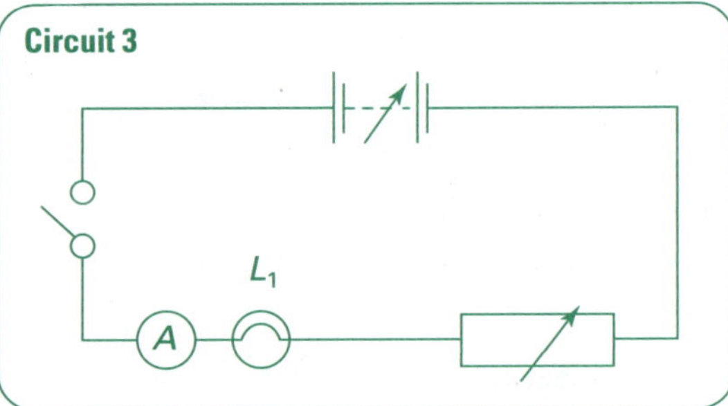

Figure 4.51 Construct these three simple circuits

6. With the power off, unscrew one lamp from its holder. Turn the power back on. Record what you observe.
7. Turn the power off and replace the lamp.
8. Connect an ammeter into the circuit as shown in Circuit 3.
9. Investigate how the current changes as the voltage is increased. Record your observations.

Analysis

1. What happened as the dial setting on the power pack was increased to higher voltages? Explain your observations.
2. Did electrical energy only produce light energy in the lamp? How do you know?
3. How did the brightness of the lamp change when another lamp was included in the circuit? Explain your observations.

4. Why is there no light when one lamp is removed from its holder? What does this tell you about what is required for an electric current to flow?
5. How did the reading on the ammeter change as the voltage increased?

Go to pp. 244–245 to check your answer.

Conclusion

Write a suitable conclusion.

Go to p. 245 to check your answer.

Circuits in the home

Power plugs

Each power device has two or three distinct wired contacts. These may be steel or brass, often plated with zinc, tin or nickel. The two slanting flat pins are the live and neutral contacts and typically carry current back and forth from the source to the load. Many power plugs also include a third contact to connect to earth ground, and is designed to protect should the insulation fail on the connected device. This earth pin is slightly longer for added safety: it is the first to make contact in the wall socket, and the last to break contact.

Figure 4.52 shows the plugs of an extension cord.

Figure 4.52 The 'female' and 'male' plugs of an extension chord

Some plugs have two angled power pins but no earthing pin. This is used with small double-insulated appliances; however, the power point (wall) outlets always have three pins, including an earth pin.

Lighting and power circuits

Parallel circuits are used for lighting and power circuits in our homes, as shown in Figure 4.53. This ensures that when the filament in one light breaks, the other lights in the house do not go out (as would happen in the case of a series circuit). In a parallel circuit, individual switches control each light or power point. In a series circuit only one switch would be required to turn all lights on or off.

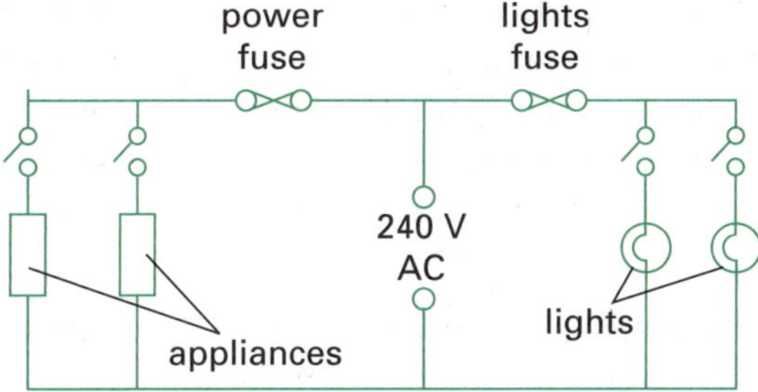

Figure 4.53 Parallel lights and switches in a home circuit

Christmas lights

Lights for Christmas trees are often arranged in series. If one bulb fuses, then it has to be replaced to allow them all to work.

Radiators and electric blankets

Radiators and electric blankets use resistors to generate heat as shown in Figure 4.54. These appliances often use combinations of series and parallel circuits for the resistors to allow variable heat settings (high, medium or low). For a high heat setting, the current flow must be high and thus the resistance must be low. Low resistance is achieved by turning switches so that the resistors are in parallel. For a low heat setting, the current flow must be low and thus the resistance must be high. High resistance is achieved by turning switches so that the resistors are in series.

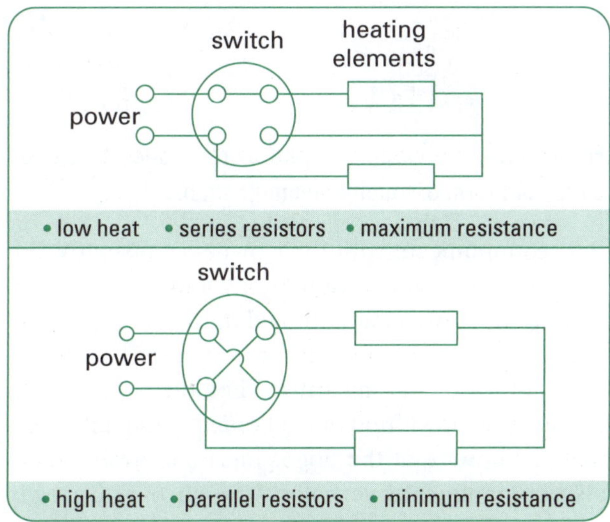

Figure 4.54 Circuits in an electric blanket

4.5 Communications

Communication is the notion of conveying meaningful information. This requires a sender, a message and an intended receiver. However, the receiver need not be present and so communication can occur across vast distances in time and space. For example, the words you are reading now were written some time ago and the author is nowhere near you, but the message is getting through.

There are a number of ways humans communicate with each other. Besides written language, as mentioned in the previous paragraph, there is oral communication (speaking), body language, facial expressions, signs, artworks, music, clothing, hairstyles, architecture and so on.

Animals also communicate with each other. Examples include distinctive threat displays that are made during competition over food, mates or territory; courtship rituals; food-related signals; and alarm calls. One of the most elaborate food-related signals is the dance of honeybees, which occurs in the dark inside the hive to inform other bees where there is a source of nectar. This is shown in Figure 4.55.

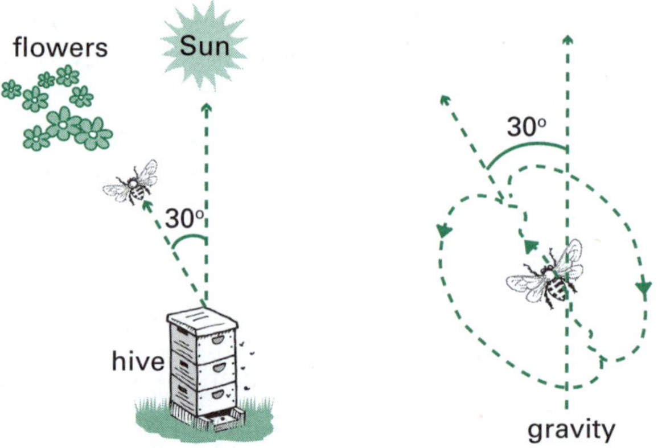

Figure 4.55 The waggle dance of the honey bee is an advanced form of touch communication.

Pets communicate with their owners constantly. For example, when your dog is happy and excited to see you, there is wagging and jiggling from his head to the tip of his tail. He may even dance around and jump into your waiting arms. When the dog tucks his tail, he is feeling afraid or frightened; if the tail is held high, it shows that the dog is alert and aware. Dogs also communicate verbally through barks, growls, whines, whimpers and howls.

Traditionally, Aboriginal peoples communicated orally and, as they didn't have a formal written language, through rock art. Message sticks contained no messages written on them, but were used to help identify the bearer and give him some authority. Sign language, too, was a common method of overcoming the language difficulty between groups. Smoke signals were pre-arranged signs useful when hunting and to indicate a camp location. Visitors also made them to announce arrival in a strange territory.

When the First Fleet made camp in Australia in 1788, the only way to communicate with England was through messages sent by sailing ships. This took many months, with replies typically taking two years. Today communication occurs almost instantly anywhere in the world, and uses many forms. Electronic communications are so integral to our daily lives and resulted from the advent of electricity in the 1800s which revolutionised their speed, distance, frequency and accuracy.

Mobile phone communication

Mobile phones have become an indispensable and useful part of our everyday lives, with the vast majority of Australians owning at least one. Studies have shown people can lose appetite and become depressed if they are deprived of their phones.

A mobile phone (or cell phone) allows for mobile telephone calls across a wide geographic area. These mobile phones link in with the public telephone network both domestically and across the world. Over recent decades mobile phones have become smaller and more versatile, supporting a wide variety of other services such as text messaging, email, internet access, GPS location, short-range wireless communications (infra-red, bluetooth), computing, games and photography. In fact it is ironic they are still called 'phones'.

An area, like a city or town, is divided into small cells with each cell covering around 20 to 30 km^2 as shown in Figure 4.56. Mobile (cell) phones operate within these cells and do not have powerful transmitters, so don't need a great range. They can switch cells as they move around. It is the cells that give cell phones their incredible ranges. This allows multiple frequency re-use across a city, so that millions of people can use cell phones at the same time. This is because cells which are not next to each other can use the same frequency without it creating a problem.

A cell phone is a full-duplex device. This means that one frequency is used for talking and another separate frequency is used for listening. So both people on the call can talk at once. On the modern digital system, each cell can carry about 168 channels.

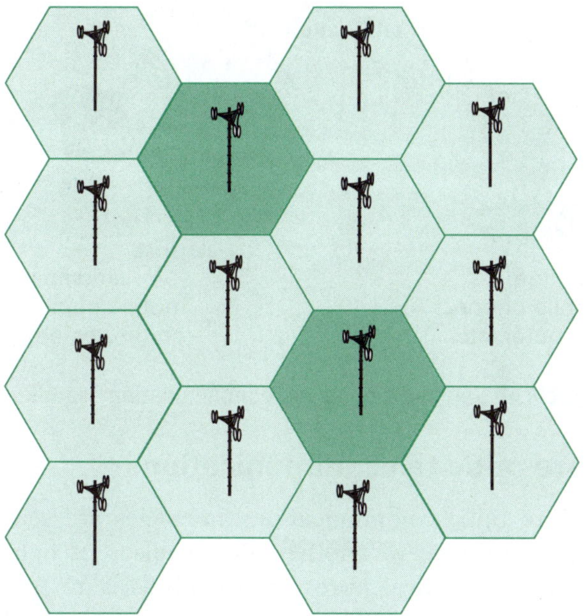

Figure 4.56 Each cell contains a base station consisting of a tower and radio equipment.

A cell phone connects as follows.

- When you turn on the phone, it listens for a five-digit code on a special channel. This control channel is what the phone and base station use to communicate about call set-up and changing channels.
- It then matches this system identification code with the one programmed in the phone.
- Your phone transmits a registration request, so that the Mobile Telephone Switching Office can electronically keep track of your phone's location.
- When this office receives a phone call, it knows in which cell you are and can divert the call to you.
- The office picks a pair of frequencies that your phone will use in that cell and alerts your phone over the control channel.
- The call is connected once your phone and the tower switch on those frequencies.
- As you move to the edge of your cell, the signal strength gets weaker but the signal in the adjacent cell becomes stronger. Eventually the base stations coordinate with each other and with the office to switch your call to that next cell (see Figure 4.57).

All this occurs electronically, automatically and so fast and seamlessly you do not realise it is happening.

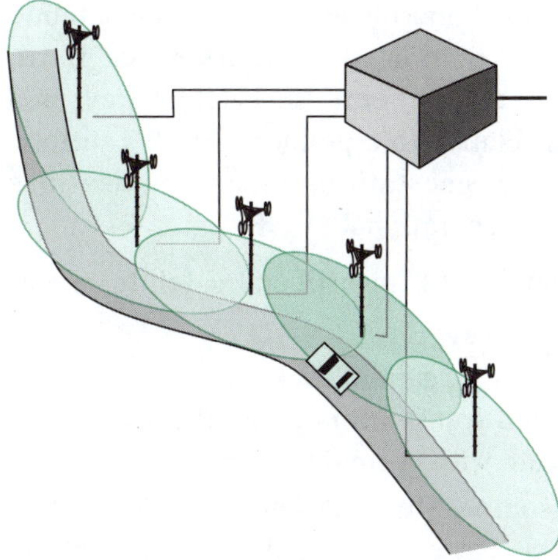

Figure 4.57 The signal passes from cell to cell as you travel.

Satellites and telecommunications

A communications satellite is an artificial satellite stationed in space to relay telecommunication messages. Modern communications satellites use a variety of orbits, among the most popular being geostationary orbits. These orbits are ideal for weather satellites and communications satellites.

A satellite in a geostationary orbit appears to be in a fixed position above the Earth, as shown in Figure 4.58. This satellite revolves around the Earth at the same angular velocity as the Earth itself. The benefit is that ground stations (also known as command centres) can operate effectively without needing expensive equipment to track the satellite's position.

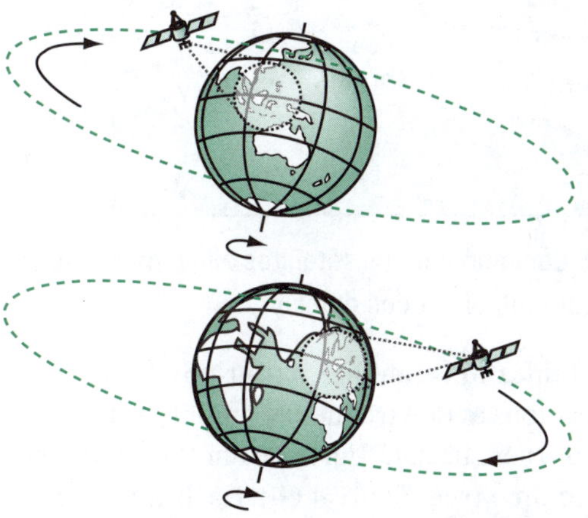

Figure 4.58 A satellite in a geostationary orbit appears to be stationary over the same part of the world.

A geostationary satellite maintains an orbit of 35 786 km above Earth, and travels at 11 300 km/h.

That orbital speed and distance allows the satellite to make one revolution in 24 hours. Since Earth also rotates once in 24 hours, the satellite stays in a fixed position relative to a point on Earth's surface. Australia's first geostationary communications satellite became operational in 1985.

The basic features of an artificial satellite are as follows.

- Communication capabilities. Antennae, radio receivers and transmitters allow the satellite to communicate with ground stations.
- A power source. Most satellites are powered by rechargeable batteries, using the ultimate battery charger, the Sun. The solar panels are prominent on many satellites.
- A control system to accomplish its function.

Radio signals around microwave frequency range are best at carrying large volumes of communications traffic. These waves are not deflected by the Earth's atmosphere as some other frequencies are. The signals travel in a straight line, known as 'line-of-sight' communication. Over short distances, microwave towers every 40 km or so act as 'repeaters' to boost and repeat the signal. Geostationary communications satellites act as repeaters, but over greater distances.

Figure 4.59 shows how messages can be relayed over long distances using satellites.

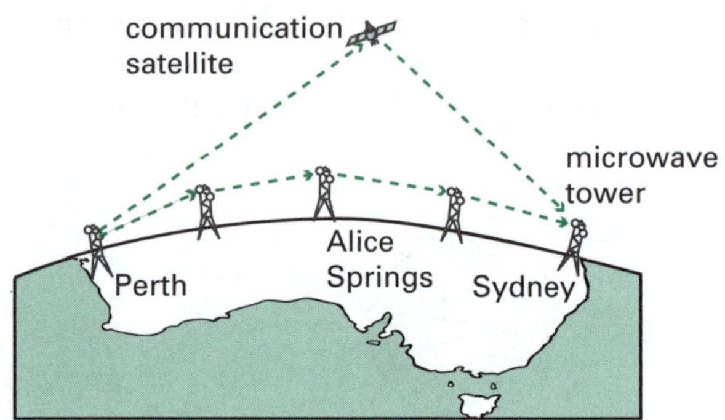

Figure 4.59 Communications satellites allow messages to be sent over long distances quickly.

In a communications satellite, a transponder receives communications at one frequency, amplifies them and re-transmits them back to Earth on another frequency at the same time (see Figure 4.60). Each transponder is a receiver and transmitter (transceiver). With digital video data compression and multiplexing, several video and audio channels can pass through a single transponder on a single wideband carrier. Communications satellites usually contain hundreds, and even thousands, of transponders.

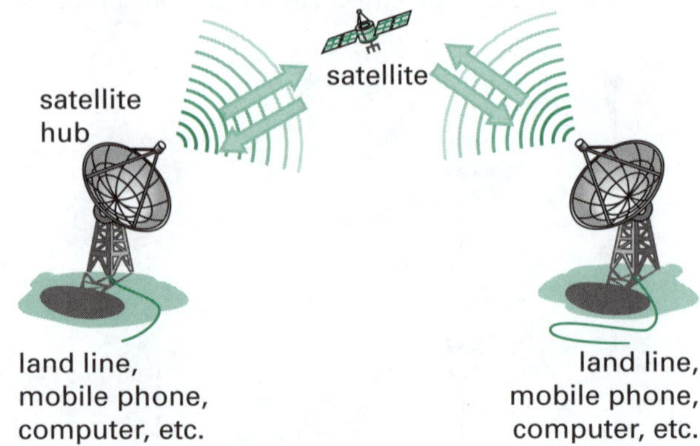

Figure 4.60 Sending signals via communications satellite

Fibre optic telecommunications

In fibre optic communications, messages are sent from one place to another along pulses of light through an optical fibre. Optical fibres were first developed in the 1970s and have revolutionised telecommunications, especially long-distance phone calls, cable TV and the internet. The many advantages of optical fibres have led to them largely replacing copper wire communications in core networks in the developed world.

A fibre optic cable is made from very pure and transparent glass that is drawn into a very thin strand, about as thick as a human hair. This glass strand is then coated in two layers of plastic. This acts as a mirror around the glass strand, allowing for total internal reflection. This means that the light is reflected back from the edge of the medium, in this case glass, it is travelling through. When light moving through the fibre hits the wall at a shallow angle, it bounces back and stays completely within the fibre, as shown in Figure 4.61.

When a telephone conversation is sent through an optical fibre, the analog voice signal is translated to a digital signal (see Figure 4.62). Then a laser switches on and off billions of times each second sending pulses of light and no light through the fibre. Newer systems use lasers with different colours to fit multiple signals into the same fibre.

Optical fibres are usually found where higher bandwidth is required or longer distances need to be covered. Advantages of optical fibres over electrical (copper) cabling include the following.

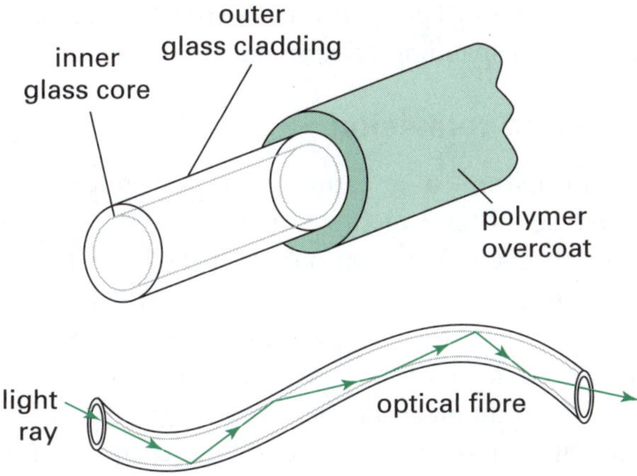

Figure 4.61 Light bounces from the edges of an optical cable as it moves through.

- Optical fibres are cheaper than copper cabling of similar length.
- Its exceptionally low loss of signal allows long distances between amplifiers and repeaters. For example, copper wire signals need repeaters every few kilometres, while optical fibre signals can go for over 100 km before needing to be boosted.
- The signal cannot be degraded or interfered with by ground currents and other external power issues.
- Optical fibres are thinner than copper wires, so more fibres can be bundled into a given-diameter cable than copper wires.
- They have a high data-carrying capacity. One thin optical fibre can replace thousands of copper links.
- There is no crosstalk between fibre cables that run alongside each other for long distances. Copper cables running parallel and close to each other can sometimes bleed (interfere with) a signal from one wire to the other.
- Fibre cabling can be installed in areas of high electromagnetic interference as these are immune from problems.

Radar

Radar was developed by the military during World War II, with the word 'radar' being an acronym for **ra**dio **d**etection **a**nd **r**anging. It uses radio waves to determine the range, altitude, direction or speed of both moving and fixed objects. This includes aeroplanes, spaceships, rainfall, hurricanes and even cars on the road. Today, radar has very diverse uses such as air traffic control at airports, radar astronomy, detecting and guiding anti-missiles, detecting landmarks, and ground-penetrating radar for geological observations. Some aircraft have a radar altimeter that bounces a signal from the terrain beneath it to give a measure of the plane's altitude.

Meteorological (weather) radars are very effective at detecting rain. Forecasters can interpret the patterns and intensity of the radar images to give warnings of major weather events such as severe thunderstorms, tropical cyclones and areas where there may be heavy rainfall. Radar does not pick up clouds, since cloud droplets are too small, but it can detect the rainfall which those clouds produce. How much reflection occurs depends strongly on the diameter of raindrops in, or falling from, the cloud and not the amount of raindrops.

A radar dish, or antenna, sends out pulses of radio waves (or microwaves) that bounce off any object in their path. Some of the wave's energy is returned to the dish which picks up the signal, as shown in

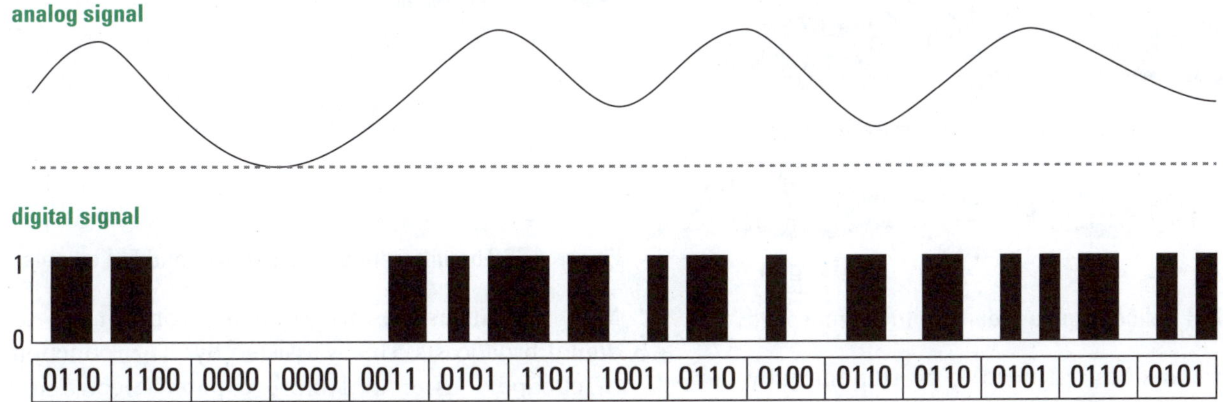

Figure 4.62 Changing an analog signal into a digital signal

Figure 4.63. How long it takes for the signal to return, and any change in its frequency, is fed into a computer that calculates how far away it is and how fast it is moving.

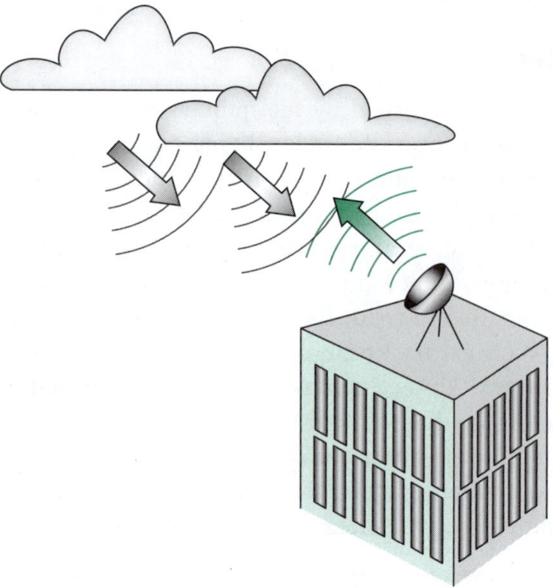

Figure 4.63 Radio waves bounce off raindrops.

Police use radar guns to detect speeding drivers. The waves bounce off a car and come back to the gun as shown in Figure 4.64. If a car is stationary, the frequency of the waves sent by the gun comes back exactly the same. If the car is moving towards the radar gun, these waves come back at a higher frequency. The faster the car is moving, the more scrunched up the returning waves are. If the car is moving away from the radar gun, these waves come back at a lower frequency. The faster the car is moving, the more stretched out are the reflected waves.

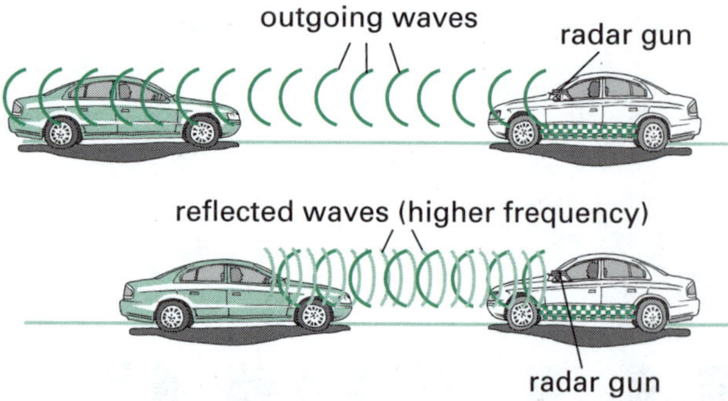

Figure 4.64 Police using a radar gun to detect speed

Police radar does not need to be stationary by the side of the road. Police now have the ability to link their radar with their police car's speedometer and so can detect speeding drivers while they are moving.

Radio and television

We can listen to a radio or watch television because sound and picture signals can be changed to electromagnetic radiation and transmitted. The Italian inventor Guglielmo Marconi (1874–1937) developed the process of wireless telegraphy and transmitted Morse code over the airwaves. This allowed communications between ships and shore. (Marconi won the Nobel Prize in Physics in 1909 for his work.) When the RMS *Titanic* sank in 1912, this new technology had proved its worth. Unfortunately, the ship's distress signals didn't save around 1500 passengers.

The first radio station in Australia broadcasted from Sydney in 1923, while mainstream professional television was launched in 1956, again in Sydney. Since those times there has been an increase in the number and quality of radio and television stations. There are now both AM and FM stations, where the signal is sent out as either amplitude modulation or frequency modulation, with television transmissions no longer in poor-quality black and white (monochrome) but high-definition colour.

Figure 4.65 compares amplitude modulation (AM) and frequency modulation (FM) signals.

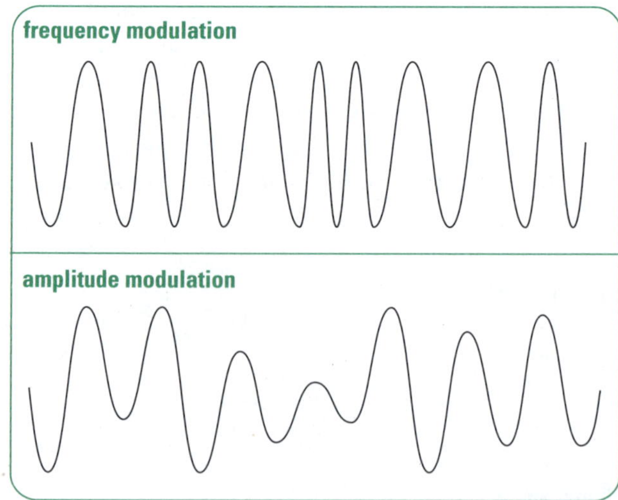

Figure 4.65 The difference between AM and FM signals

Many countries are now moving from analog to digital broadcasts. This is assisted by the production of cheaper, faster and more capable electronics. The main advantage of digital broadcasts is that they eliminate a number of complaints common to traditional analog broadcasts, such as distortion.

Changes in speed and type of news coverage as technology changes

Telecommunications play an important role in the world economy. Events from around the world that used to take hours or days to become common knowledge can now be brought instantaneously to consumers using radios, televisions, mobile phones and the internet. Commercial broadcasting is available 24 hours a day and uses satellite technology to bring current events into consumers' homes and offices as the event occurs. Major organisations have journalists on the ground where news is breaking and can bring that information immediately (in real time) to consumers. And, with the ability to send multiple streams of information simultaneously, more details of events can be communicated instantly. This includes the ability to carry on two-way communications with people across the world with ease.

Advertising in the media

The broad reach of radio and television makes them powerful for and attractive to advertisers. Advertisers can get their information to many hundreds of thousands, if not millions, of potential customers using a single message. Many radio and television networks sell blocks of broadcast time to advertisers (sponsors) so they can fund their programming. What these broadcasters can charge depends on the time period and audience reach; hence, it is crucial to have rating systems in place.

Using science to test claims made in the media

The media has the ability to make claims believable. In advertising it is common to hear expressions like '9 out of 10 dentists recommend' or 'a leading university has found'. Sometimes spurious science claims are made to back up a product: 'it contains polycyclocrapinol so you know it must be good'. Most people have no idea what this chemical is, but at least it sounds authoritative. This information can also be enhanced by having personalities (e.g. singers, actors and sportspeople) deliver the message. What makes a celebrity an authority about a certain product, especially when they are being paid to make that claim, over anyone else?

In logic, the expression *argumentum ad populum* ('appeal to the people') is an incorrect argument that states that something is true or correct because many or most people believe it. Children use it when they say 'but all of my friends are doing it (or going)'. During the Middle Ages most people thought the world was flat. While this notion was widely popular, it didn't make it true. So saying 'product X is the most popular brand, so buy brand X' or 'you must watch it as it is the most watched show on TV' doesn't necessarily make them the best products or shows.

Another logical argument is *argumentum ad verecundiam* ('arguing from authority') and can be a powerful appeal if the authority is a legitimate expert on the subject. But the argument breaks down when an inference relies on individuals or groups without relevant expertise or knowledge. Consider, for example, how footballers, models or actresses are used to advertise clothes or facial products. These people are no more believable than any other person in the community. There is currently strong debate on climate change and even scientists expert in the field are not all in total agreement. It is therefore easy to find some authority which agrees with the presenter's point of view and have their argument dominate the debate.

Argumentum ad ignorantiam ('appeal to ignorance') asserts that some proposition is true because it has not been proven false. The opposite also works; it is false because it has not been proven true. For example, some have argued that there is intelligent life elsewhere in the universe since there is no compelling evidence that UFOs have not visited Earth and so they must exist. This can form the basis of some television documentaries.

Careers in telecommunications

Telecommunications is a rapidly growing and changing field. There is strong demand for people trained in information technology, engineering, networks and technicians. However, only a small percentage of occupations in the industry relate to science and research. Figure 4.66 compares the employment in the telecommunications industry by occupation.

Telecommunications equipment installers and repairers install, repair and maintain the complex array of sophisticated communications equipment and cables. Line installers and repairers connect central offices to customers' buildings. They also install poles and terminals, and place wires and cables leading to the consumer's premises.

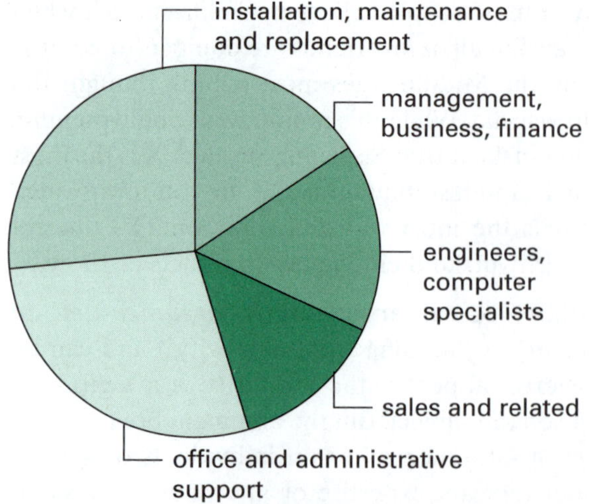

Figure 4.66 Current employment in the telecommunications industry by occupation

Around 10 to 15% of occupations in the industry are scientific and technical personnel such as engineers and computer specialists. Engineers plan cable and microwave routes, central office and PBX equipment installations, and the expansion of existing structures, as well as solve other engineering problems. There are some engineers who engage in researching and developing new equipment. Many specialise in telecommunications design or voice, video or data communications systems, and link communications equipment with computer networks. They may work closely with clients, who may not understand sophisticated communications systems, and design systems that meet their needs. Computer software engineers and network systems and data communications analysts design, develop, test and debug software products. These include computer-assisted engineering programs for schematic cabling projects; modelling programs for cellular and satellite systems; and programs for telephone options, such as voicemail, e-mail and call waiting.

In this growing and changing field, new occupational specialties are emerging based on innovations and new technologies. For example, some engineers research, design and develop gas lasers and related equipment needed to send messages through fibre optic cable transmission. They study the limitations and uses of lasers and fibre optics, find new applications for them and oversee the building, testing and operations of the new applications. Job prospects for telecommunications engineers are good, with employment for the next few years expected to grow very strongly.

Test yourself 2

Part A: Knowledge

1. Which of the following materials would make the best connecting wires in an electric circuit? *(1 mark)*

A plastic
B glass fibres
C iron
D copper

2. A speeding motorist is travelling towards a police radar gun. Which of these changes would show the motorist is speeding? *(1 mark)*

A an increase in the returning wave's frequency
B an decrease in the returning wave's frequency
C an increase in the returning wave's amplitude
D an decrease in the returning wave's amplitude

3. A suitable device to measure the current flow in a circuit is *(1 mark)*

A an ammeter.
B a voltmeter.
C an ohmmeter.
D a battery.

4. Five resistors are arranged in parallel with a 12-V battery. Three resistors each have a resistance of 1 Ω and the other two each have a resistance of 2 Ω. Select the correct statement about this circuit. *(1 mark)*

A The voltage drop across the 2-Ω resistors is greater than that across the 1-Ω resistors.
B The current flow is the same through all resistors.
C Each resistor has the same voltage drop across it.
D The current is the same in all parts of the circuit.

5. In transmitting radio waves, AM means *(1 mark)*

A ante meridiem.
B alternating modulation
C ampere.
D amplitude modulation.

6. Complete the following restricted-response questions using the appropriate word. *(1 mark for each part)*
   a) An electric current is actually due to the flow of ....
   b) A .... is used to measure the drop in electric potential across a resistor.
   c) Ammeters are placed in .... in a circuit to measure the current flowing in the circuit.
   d) Two resistors in series have a .... resistance than each individual resistor.
   e) The unit of electrical current is the ....

7. Use the code letters to match the terms or phrases in each column. *(1 mark for each part)*

Column 1	Column 2
A series circuit	F same drop in voltage across each resistor
B conductors	G resistance
C conventional current	H metals
D Georg Ohm	I same current through each resistor
E parallel circuit	J flow of positive charge

Part B: Skills

8. Figure 4.67 shows a water circuit containing a water pump, filter, water flow meter, pressure gauge and a tap. Draw this as an electric circuit using equivalent symbols for the devices as in the water flow circuit. *(2 marks)*

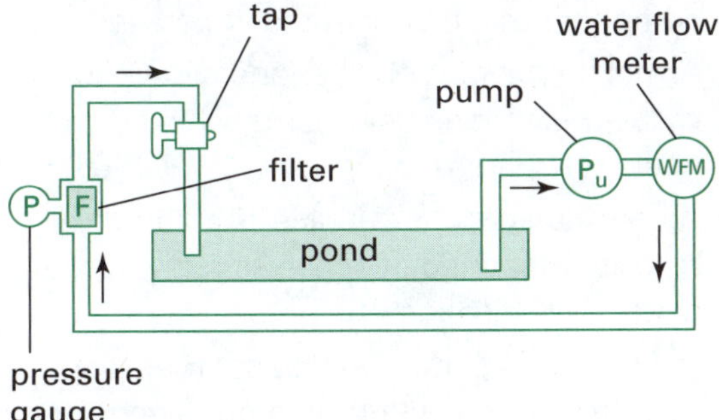

Figure 4.67 Water flow circuit

9. Figure 4.68 shows some different models for charge flow in a metallic conductor. Explain which model correctly shows the flow of charge in the conductor. *(2 marks)*

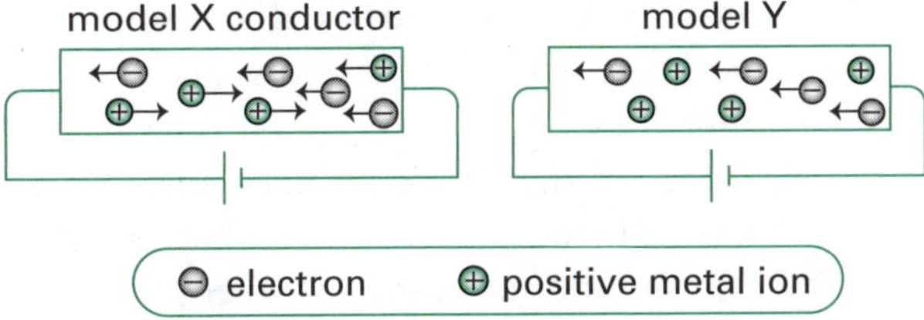

Figure 4.68 Models for charge flow in a conductor

10. a) How does the cellular telephone gets its name? *(1 mark)*
    b) A cell phone is basically a sophisticated two-way radio. Explain. *(1 mark)*
    c) Cell phones use two frequencies, one for talking and the other for listening. What type of communication is this? *(1 mark)*

11. A student is asked to set up a circuit in which three identical lamps are arranged so that two lamps have the same brightness but the third lamp is the brightest. The equipment available is three identical lamps, a 12-V battery, a switch and connecting leads. Draw a circuit diagram for the circuit constructed by the student. *(2 marks)*

12. Since the earliest times, people have wanted to send messages to each other over large distances.
    a) What were some of the ways people tried to communicate over long distances in past centuries using light? *(2 marks)*
    b) What are some of the limitations of these methods? *(2 marks)*
    c) Describe how an optical fibre can be used to send messages over very long distances. *(5 marks)*

13. a) The following statements explain how radar works, but the steps are out of order. List the correct sequence. *(3 marks)*
    A. Only a very small fraction of the original signal is transmitted back in the direction of the receiver.
    B. Just after the pulse is transmitted, a switch switches control to the receiver.
    C. The switch may toggle control between transmitter and receiver up to 1000 times each second.
    D. This allows the antenna to receive echoed signals.
    E. A transmitter transmits a high-power pulse out of the radar antenna.

F. If the pulse hits a surface that is not perfectly flat, it is reflected in all directions.

G. Once the signals are received, the switch then transfers control back to the transmitter to transmit another pulse.

b) How does a radar determine how far away an object is? *(2 marks)*

14. A 24-V battery is connected in series with three known and one unknown resistors (R), a switch and an ammeter. The three known resistors have the following resistances: $R_1 = 3\ \Omega$; $R_2 = 4\ \Omega$; $R_3 = 1\ \Omega$.

a) Draw a circuit diagram. *(2 marks)*

b) *Mathematical extension:* If the ammeter reads 1.5 A, calculate the value of resistance R. *(2 marks)*

c) A voltmeter is placed in turn across each resistor. In which case will the lowest voltage drop reading be recorded? *(1 mark)*

Go to pp. 245–247 to check your answers.

Summary

1. Heat energy is transferred by conduction, convection and radiation.
2. Sound energy is a mechanical wave that can be transferred through solids, liquids and gases.
3. Our ears detect sound waves and loud sounds can cause deafness.
4. Light is an electromagnetic wave which can propagate through a vacuum as well as matter.
5. Our eyes detect visible light waves.
6. Energy can be transmitted through an electric circuit.
7. Electric circuits can be arranged in series or parallel.
8. Technology has led to the development of the cochlea implant.
9. The bionic eye is under development.
10. Mobile communication technologies have accelerated the rate of information transfer.
11. Electrical energy is the flow of electric charges along a conductor.
12. Conductors allow an electric current to flow; insulators prevent an electric current from flowing.
13. There are two important forms of electric current: DC (direct current) and AC (alternating current).
14. An electrical circuit is a complete conducting pathway for an electric current.
15. Ammeters measure electrical current; voltmeters measure potential difference.
16. An electrical current is the flow of positive charge, but electricity is the flow of electrons.
17. Voltage is a measure of the potential energy differences between any two points in a circuit.
18. Resistance is a measure of the electrical conductivity of the conductor.
19. Ohm's Law can be expressed as $R = V/I$. That is, resistance = voltage ÷ current.
20. In a series circuit, components are placed one after the other so that the electrical current passes in turn through each of them.
21. In parallel circuits, the voltage drop across each component is the same.
22. Circuits in the home use AC current and components are connected in parallel.
23. A mobile phone or cell phone allows for mobile telephone calls across a wide geographic area.
24. A communications satellite is an artificial satellite stationed in space to relay telecommunication messages.
25. In fibre optic communications, messages are sent from one place to another along pulses of light through an optical fibre.
26. Radar is used to detect the presence, direction, distance and speed of moving objects by sending out pulses of high-frequency electromagnetic waves which are

Summary cont.

reflected off the object and return back to the source.

27. In television, transmitted sound and visual images are displayed for entertainment, information and education.

28. Telecommunications play an important role in the world economy, with information able to be accessed from around the world almost instantly.

Syllabus checklist

Are you able to answer every syllabus question in this chapter? Tick each question as you go through the list if you are able to answer it. If you cannot answer it, turn to the appropriate page in the guide as is listed in the column to find the answer.

For a complete understanding of this topic		Page no.	✓
1	Can I explain how heat energy can be transferred by conduction, convection and radiation?	129–131	
2	Can I state the wave type of a sound wave and name the materials through which sound travels?	132–134	
3	Can I explain how our ears detect sound waves and how loud sounds can cause deafness?	135–137	
4	Can I explain the principles behind cochlea implant technology?	136–137	
5	Can I state the wave type of a light wave and explain the types of media through which it travels?	137–140	
6	Can I describe how our eyes detect visible light waves?	141–142	
7	Can I explain the issues in the development of the bionic eye?	142–143	
8	Can I explain the difference between conductors and insulators?	145–146	
9	Can I explain what an electrical current is?	146	
10	Can I explain the difference between AC and DC?	146–147	
11	Can I explain how energy can be transmitted through an electric circuit?	146–147	
12	Can I explain what ammeters and voltmeters measure?	147	

For a complete understanding of this topic		Page no.	✓
13	Can I describe and draw an electrical circuit?	147–148	
14	Can I explain the terms electrical current, voltage and resistance?	148–149	
15	Can I explain and use Ohm's Law?	149	
16	Can I explain the difference between arranging electric circuits in series and arranging them in parallel?	150–151	
17	Can I explain the nature of circuits used in the home?	153	
18	Can I explain why mobile communication technologies have accelerated the rate of information transfer?	154	
19	Can I explain how mobile phones work?	154–155	
20	Can I explain how communications satellites work?	155–156	
21	Can I explain how fibre optics works?	156–157	
22	Can I explain how radar works?	157–158	
23	Can I explain the influence of radio and television in our modern lives?	158	
24	Can I appreciate the importance of modern communications technology?	159–160	

Go to p. v for *Tips for tests and examinations* 80 MIN

Part A: Multiple-choice questions

(1 mark for each)

1. Which statement is true of waves in deep ocean water?
   - A The water particles in surface water waves do not progress but simply oscillate up and down.
   - B Water waves are examples of compression or longitudinal waves.
   - C Water particles oscillate along the axis of propagation of the wave.
   - D Water waves have wavelengths similar in magnitude to infra-red rays.
2. Water waves near a beach have a frequency of 0.4 Hz. Which statement is true about these waves?
   - A In one second the wave will travel a distance of 40 cm.
   - B The distance between adjacent crests is 0.4 m.
   - C These waves have similar frequencies to microwaves.
   - D In 10 seconds, four waves will pass a fixed point.
3. A piece of shiny black cardboard is placed under a fluorescent light. Which of the following best states what you would see?
   - A The light is partly absorbed and partly reflected.
   - B The light is completely absorbed as the cardboard is opaque.
   - C The light is absorbed and refracted.
   - D Most of the light is scattered after reflection and little light is absorbed.
4. A correct practical application of the reflection of light from a convex mirror is
   - A the passenger side mounted rear-vision mirror.
   - B the mirror in the Hubble space telescope used to capture starlight.
   - C a shaving mirror.
   - D a car headlight reflector.
5. A candle on a birthday cake is lit, glowing and hot. Yet the end pushed into the cake does not melt the icing. Why is this?
   - A The candle is a good conductor of heat and radiates heat into the surroundings.
   - B The candle is a poor conductor of heat, so the other end of the candle is not hot at all.
   - C The heat is completely lost upwards due to a convection current.
   - D Icing has a very high melting point.
6. Select the correct statement about an electric circuit.
   - A Currents flow through the circuit when it is open.
   - B Current flows around the connecting wires of a circuit from the positive terminal of the battery towards the negative terminal.
   - C The filament in an electric light bulb is chosen so that little electrical energy is wasted as light and heat.
   - D When two equal resistors are arranged in parallel, the total resistance is higher than when they are arranged in series.
7. An electric blanket contains two resistors. One resistor has a higher resistance than the other. A variable selector switch allows different combinations of resistors to be combined in a circuit with the power supply. What is needed to obtain the greatest heat from the blanket?
   - A Arrange the two resistors in parallel.
   - B Arrange the two resistors in series.
   - C Use only the resistor with the greater resistance.
   - D Use only the resistor with the lower resistance.
8. Which of these statements about geostationary satellites is correct?
   - A They remain stationary at a fixed point in space.
   - B They are located directly above the equator.
   - C Their orbit can be at any altitude.
   - D All geostationary satellites are used for communications.
9. In the last 60 years radar has been used for
   - A speed traffic control.
   - B weather monitoring.
   - C air traffic control.
   - D all of the above.

10. Signals sent or received by fibre optics, mobile phones, radar, radio and television
   A are often visible.
   B are transmitted at the same frequency.
   C all travel at the speed of light.
   D are not always forms of electromagnetic radiation.

Part B: Short-answer questions

11. Figure 4.69 shows two mirrors (A and B) at right angles to each other. A ray of light strikes mirror A. Complete the diagram to show the subsequent path of the light ray. *(2 marks)*

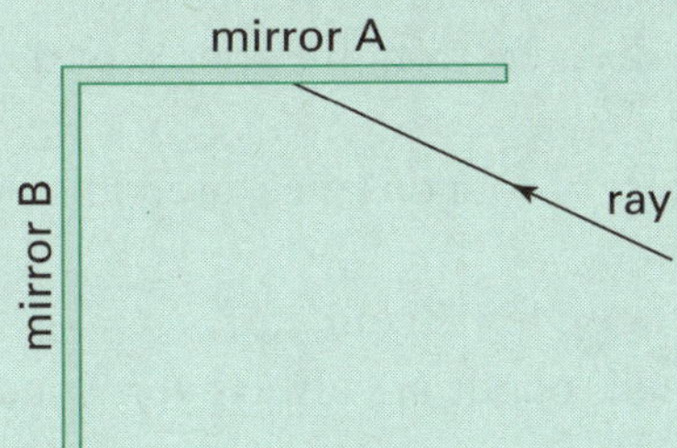

Figure 4.69 Double reflection

12. Figure 4.70 shows a ray of light in air that is about to enter a semicircular block of plastic. The angle between the ray and the plastic block is 30°.

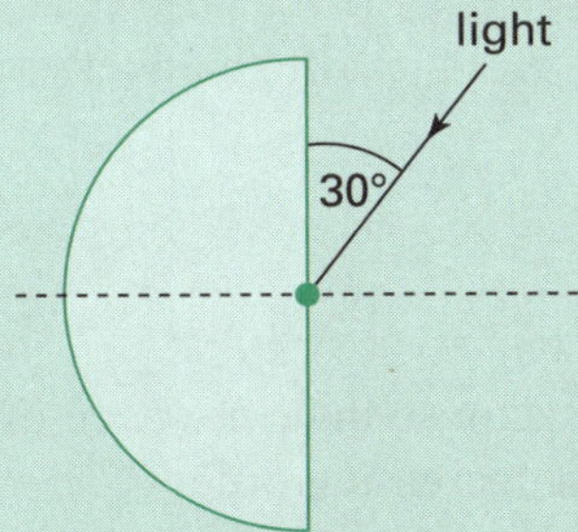

Figure 4.70 Refraction

   a) State the angle of incidence. *(1 mark)*
   b) i) Copy and complete the diagram to show the subsequent path of the ray into and out of the plastic block. *(1 mark)*
   ii) On your diagram, label the angle of incidence, the angle of refraction and the refracted ray. *(3 marks)*
   c) When a student conducts this experiment, she observes that the refracted ray has a lower brightness than the incident ray. Account for this observation. *(2 marks)*

13. Geologists have placed a sensitive detector in a shaft drilled into a granite rock as shown in Figure 4.71. An explosive charge is set off in the same rock layer 60 km away. How long will it take for the detector to register the arrival of the sound energy from the explosion if the speed of sound in granite is 6000 m/s? *(2 marks)*

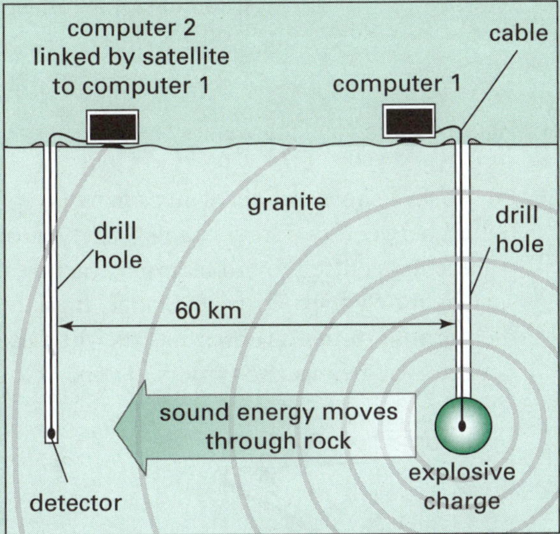

Figure 4.71 Sound energy moving through rock

14. Figure 4.72 shows an experiment to measure the speed of sound in air using timers and a starting pistol. Use the information to calculate the speed of sound. *(1 mark)*

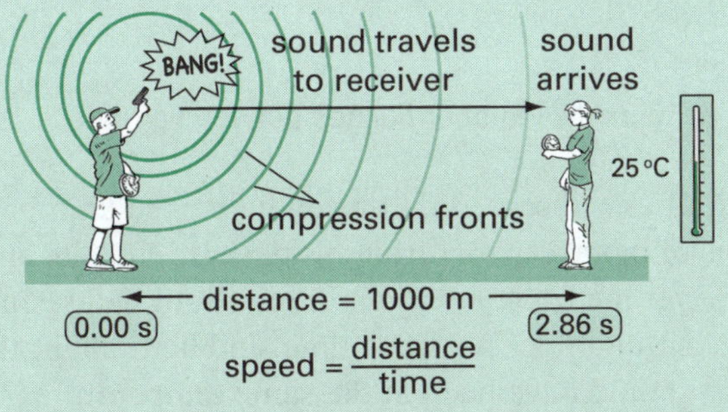

Figure 4.72 Measuring the speed of sound

15. Figure 4.73 shows a boat using sonar to determine the depth of water. Sound waves are produced from a transmitter at the bottom of the boat. These reflect off the ocean floor and are detected by the boat. The time delay allows the depth of the water to be measured. Assume sound waves travel at 1500 m/s. Use the information in the diagram to estimate the depth of water under the boat. *(2 marks)*

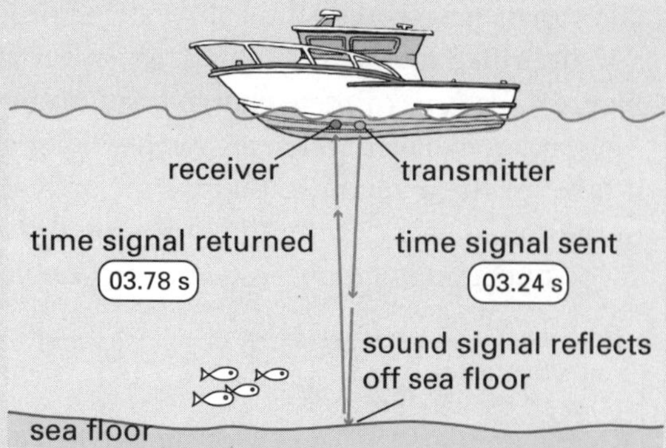

Figure 4.73 Depth of the water

16. Figure 4.74 shows rubber flanges (sheets) bolted between two steel beams. This type of connection is often found in large structures such as the Sydney Harbour Bridge. Explain the purpose of these rubber inserts when used between the beams and girders. *(1 mark)*

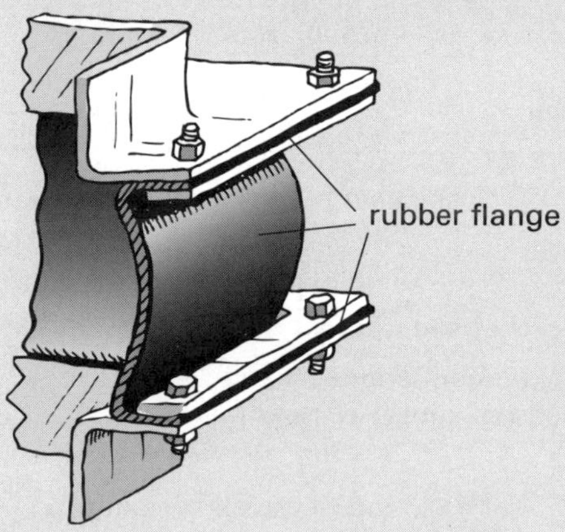

Figure 4.74 Rubber flanges between girders

17. Ricky observed that when he picked up a pair of metal scissors from his desk it felt cold, and yet the book next to it didn't. Both had been sitting there for some time, and he thought they should have been at the same temperature. If the scissors and the book are both at room temperature, why does one feel colder? *(2 marks)*

18. A science teacher took a musical bell and showed the class it made a sound by striking the hammer against the side of the bell. Next, she filled a bucket with water and immersed the bell in it. She then shook the bell to make it ring.
    a) The students heard a dull ringing sound which was not as clear as before. What does this experiment tell us about the transfer of sound energy between different media? *(1 mark)*
    b) If the experiment was repeated in a large swimming pool with the experimenter and observers underwater, predict whether a sound will be heard. Justify your answer. *(2 marks)*

19. Which of these statements are true and which are false? *(6 marks)*
    a) The pitch of a sound is related to its frequency.
    b) The higher the pitch of a sound, the lower is its frequency.
    c) The loudness of sound decreases with increasing distance from the source of the sound.
    d) The loudness of sound is measured in hertz (Hz).
    e) A 100-dB noise hurts your ears.
    f) High-pitched sounds are always very loud.

20. Table 4.8 lists the speed of sound through some gases at 0 °C.

Table 4.8 Speed of sound through different types of gas at 0 °C

Gas	Speed (m/s)
air	331
carbon dioxide	259
helium	965
nitrogen	334
oxygen	319

    a) Re-organise this data so that the speed of sound is listed in decreasing order. *(1 mark)*
    b) In which gas it the speed of sound the highest? *(1 mark)*
    c) In which gas is it the lowest? *(1 mark)*
    d) Plot a column graph of this data. Give your graph a heading. *(4 marks)*
    e) What evidence is there to show air is made up mainly of nitrogen, with some oxygen? *(1 mark)*
    f) How can you tell there is very little carbon dioxide in our atmosphere? *(1 mark)*

21. Sometimes when a cool or cold drinking glass is put into hot water it cracks. Why is this? *(1 mark)*

22. Jim picked up an iron bar, which had been lying on the floor for some time, and immediately

realised it was cold. He explained that the coldness from the iron moved onto his hand, and that's why it was cold. Do you agree with Jim's explanation? Why? *(2 marks)*

23. Diamond has a much greater ability to refract light than glass. Figure 4.75 shows a slab of glass sandwiched between two diamond slabs. A ray of light in air is incident on one diamond surface as shown. Copy and complete the diagram to predict the subsequent path of the light ray. *(4 marks)*

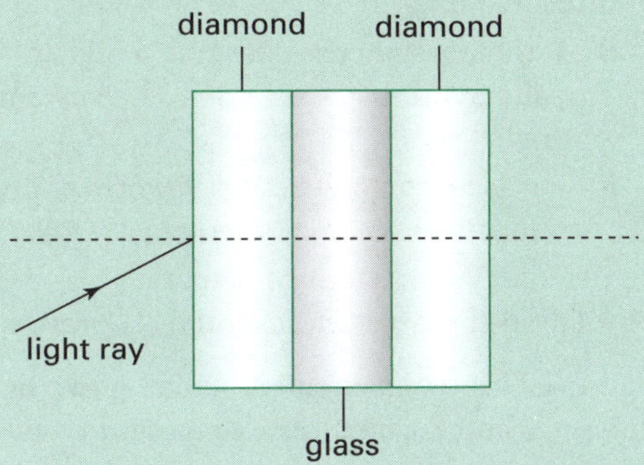

Figure 4.75 Glass and diamond slabs

24. The hearing and sound production ranges were measured for several cats and dogs. The results are in Table 4.9.

Table 4.9 Hearing and sound ranges for cats and dogs

Animal	Frequency range of hearing (Hz)	Frequency range of sound production (Hz)
cat	60–65 000	700–1500
dog	15–50 000	450–1000

a) Present this information in appropriate graphical formats for hearing and sound production. *(2 marks)*

b) Which of these animals can:

i) hear sounds of the higher pitch? *(1 mark)*

ii) produce sounds of the higher pitch? *(1 mark)*

c) A bat emits a sound of frequency 90 000 Hz. Can a cat or dog hear such a sound? *(1 mark)*

25. Mirrors are made from pieces of plane glass by silvering the back of the glass. The mirrors have various uses, including as a periscope. Periscopes are used in submarines to observe surface boats while the submarine is still submerged.

a) Draw a diagram to show how two mirrors and various pieces of tubing or cardboard can be used to make a periscope. *(2 marks)*

b) Explain why the periscope will not work well if the glass in the mirrors is too thick. *(1 mark)*

26. Copy and complete Table 4.10 by using some of the words and phrases from the following list: yes; no; travel as invisible rays; particles move from one place to another; particles vibrate and bump into their neighbours. *(6 marks)*

Table 4.10 Conduction, convection and radiation

	Conduction	Convection	Radiation
How heat travels			
Can heat travel through empty space?			

27. A student buys a pack of four AA dry cells, each with a voltage of 1.5 V.

a) If one dry cell is connected into a circuit, what type of current flows in the circuit? *(1 mark)*

b) The student connects all four cells in series with a small lamp. Explain why the lamp glows more brightly with this arrangement than with one cell alone. *(2 marks)*

28. A student is supplied with a lamp, a 12-V battery, connecting leads and two switches.

a) Design a circuit so that the lamp will light when either switch is closed (on) but does not light when both switches are open (off). *(2 marks)*

b) Name one practical application of this type of circuit. *(1 mark)*

29. *Mathematical extension:* A simple series circuit was set up as shown in the Figure 4.76. A 6-V battery is in series with an ammeter and two resistors (AB and CD). Two voltmeters are connected as shown. The voltage drop across AB was 2 V. The ammeter reads 1.0 A.

a) What will the voltmeter across CD read? *(1 mark)*

b) What is the resistance of AB? *(1 mark)*

c) What is the resistance of CD? *(1 mark)*

d) What is the total resistance of the circuit? *(1 mark)*

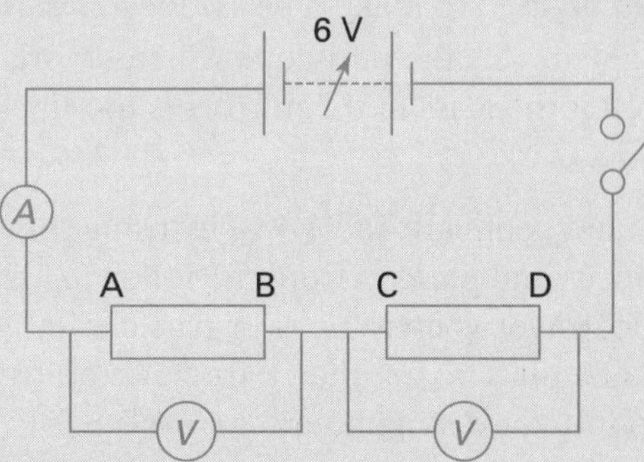

Figure 4.76 Simple series circuit

30. a) The range served by each cell phone base station is about 25 km^2. State one advantage and one disadvantage of this small coverage area. *(2 marks)*
    b) How do cell phones allow you to keep talking as you move from one cell to the next? *(2 marks)*
    c) Why do cell phones use lower-power radio signals? *(2 marks)*

31. Water flows around a series of pipes that form a complete circuit. The circuit also includes a pump, tap and water-flow meters in each section of pipe. There are three types of pipes with different internal diameters (1 cm, 2 cm and 3 cm) which are arranged in series in the circuit. This water-flow circuit is compared to the operation of an electric circuit involving a three-speed electric fan.
    a) Which component of the water-flow circuit corresponds to the:
       i) switch? *(1 mark)*
       ii) power supply? *(1 mark)*
    b) Compare the water flow rate in each of the three sections of pipe. *(1 mark)*
    c) In which pipe section is the resistance to flow greatest? *(1 mark)*

32. a) State two advantages of using geostationary satellites. *(2 marks)*
    b) State two disadvantages of using geostationary satellites. *(2 marks)*

33. Table 4.11 shows the results of an experiment in which a student measured the current and voltage in various simple circuits involving different known resistors.

Table 4.11 Current and voltage in simple circuits

(*R*) Resistance (ohms)	(*V*) Voltage (volts)	(*I*) Current (amps)
4	2	0.5
4	4	1
2	6	3
4	8	2
4	12	3

a) Complete the following sentences by stating a conclusion that is consistent with the tabulated data.
   i) If the resistance is constant, a higher voltage produces a .... current flow. *(1 mark)*
   ii) For a constant current, a lower resistance requires a .... voltage. *(1 mark)*

b) Use the data to state an algebraic relationship between *R*, *V* and *I*. *(1 mark)*

34. Figure 4.77 shows a circuit containing two light globes (Y and Z), a battery, ammeter and two switches (A and B).

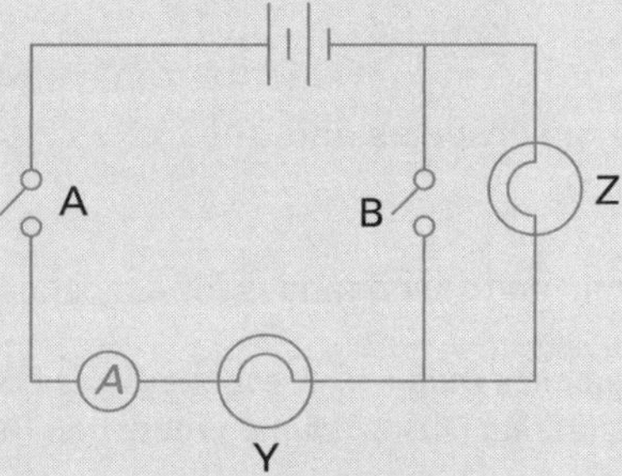

Figure 4.77 Simple electric circuit

a) If both switches are closed, which globes will light up? Why? *(2 marks)*
b) If only switch A is closed, which globes (if any) will light up? *(1 mark)*
c) Switch B is moved so that it is in series with Y and Z. Switch B is closed and switch A is open. Will any of the globes light up? Explain. *(2 marks)*

35. Figure 4.78 shows two resistors (40 Ω and 80 Ω) in parallel with a lamp (10 Ω) and a constant DC power supply. A selector switch can be used to allow the current to flow through either resistor via the connecting points X and Y in the circuit. Explain how the brightness of the lamp will change as the selector switch is moved from Y to X. *(2 marks)*

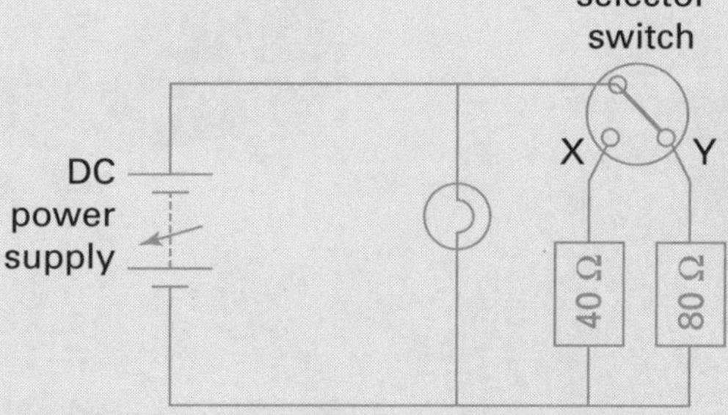

Figure 4.78 Circuit with selector switch

36. **a)** The simplest function of radar is to determine how far away an object is. Explain how radar is used for this. *(3 marks)*

b) Radar can also be used to measure the speed of an object. Explain how radar can do this. *(5 marks)*

37. 'A geostationary orbit can only be achieved at an altitude of 35 786 and directly above the equator.'

a) What is a geostationary orbit? *(1 marks)*

b) Why must it be at this altitude? *(2 marks)*

c) Why must a satellite in a geostationary orbit be located directly above the equator? *(2 marks)*

d) A geostationary satellite has an orbital speed of 3.07 km/s. It takes 23 hours, 56 minutes and 4 seconds to orbit once around the Earth; this is called a sidereal day.

i) How many seconds in 23 hours, 56 minutes and 4 seconds? *(2 marks)*

ii) What distance does a geostationary satellite cover in one orbit around the Earth? *(1 mark)*

38. Use the following words to complete this cloze passage: base, connected, communicates, geographic, network, transceivers, transmission, phone, radios, moving, efficient, data. *(12 marks)*

Mobile phones are sophisticated two-way .... These devices use high-frequency signals to transmit and receive voice and other .... When you make or receive a call, your mobile phone .... with a network of low-powered radio .... These are known as base stations. Each base station covers a small .... area, often called a 'cell'. Cells are interlinked to create the cellular .... A base station consists of antennas and .... dishes connected to an equipment cabin. The base station collects .... signals and passes them on. A mobile phone may communicate with several different .... during a particular call, and this makes it possible for you to continue your call while .... When a call is ...., both the mobile phone and the base station automatically adjust to the minimum power level needed to maintain a quality call. This makes the mobile phone one of the most .... communication devices in the world.

39. A 3-Ω resistor and a 4-Ω resistor are arranged in parallel with a 12-V battery and three ammeters as shown in Figure 4.79. Three voltmeters are also connected: one across each resistor and across the battery terminals.

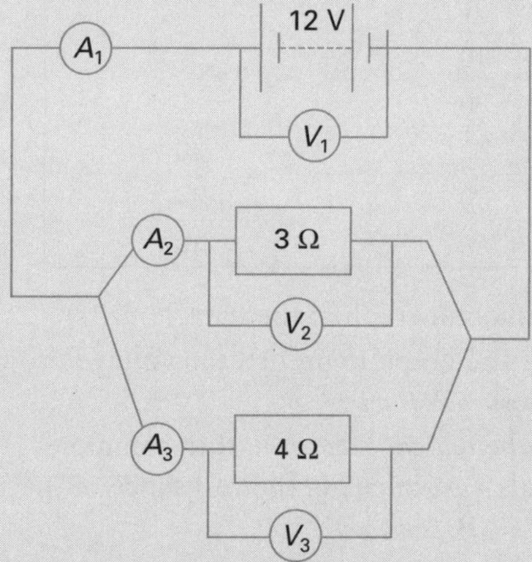

Figure 4.79 Parallel circuit

a) *Mathematical extension:* Which ammeter will have the **i)** highest and **ii)** lowest, reading? Explain. *(2 marks)*

b) Compare the readings on the voltmeters. *(2 marks)*

c) *Mathematical extension:* A switch is inserted in the circuit to turn off the branch to the 3-Ω resistor. How will this affect the readings on each ammeter? *(1 mark)*

40. Figure 4.80 shows a 100-cm resistance wire AB connected in a series circuit with a 20-V DC transformer and an ammeter. Two wires allow a voltmeter to be connected at any two points along the resistance wire AB. One end is permanently attached to A and the other slid along to a point X. Readings of the voltmeter are made at each position of X.

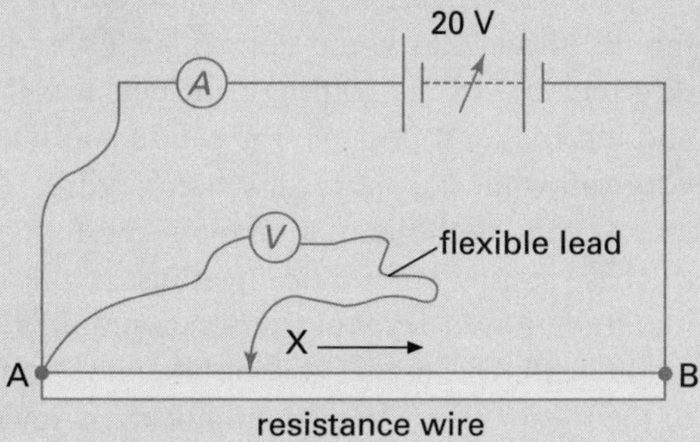

Figure 4.80 Resistance wire experiment

Table 4.12 shows the results of the experiment.

Table 4.12 Results of resistance wire experiment

Distance AX (cm)	Voltmeter reading (V)
0	0
10	2
30	6
50	10
70	14
90	18

a) Plot a line graph of the data. *(3 marks)*

b) Use the graph to predict the voltage drop across AB. *(1 mark)*

c) *Mathematical extension:* If the ammeter reads 4 A, calculate the resistance of the wire AB. *(2 marks)*

Go to pp. 247–251 to check your answers.

CHAPTER 5

Investigations and problem solving

Overview

In this chapter you will learn about:

- first-hand investigations
- inferences, predictions and generalisations
- hypotheses and theories
- conducting background research
- working individually and in a team
- selecting equipment
- variables and fair testing
- experiments with animals
- making and testing a hypothesis
- designing an experimental procedure
- safety during a first-hand investigation
- accuracy and reliability
- writing scientific reports
- graphing and drawing diagrams
- random and systematic errors
- explanations and conclusions
- cause and effect relationships
- report presentation.

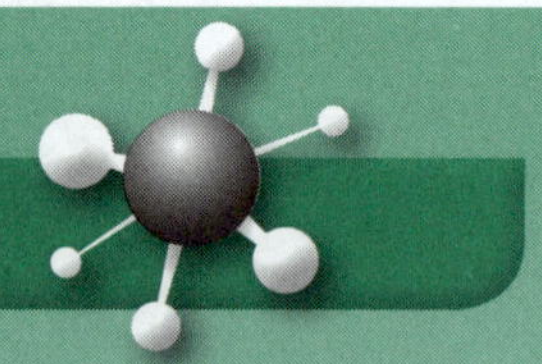

Glossary

Continuous data—data that can take any of an infinite number of values between whole numbers, such as measuring

Controlled variable—a factor that is held constant during an investigation

Dependent variable—a factor that the investigator is observing or measuring in response to the independent variable

Discrete data—data where there is only a finite number of values possible, such as in counting

Extrapolation—extending a graph beyond the plotted values to an unknown situation by assuming that existing trends will continue or similar methods will apply

Generalisation—a statement that is true in the majority of cases

Hypothesis—a logical explanation of an observation or solution to an investigation that is made after all available evidence and information has been collected

Independent variable—a factor that the investigator selects to change or vary

Inference—a possible and reasonable explanation of an individual observation

Interpolation—constructing new data points within the range of a discrete set of known data points

Observation—something that can be seen, heard, felt, smelt or measured using a device

Placebo—a substance that does not contain the active material to be tested

5.1 First-hand investigations

The main objective of all scientists is to get the correct answer to the problem they are working on. Some would call this the pursuit of truth. And the very meaning of truth implies a way of checking and verifying the results. This checking needs to be exhaustive, as the truth of a general proposition can be disproved by a single exception.

The first step of a scientific investigation takes place when an observation is made. This might be an observation of an event such as a pine cone falling from a tree. It might be an observation of some feature of an animal such as the presence of webbing between the toes of a bird. These observations may then lead you to think about the cause or reason behind the observation.

Louis Pasteur wrote: 'In the field of observation, chance favours only the prepared mind'. This means that you will only discover something new if your mind is prepared to understand what you observe. This is done by reading and planning.

The scientific method is generally described as a series of steps in which an observation or problem is investigated. These steps are:

- making observations (using your senses or with the aid of instruments)
- making inferences
- reviewing the known facts using data sources
- making predictions
- planning, designing and experimenting to test predictions
- formulating a hypothesis based on an analysis of collected data
- testing and modifying the hypothesis by further investigations
- discussing collected results
- making conclusions
- planning for further investigations about the problem.

Identifying data sources

In order to prepare for an investigation of some problem or observation you have made, you need to review what is currently known about this topic. This involves identifying:

- what information or data needs to be collected to help you understand the issues
- the possible sources of this information or data (e.g. library books, textbooks)
- dictionaries, encyclopaedias, periodicals and multimedia resources including the internet

- the relevant parts of the gathered information by doing some background reading, highlighting key ideas and summarising.

In this way you will develop an understanding of the problem to be investigated.

Inferences, predictions and generalisations

Some problems will be investigated by conducting research in the laboratory or in the field. Some problems will be investigated by second-hand data research. Second-hand data is work that has been published by others.

We will now look at examples of three of the steps of the scientific method.

Inferences

Following an observation, a scientist may make an inference (see Figure 5.1). This is a logical explanation of the observation.

Figure 5.1 Scientists make inferences about the observations they make.

Predictions

The next step is to test the inference by making a prediction and performing an experiment to test the prediction. These experiments may support the inference or they may not. A new explanation may need to be sought.

Generalisations

As more supporting data is collected, the investigator may see a pattern emerging. At this point the investigator may be able to make a generalisation about the topic under investigation. Generalisations are statements that are true in the great majority of cases.

Examples

1. After the last lesson was over and the students had gone home, the teacher observed that the last sink in the laboratory had chocolate wrappers in it. She decided that the three boys sitting there had been eating chocolates during the lesson.

   This is an inference based on her observation. The inference may not be correct as the wrappers may have been placed there in a previous lesson or other students placed the wrappers in the sink.
2. If the temperature of the acid is raised, the speed of the reaction will increase.

   This is a prediction and suggests that the student has done some preliminary experiment on the rate of the reaction and he is now predicting that heating the acid will make the reaction go even faster.
3. A senior student used a chemical data book to record the colours of a wide variety of different ionic compounds. In his report he wrote that all nickel compounds were coloured but all sodium compounds were white.

   This is a useful generalisation about the compounds of nickel and sodium.

Hypotheses and theories

The next step in the scientific method is to generate a hypothesis. What is a hypothesis? A hypothesis is a statement concerning the observation(s) or solution to a problem being investigated. A hypothesis is *not* a wild guess. Hypotheses are temporary statements that lead towards the development of scientific theories.

Hypotheses are only made after looking at *all* the available evidence and information that has been collected. This is why background reading of identified data sources is so important. The work of others may help you to produce a good hypothesis. Hypotheses are the result of a scientist making inferences, making predictions and developing generalisations.

The hypothesis must be able to:

- predict the results of future investigation about the problem
- suggest ways in which it can be tested to determine whether or not it is true.

Hypotheses are not always correct. If the data collected during the investigation does not support the hypothesis, a new hypothesis must be developed.

Examples

1. A student wrote in her notebook: 'Based on our bushwalk, this plant species appears to grow best in moist, rich soils and a shaded environment rather than dry, sandy soils and a sunny, exposed environment'.
   This is a hypothesis that can be tested. The student can now perform an experiment in which this plant species is grown under different conditions.
2. The student read the following comment on a social web site: 'Most Australians prefer hip-hop music to rock music'.
   This is *not* a hypothesis as it involves a personal opinion and is not scientifically based. This opinion will vary from one group to another.
3. A student read on a blog that human ancestors were tree dwellers in the African plains and that earlier hominid species were exclusively cave dwellers.
   This is *not* a hypothesis. It does suggest an answer to a problem. It cannot, however, be tested directly as there are no living human ancestors or hominids. The finding of hominid bones in caves does not provide sufficient evidence that they exclusively lived in caves.

A scientific theory is the result of very extensive research that supports a hypothesis, or a group of related hypotheses. Over a longer period of time the theory may become so universally accepted that it becomes known as a scientific law. Sometimes old theories are replaced by new theories. The law of mass–energy conservation is an example of an old theory (the law of energy conservation) being replaced by a newer law when Albert Einstein showed the equivalence of mass and energy. No experiments have ever been able to disprove the conservation of mass–energy theory and so it has been elevated to the status of a law of science. Many scientists believe that the theory of evolution should now be called the law of evolution as ongoing research has always supported the theory for well over a hundred years.

Background research and reliability of secondary data sources

The following points should be noted when gathering information from secondary sources.

- Use a wide range of resources to gather secondary information. Do not just rely on the internet. Not all the information on the internet is accurate or correct. Much of it is not peer reviewed. CD-ROMs, periodicals, library books and newspapers are also valuable sources of information. Ensure that you reference sources from which you have collected information in the appropriate manner (e.g. Author, Year, Title, Publisher, Place). For articles on the internet, record the title of the article, author (if known) and the URL.
- When using library books or textbooks to locate information, you should look for key words in the index. Look at the table of contents at the front of the book to locate relevant chapters. There may also be a glossary of terms to help you understand difficult technical words.
- Once you have gathered the second-hand data, you will need to collate and summarise it. You must not plagiarise the work collected. You must read and make notes from this collected information, placing your notes into a scaffold/table which you have constructed earlier. You can then use these note summaries to construct your own text.
- Information from secondary sources may contain various types of graphs (e.g. histograms, sector graphs, line graphs, divided bar graphs), flow charts and diagrams. You can use this type of information to extract data.
- As you gather second-hand data, you must decide whether the information is relevant or irrelevant. For example, if you are gathering information on water movement in flowering plants, any data on non-flowering plants would be irrelevant.
- Not all the gathered data will be reliable. You need to check its reliability by comparing it with other sources. Science articles that appear in respected journals and that have been peer reviewed are more likely to be reliable than an article that is self-published. Articles written by creation scientists are not reliable as they do not follow the scientific method.
- Your collected data needs to be organised for easy analysis. Tables, proformas, scaffolds and diagrams are commonly used to organise information. Computer databases and spreadsheets are becoming increasingly important ways of organising collected data.

Using science to make claims about commercial products and practices

Most companies that produce commercial products behave ethically but there are some companies which may be unethical in that they only keep data that is favourable to them. They may employ scientists to monitor quality control or to do research with the aim of producing new products. When these products are advertised, the advertising company may present only the positive data. It is only when independent tests are done that the consumer finds out the truth of the claims made about a product. For example, a company may produce a skin cream that claims to reduce face wrinkles. Independent testing, however, may show that the reduction in wrinkles is minimal and that the product simply fills in the wrinkles and makes the skin look smoother. Such companies often do insufficient testing to ensure reliability of the conclusions.

The labels of bottles contain information about the product as well as various claims about it (see Figure 5.2). All commercial products must be properly tested to meet government regulations before they can be sold.

Figure 5.2 Bottle labels contain information and claims about the product.

There are laws in Australia which insist that organisations and individuals that are making claims behave ethically. While most do, there will be instances that come to the attention of government authorities responsible for codes of conduct and behaviour. However, this might be some time after misleading claims have been widely communicated.

5.2 Working individually and in a team

Students need to be able to work individually on some occasions and in teams on others. There are advantages to both ways of working. There are also responsibilities that must be considered.

Working individually requires students to:

- appreciate the importance of taking personal responsibility in planning and performing a task
- work to realistic timelines and goals
- persevere with an activity to achieve a reasonable endpoint
- accept personal responsibility for a safe working environment for themselves and others in the laboratory
- evaluate the effectiveness of their work in completing the task.

Working in teams requires students to:

- identify the different roles of each team member
- negotiate and allocate responsibility as a team member
- accept and perform the roles decided by the team
- collaborate with others in setting and working to agreed timelines and goals
- accept personal responsibility for maintaining a safe working environment for all team members
- work cooperatively to monitor the progress of the investigation
- evaluate the process used by the team and the effectiveness of the team in completing the task.

5.3 Selecting equipment

The equipment chosen for a practical investigation must be appropriate for the task. You need to ensure that the experiment can be done safely using the correct equipment.

Examples

1. A digital temperature probe is often more accurate than a simple alcohol thermometer. A mercury thermometer is more accurate than an alcohol thermometer. A thermometer whose scale is divided into fifths or tenths of a degree is more suitable to determine the variation in

body temperature of a student over 2 hours than a thermometer marked in whole degrees.

2. A 100-mL measuring cylinder marked with 1-mL scale divisions is more suitable to measure exactly 73 mL of water than a 100-mL beaker with 10-mL divisions.
3. The reaction of solid potassium sulfite with dilute hydrochloric acid should be performed in a beaker in a fume cupboard, as the evolved sulfur dioxide is very poisonous. The student should also wear safety glasses to avoid the splashing of acid into the eyes.
4. An ammeter is used to measure the current flowing in a circuit. If the current is very small then a milliammeter or a microammeter should be used.
5. Distillation of volatile, flammable liquids is best performed using an electric heater rather than a Bunsen burner. The absence of a flame makes it more likely that accidental fires will not occur.
6. Using an electronic balance which weighs to 2 or 3 decimal places is more accurate than using a sliding poise balance when measuring the weights of small quantities of materials.
7. Quantitative filter paper has finer pores than qualitative filter paper and so it is better to use when filtering very fine precipitates.

5.4 Fair testing and variables

Many factors can influence the results of an experiment. These factors are called variables. In order to investigate a problem experimentally, all the variables must be controlled so that only the variable being investigated can affect the result. This is known as fair testing.

Variables

During an investigation, the experimenter tests what happens when he or she alters one variable. The effect of a change in this factor (or independent variable) on another variable (the dependent variable) is investigated. All other variables are kept constant (i.e. controlled).

- The independent variable is the one that is systematically allowed to change by the researcher. It is the 'change' that you allow to happen when conducting an experiment.
- The dependent variable is affected by the change in the independent variable. It is the 'what happens' part of the investigation.
- The controlled variables are the other factors that can alter the result if they are not held constant.

Example 1: Battery life

- Hypothesis: dry cell batteries used in a CD player have a longer life if kept in the refrigerator when not in use
- Independent variable: temperature
- Dependent variable: battery life time
- Controlled (constant) variables: same brand of batteries; batteries used in the same CD with the same settings on the player; same refrigerator and temperature settings

Example 2: Rate of reaction

- Hypothesis: magnesium metal dissolves faster in hydrochloric acid if the acid is more concentrated
- Independent variable: concentration of hydrochloric acid
- Dependent variables: rate at which the magnesium dissolves
- Controlled (constant) variables: same mass and shape of magnesium pieces; same volume of acid; same temperature of the acid; same vessel

Using a control

Measurements may only be reliable in some experiments if a control is used. This is particularly true in biological or biochemical systems where many factors can alter a result. A control is an experiment that is performed as a comparison with those in which the independent variable is allowed to change.

Examples

1. A new plant fertiliser made from seaweed is to be tested to see if it improves the growth rate of geraniums. Ten plants are tested. Five plants receive the dose of fertiliser recommended on the packet. Five plants are used as controls. They do not receive any fertiliser. The growth rate of the two groups is compared in order to determine whether the fertiliser makes any difference to the growth of the geraniums.

2. A new petrol additive claims to give improved distance travelled per litre of fuel. Using the same car in each run, the car is filled with normal petrol and driven 200 km and then the amount of fuel required to refill the tank is determined. This is repeated five times and the results averaged. These results are the control group. The experiment is repeated five times over the same route with the same amount of fuel additive added to the tank. These results are averaged and compared with the average of the control group to determine whether there has been an improvement with the fuel additive.

Experiments with living things

Animals cannot be used in laboratory experiments unless permission is sought from the appropriate governmental authorities. Supporters of the use of laboratory animals in university research believe that such research is essential to discover the possible toxicity of drugs as they are developed. Mice, rats, fish and birds have commonly been used in such experiments. Many animal rights groups believe that animal testing of drugs and other medications should be banned altogether and that only human volunteers should be used. In schools, no live animal testing is allowed.

In many experiments involving human subjects, scientists adopt special methods to ensure that bias does not influence the result. In addition, all living things of the same species have such variability that it is difficult to control the variables. Two common methods used to overcome these problems are the blind and double-blind experiments.

Blind experiments

When testing whether a particular medication works on humans, it is important that the subjects do not know whether they are the test group or the control group. The control group receives a placebo. This is a tablet, capsule or other medication that looks the same but has no active or test ingredient. Only the test group receives the medication with the active agent being tested. This procedure overcomes problems in attitude and wishful thinking.

Double-blind experiments

This technique goes one step further than the blind experiment. In this case the possible bias of the scientist (or, more likely, the person administering the test) is removed because he/she does not know which medication samples contain the test ingredient. Only the person preparing the samples knows the code and that person does not communicate with the scientist (see Figure 5.3).

You can grow grass on a busy street.

Are you male pattern bald?
Then this tablet is for YOU!

It will:

- grow your existing hair faster
- grow new hair on your head
- grow your existing hair thicker
- give you positive proof within 6 months, provided you remain on the course.

Figure 5.3 This hair product makes claims that would need to be tested by a double-blind study.

5.5 Designing an experimental procedure

It is important when undertaking a practical first-hand investigation that the procedure adopted is safe and that modifications are made as problems arise. It is important to develop a planned procedure before undertaking experimentation. The planned procedure will:

- identify the variables that need to be kept the same
- specify the independent variable
- specify the dependent variable
- ensure that the selected procedure is valid
- ensure that the measurements are valid and accurate
- ensure that the measurements are reliable.

Example: Rate of cooling of hot water

Hypothesis: hot water cools down faster in darker coloured vessels

Experimental procedure:

1. Use two identical metal cans (e.g. drink cans). One is painted white and the other is painted black.

2. Pour 200 mL of 90 °C water carefully into the white can.
3. Measure the temperature of the water every 2 minutes for 20 minutes, as shown in Figure 5.4.
4. Repeat the experiment with the black can.
5. Repeat the experiment on each can five times.

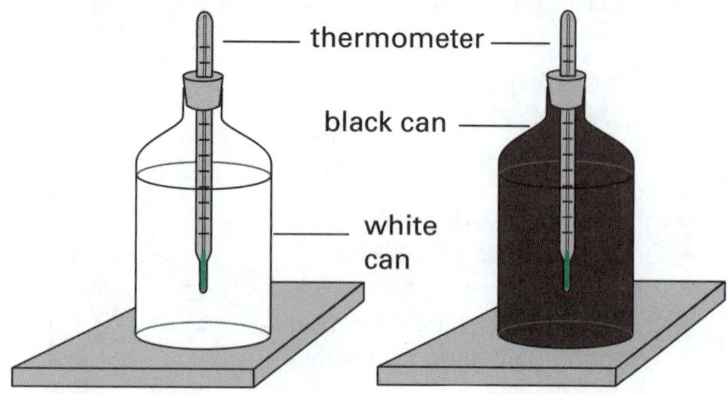

Figure 5.4 White and black can experiment

Independent variable: colour of can

Dependent variable: temperature

Constant variables: same volume of water; same initial temperature of water; same size vessels; same material of each vessel; same external temperature; same thermometer; thermometer bulb at the same depth in the water

Procedure and measurements are valid and accurate: The procedure is valid as it has been verified by many tests carried out by scientists. The measurements are valid if they are 'trustworthy' and the method is one that is commonly used by many scientists. The accuracy of the measurements depends on the equipment selected and how carefully the experiment is conducted. The accuracy in determining the temperature of the water can be improved by using an electronic thermometer connected to a data logger. The accuracy of measuring volume can be improved by measuring volumes from a pipette or burette rather than a measuring cylinder or beaker.

Measurements are reliable: The measurements are reliable if repeated experiments lead to consistent results. Sometimes in a large set of repeated measurements, an individual measurement may differ widely from the other measurements. You may be justified in ignoring this measurement before calculating an average. Measurements cannot be reliable if they are not performed with high precision and accuracy.

Experimental results: The results of each five runs are averaged and compiled into Table 5.1.

Table 5.1 Experiment results (averages of five runs)

Time (min)	Temperature of white can (°C)	Temperature of black can (°C)
0	90	90
2	86	84
4	82	78
6	78	73
8	74	67
10	70	62
12	67	57
14	63	52
16	59	47
18	56	43
20	52	39

Graphical results: Figure 5.5 shows the results of the experiment.

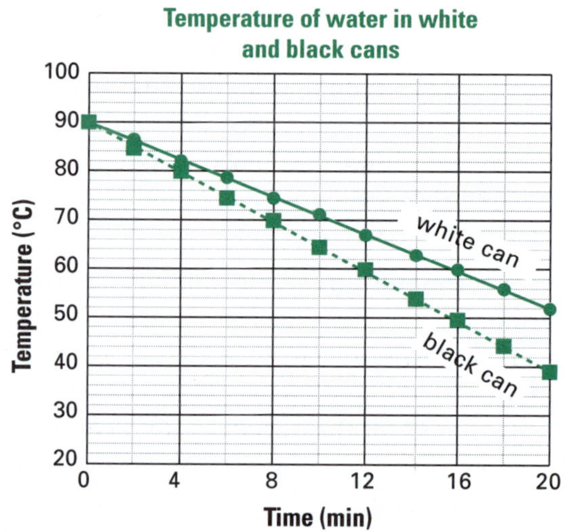

Figure 5.5 Graph of temperature versus time

Conclusion: The results of the experiment show that the hot water cools down faster in a black can than a white can.

5.6 Lab rules and safety in a first-hand investigation

A laboratory is a potentially dangerous place. In order to work safely in a laboratory, it is vital that you follow the safety rules. Now it is time to remind you of these important rules. Follow them and accidents will be minimised.

These rules are not in any order because they are all equally important.

Rule 1: Follow the instructions! Ask if you do not understand what to do.

You may be doing an experiment with many steps. Listen to the teacher's instructions. Follow the demonstration of how to use the equipment safely. Read the steps of the experiment. If you still do not understand, ASK!

Rule 2: Never run or fool around! You may damage yourself or someone else.

This rule is common sense. You should always stay with your work group. Don't wander off to visit friends. Keep your mind on the task. If you muck about, you may burn or spill chemicals on yourself or others.

Rule 3: If an accident occurs, stay calm and get help.

If you cut yourself, burn your skin or clothes, or spill chemicals on your skin or clothes, it is important not to panic. Ask others to help you. Someone else in the group can alert the teacher. Remember the following.

- Always wear your safety glasses.
- If you cut yourself, wash the wound and stop the bleeding by wrapping a clean cloth around the wound. Seek help and follow the teacher's instructions.
- If you burn your skin, run cold water over the burn. If necessary, use the shower. Keep the cold water treatment up for 15 to 20 minutes. Follow the teacher's instructions.
- If your clothes catch fire, roll onto the floor and smother the flames. Your classmates should wrap you in the fire blanket. One member of the group should immediately fetch the teacher.
- If you spill chemicals on your skin or clothes, wash the chemical off immediately and seek the teacher's advice.
- If you get chemicals in your eye (this probably means you aren't wearing the safety glasses!), wash your eye with large amounts of water. Keep doing this while others seek the teacher's assistance. (See Figure 5.6.)

It is important that you know where emergency equipment, such as fire blankets or the first aid kit, is kept in the laboratory so that in an emergency you can access them immediately.

Figure 5.6 Always wear safety glasses when doing experiments.

Rule 4: Never eat or drink in a laboratory.

Your food may become contaminated if left on the tables. Your hands will not be clean and so food should not be consumed. In any event, laboratories are for experiments. There are other, more suitable, places for eating.

Rule 5: Do not sit on the experimental benches.

When an experiment is being conducted, you might knock it over if you sit on the bench. You will also take longer to react and move away from any spills which may occur. Chemical spills often occur on benches. While the spill may have been cleaned up, there is no guarantee this was done satisfactorily and so your clothes can be damaged by stains or acid spills.

Assembling equipment

In a first-hand investigation, the selected equipment needs to be assembled to produce an apparatus that will perform the required task. Assembling pieces of equipment and using the apparatus requires good manipulation skills and attention to safety.

Example 1: Evaporation of a solution

The process of evaporation involves heating a solution in an evaporating basin using a Bunsen burner.

- Care must be taken not to touch any hot equipment until it is cooled down.
- You must wear safety glasses to ensure that your eyes are protected in case hot liquid is ejected from the flask.
- Heating should be done slowly to avoid possible ejection of hot liquids. Whenever you are heating

anything, you should always monitor the situation. Don't leave apparatus to boil over while you leave the area to do something else.

- When a Bunsen burner is not in use, it should be turned off or adjusted to give a yellow safety flame, and removed from under any glassware.

Example 2: Reactions with acids

- Acids are corrosive and can damage the skin and eyes.
- Safety glasses must be worn at all times.
- Never use concentrated acids as their reactions can lead to excessive heat production.
- Use only small volumes of acids at any time.
- When pouring liquids from reagent bottles (e.g. acids), it is good practice for you to pick up the bottle with the label facing the palm of your hand. There are two important reasons for this. If any spills do occur, they will run down the side of the bottle away from where your fingers are. This will avoid damaging the label because if any spills do occur, they will consistently be away from it.
- Always put the lid back on the bottle of acid immediately after use so it doesn't get mixed up with other reagents you might be using.

Example 3: Electrical circuits

- Ensure all electrical leads are free of corrosion.
- Make sure any leads and power cords do not trail over the edge of a bench or near the sink.
- Ensure the power is always turned off when not in use.
- Use a tapping key to turn on the current and collect your measurements rapidly. Do not leave the current on when no measurements are being made.

Experiment 1

Compression of air

Aim

- To determine the change in volume of air as it is compressed by known weights

Materials

Plastic gas syringe; steel masses of the same weight; stopper

Method

1. Use the stopper to seal one end of the plastic gas syringe after the plunger has been pulled out so that the volume of air is 100 mL.
2. Progressively add masses and record the volume of air after each mass is added (see Figure 5.7).

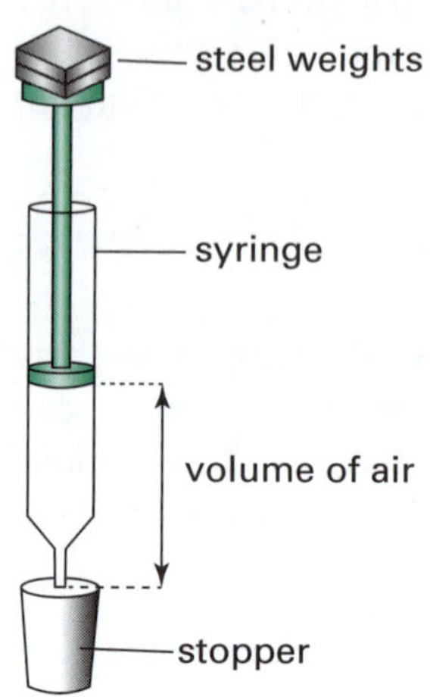

Figure 5.7 Experiment to measure volume of air as it is compressed

Results

The results of this experiment are shown in Table 5.2.

Table 5.2 Volume of air versus applied pressure

Number of added lead weights	Volume of air (mL)
0	100
1	92
2	84
3	76
4	68
5	60

Analysis

Analyse the table of results by drawing a line graph. Predict the volume of the air if eight weights were added. Go to pp. 251–252 to check your answer.

Conclusion

Write a suitable conclusion.

Go to p. 252 to check your answer.

Experiment 2

Forensic analysis of soil

Aim

- To obtain evidence of a crime using soil analysis

Background

Soils vary from one location to another. They can vary within a garden or across a larger area. One way to compare soils is to measure the size of the particles and work out what proportion of the sample is made up of each size. This data can be used in forensic analysis.

Method

1. Samples of soil were collected from different parts of a victim's garden.
2. Soil was also removed from the soles of a suspect's shoes.
3. The size of the particles was measured by using sieves with different mesh sizes. After sieving, the separated samples were weighed and their proportions calculated.

Results

Figure 5.8 shows graphical data from zones A, B and C in the victim's garden.

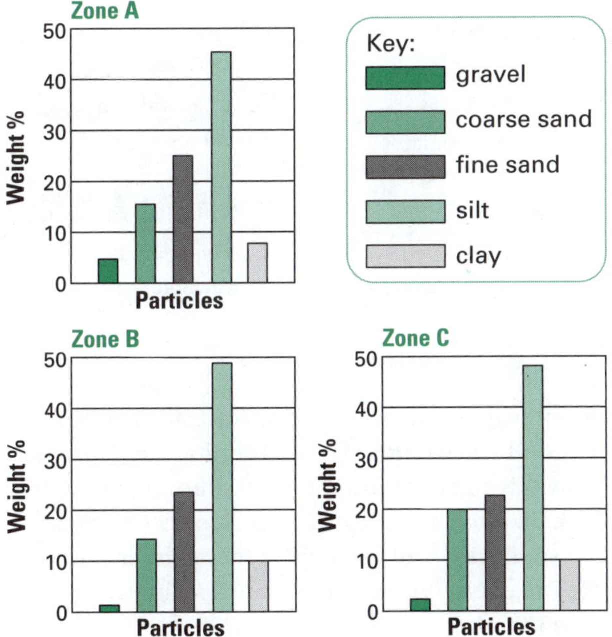

Figure 5.8 Soil analysis

Table 5.3 shows the results from the suspect's shoe.

Table 5.3 Particle weight % of total soil sample

Type of particle	Weight % of total soil sample
gravel	0
coarse sand	15
fine sand	25
silt	50
clay	10

Analysis

Plot a column graph of the analysis of soil from the shoe of the suspect. Analyse the data and explain whether or not there is a match between the suspect's samples and the garden samples.
Go to p. 252 to check your answer.

Conclusion

Write a suitable conclusion.

Go to p. 252 to check your answer.

Test yourself 1

Part A: Knowledge

1. Generalisations are *(1 mark)*
   A statements that are true in the great majority of cases.
   B statements that are used to test inferences.
   C logical explanations of an observation.
   D theories developed to explain an observation.

2. If you accidently burn yourself while conducting an experiment, you should immediately *(1 mark)*
   A wrap your skin in cloth to exclude the air.
   B run cold water over the burn for 20 minutes.
   C ask the teacher what you should do.
   D ask your friends to tell the teacher.

3. Bert is required to heat a test tube containing an acid as part of an experiment. Identify the correct procedure he should use. *(1 mark)*
   A Hold the test tube base a few millimetres above the top of the Bunsen burner.
   B Hold the test tube with metal tongs when heating the acid.
   C Point the test tube being heated towards the window and away from other students.
   D Fill the tube with acid and then heat the tube with a yellow Bunsen flame.

4. Sally's research project involved experiments to test whether the roots of a plant always grow downwards. She filled glass test tubes with a nutrient aquaculture solution and then 1-week-old seedlings are placed into the tubes and held in place with cotton wool. She clamped the tubes at different angles from 10° to 90°

from the horizontal. She observed the direction in which the roots grew. Select the correct statement about the experiment. *(1 mark)*

A The dependent variable is the angle of the tubes.

B A mature plant should be used as a control.

C The temperature of the nutrient solution should be controlled.

D The independent variable is the direction in which the roots grow.

5. Gianni wanted to test the effect of temperature on the growth of plant seedlings. In each set of experiments only one condition was altered. All other conditions remained the same. The seedlings were allowed to grow from 1-cm plants for 7 days. Gianni's results are shown in Figure 5.9. What conclusion can be made about the effect of temperature change? *(1 mark)*

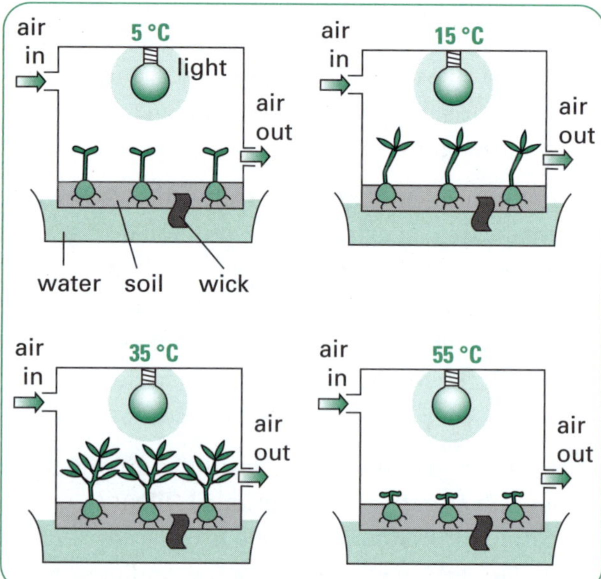

Figure 5.9 Effect of temperature on plant growth

A The higher the temperature, the better the plants grow.

B Some plants grow best at low temperatures and some grow best at high temperatures.

C No plants grow if environmental conditions exceed 35 °C.

D Plants grow best when the temperature is between 15 and 35 °C.

6. Complete the following restricted-response questions using the appropriate word. *(1 mark for each part)*

a) When you reference articles on the internet, record the .... of the article, the author (if known) and the URL.

b) When experimenting, you should appreciate the importance of taking .... responsibility in planning and performing a task.

c) Measurements are .... if repeated experiments lead to consistent results.

d) If your clothes catch fire during a laboratory activity, roll onto the .... and smother the flames.

e) A student watched a video set in North America. She noticed that the leaves on the trees had turned orange and red and she .... that the video was shot in autumn.

7. Use the code letters to match the terms or phrases in each column. *(1 mark for each part)*

Column 1	Column 2
A glossary	F repeated experiments
B accuracy	G technical words
C manipulated variable	H held constant
D reliability	I digital thermometer
E controlled variables	J independent variable

Part B: Skills

8. Fires in eucalypt forests in Australia can lead to great devastation. The convection currents and winds can carry fragments of burning material high into the air, causing outbreaks known as 'spot fires' well away from the main fire front. The energy produced by a forest fire depends on the:

- availability of fuel (t/ha)
- rate of spread (km/h).

Examine Figure 5.10 and answer the questions that follow.

a) Determine the intensity of the fire in an area where the fuel availability is 5 t/ha and the rate of spread is 2 km/h. *(1 mark)*

b) A fire has an intensity of 12 kW/km in a forest where the fuel availability is 15 t/ha. What is the rate of spread of this fire? *(1 mark)*

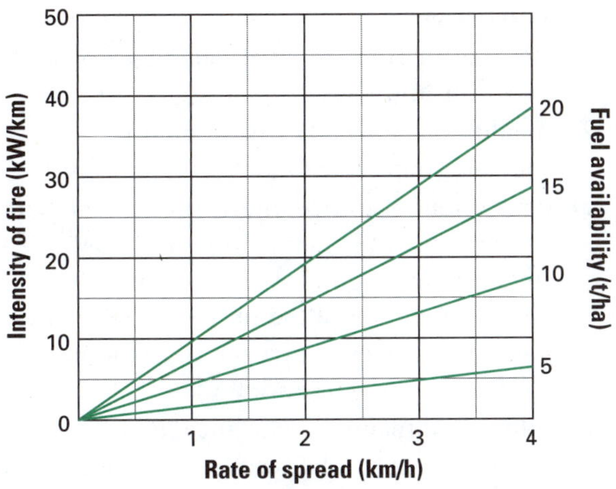

Figure 5.10 Fire intensity versus spread rate

c) It is mid-summer and you are camping in the bush. What responsibilities do you have to avoid starting a bushfire? *(2 marks)*

9. Table 5.4 provides information on a variety of indicators which change colour in acidic and alkaline (basic) solutions. Use this information to answer the questions that follow.

Table 5.4 Colour of indicators in acidic and alkaline solutions

Indicator	Colour in acidic solutions	Colour in alkaline solutions
litmus	red	blue
methyl orange	red	yellow
phenolphthalein	colourless	crimson (strong alkali)
universal indicator	red (very acidic) orange (moderately acidic) yellow (weakly acidic)	pink (weak alkali) purple (very alkaline) blue (moderately alkaline) blue-green (weakly alkaline)

a) A drop of universal indicator is added to a sample of oven cleaner in a test tube. The indicator turned purple. What can you infer about the oven cleaner? *(1 mark)*

b) Predict the colour of phenolphthalein indicator if it was added to another sample of the oven cleaner. *(1 mark)*

c) Some drain cleaner and baking soda were tested with two indicators. The results of these experiments are shown in Table 5.5. In what way are these two chemicals similar and in what way are they different? *(2 marks)*

d) Make a generalisation about oven cleaner, drain cleaner and baking soda. *(1 mark)*

Table 5.5 Testing chemicals with indicators

Chemical	Methylorange	Universal indicator
Drain cleaner	yellow	purple
Baking soda	yellow	blue-green

10. Comment on why the following contains many statements that are not considered scientific. *(1 mark)*

In justifying why the Earth is only about 6000 years old, a student stated that in the 1650s Archbishop James Ussher did extensive chronological searches through the Bible and found that the world must have been created in 4004 BC. Ussher was a learned and well-respected scholar who was appointed professor and vice-chancellor at Trinity College, Dublin. Therefore, we should believe this serious attempt at dating the Earth. Although scientists think they have found rocks that are far older, they are mistaken. Just like you can stonewash a pair of jeans to make them look older than they really are, God, in creating the Earth, made it look older than it really is.

11. The great English biologist Charles Darwin once wrote:

… From my early youth I have had the strongest desire to understand or explain whatever I have observed … (and) … reflect or ponder … over any unexplained problem. … I have steadily endeavoured … to give up any hypothesis, however much beloved … as soon as facts are shown to be opposed to it.

How difficult do you think it is for a scientist to do this? *(1 mark)*

12. Over the years many people have written about science. Here are three such quotes. Comment on each, indicating what its author means and whether you agree.

a) 'The essence of science: ask an impertinent question, and you are on the way to a pertinent answer.' Jacob Bronowski, *The Ascent of Man*, 1973. *(1 mark)*

b) 'Science is built up of facts, as a house is built of stones; but an accumulation of facts is no more a science than a heap of stones is a science.' Henri Poincaré, 1905. *(1 mark)*

c) 'Science is nothing but trained and organised common sense, differing from the latter only as the veteran may differ from a raw recruit: and its methods differ from those of common sense only as far as the guardsman's cut and thrust differ from the manner in which a savage wields his club.' TH Huxley, *Collected Essays*, 1893–94. *(1 mark)*

13. Two researchers followed the procedures below in conducting an experiment on herbicides with eggs. Read this information and answer the questions that follow.

Researcher A

1. Under sterile conditions, a fertilised egg is pierced (see Figure 5.11, part A)
2. A specific quantity of herbicide (dissolved in alcohol) is injected into the air space inside the shell.
3. The small hole is sealed with sealing wax.
4. The egg is incubated.

Researcher B

1. Researcher B repeated Researcher A's steps 1 to 4 for four eggs.
2. In addition, researcher B set up three control groups (see Figure 5.11, part B).

a) One potential source of error in Researcher A's experiment is that he doesn't know whether piercing the shell alone will affect the development of the bird. List three other potential sources of error in the experiment by Researcher A. *(3 marks)*

b) Why is it difficult to draw meaningful conclusions from the result of one embryo? *(1 mark)*

c) How did Researcher B improve on the experiment done by Researcher A? *(1 mark)*

d) What is the purpose of control group 1? *(1 mark)*

e) What is the purpose of control group 2? *(1 mark)*

f) What is the purpose of control group 3? *(1 mark)*

g) Why were several eggs used in each group? *(1 mark)*

h) Why can Researcher B be confident that any difference in results can be attributed to the presence of herbicide? *(1 mark)*

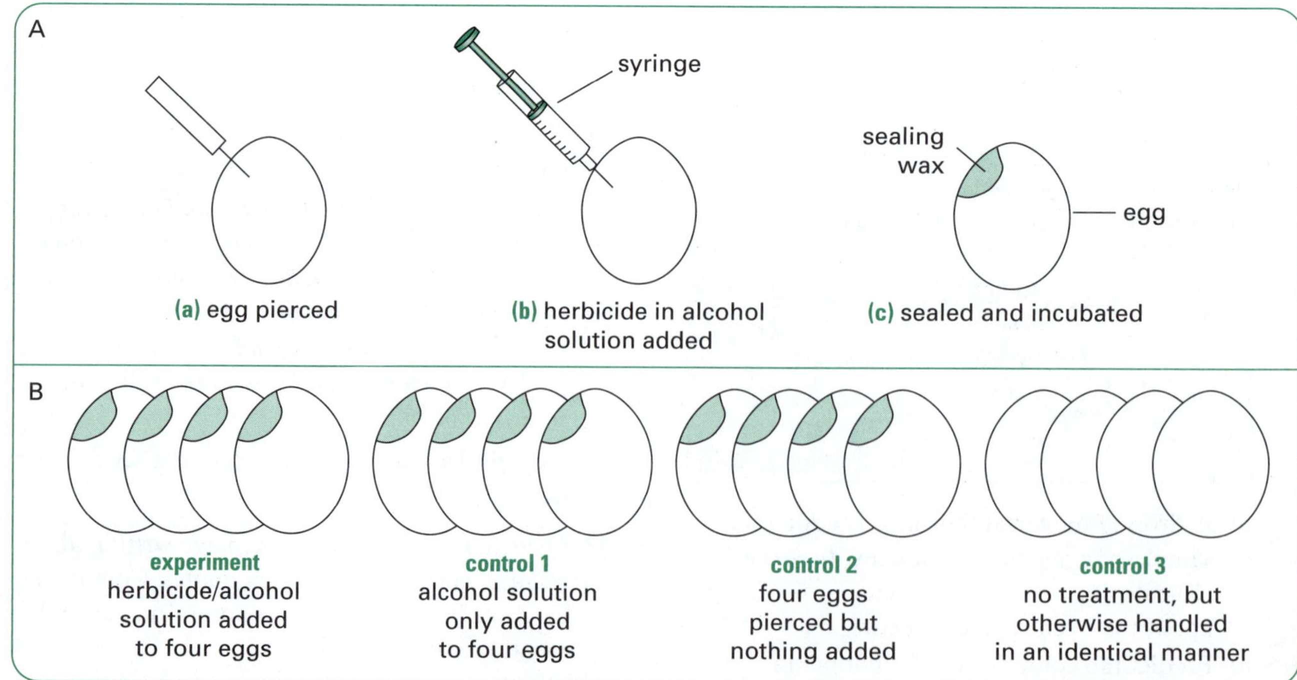

Figure 5.11 Herbicide and eggs

14. Janet was experimenting with bees in an attempt to work out if they were attracted to certain coloured flowers more than others. She had noticed that the bees seemed to like the yellow daisies in her garden more than the white ones. She set up the two situations shown in Figure 5.12. Part (a) shows the experiment using artificial light and part (b) shows sunlight.

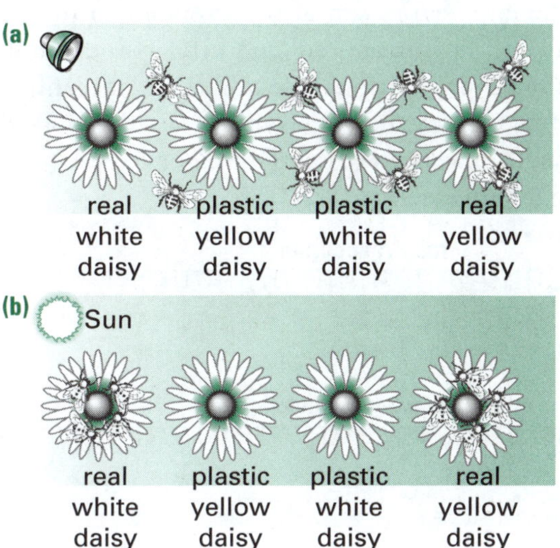

Figure 5.12 Daisies in artificial light and sunlight

From her observations, Janet hypothesised that something in the sunlight was important in attracting the bees to the yellow daisies rather than the white daisies. She decided to investigate the daisies in artificial light and in sunlight. She researched the wavelengths of the light produced from both sources (see Figure 5.13) and she used a special type of photographic film from her local scientist to photograph her daisies.

The results are shown in Figure 5.14.

a) What was Janet's original hypothesis? *(1 mark)*

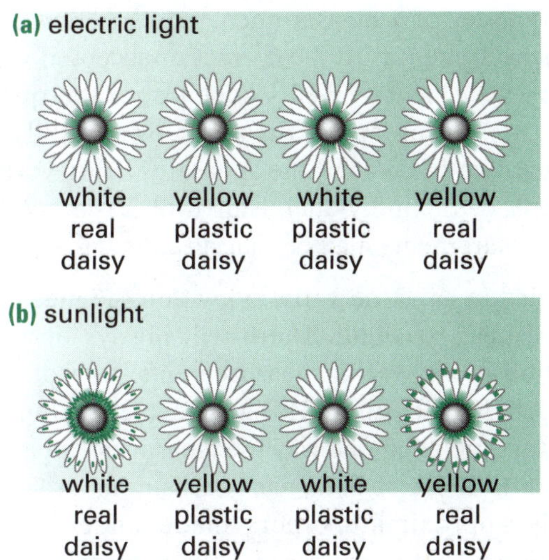

Figure 5.14 Appearance of daisies in electric light and sunlight as taken with the special film

b) Do her results support her hypothesis? *(1 mark)*

c) What did her experiment indicate? *(1 mark)*

d) What will Janet do now to further investigate the problem? *(2 marks)*

Go to pp. 253–254 to check your answers.

5.7 Improving accuracy and reliability

In science, data is obtained by taking measurements. This can lead to an explanation of why substances have different properties, or how objects behave. During an experiment other factors, which may affect the outcome, need to be controlled.

Accuracy and reliability are important in science, as they are in any research activity. If your results are not accurate or reliable, it will be difficult for anyone to take them seriously.

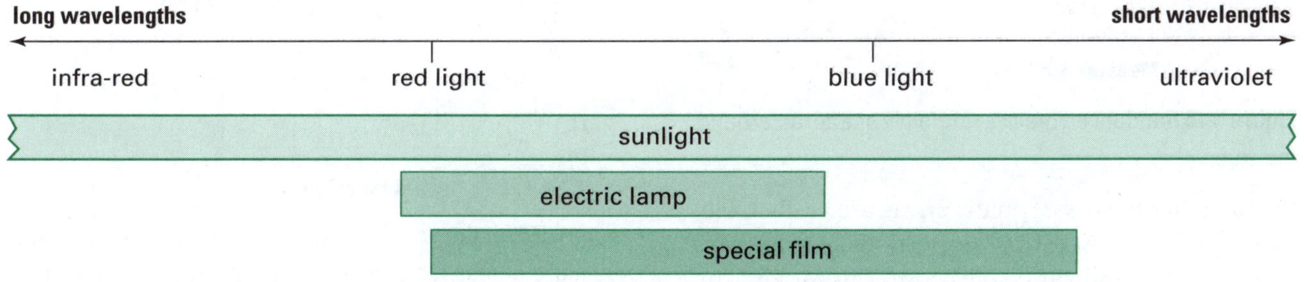

Figure 5.13 Wavelength range of sunlight, electric lamps and special film

The accuracy of a measurement refers to how close the measurement is to the correct or accepted value. Accuracy of measurements depends on the quality of the measuring apparatus and how carefully the measurement is taken. If the apparatus is not working properly, or the individual using it makes a mistake, any measurements may be inaccurate.

For reliable data, the variation within the measured values must be small. There will always be some variation in any set of measurements, regardless of what is being measured. This will be due to a number of factors, including the way the measuring apparatus is used. In a set of data, each measurement should closely approximate others made under similar conditions. That is, the results are repeatable; each time a measurement is taken it has approximately the same value. In laboratory experiments, you should perform at least five repeats. This makes the set of data more reliable as long as the variation between the repeats is small.

Random and systematic errors

All measurements are prone to random errors. Random errors in any measurement can lead to values being inconsistent when the experiment is repeated. Random errors, also called experimental errors, affect each measurement differently. Being random, they are inherently unpredictable and are usually scattered about the true value (see Figure 5.15). These errors are caused by unforeseen fluctuations in the readings of the instrument being used, or in how the experimenter interprets these readings.

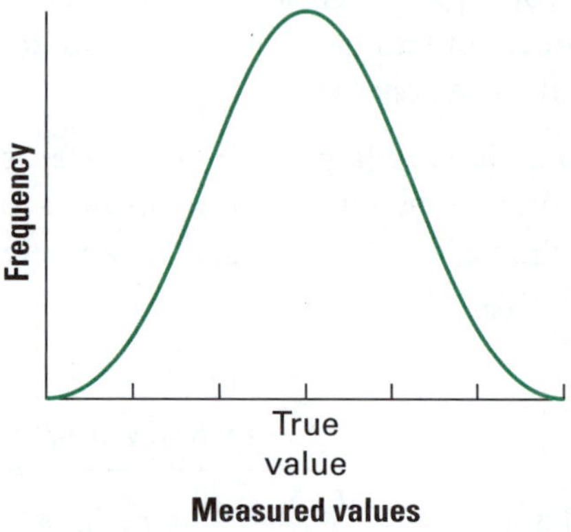

Figure 5.15 Random errors are randomly scattered around the true value.

On the other hand, systematic errors are predictable, and typically constant or proportional to the true value. One important source of this error, for example, may be the imperfect calibration of the measuring instrument (zero error). This is a constant systematic error. For example, if bathroom scales are not zeroed and show 2 kg before you step on them, then your mass will be measured 2 kg more than what it ought to be. Constant errors may be simply due to incorrect zeroing of the instrument. These errors can be very difficult to deal with because they will only be removed if their effects can be observed.

Systematic errors can also be related to the actual value of the measured quantity. For example, it could be proportional or a certain percentage of the measurement. For example, a ruler may show a length that is consistently greater or less than the true length (see Figure 5.16).

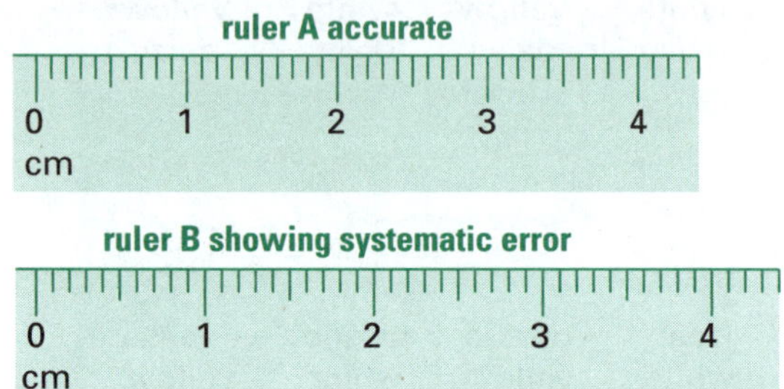

Figure 5.16 Ruler B will measure lengths a certain percentage less than the actual value.

A common way to remove systematic errors is through calibrating the measurement instrument. In voltmeters or ammeters, for example, the instrument error is very difficult to remove. These devices have built-in resistances, which can't be removed. But it can be minimised, such as providing a chart or graph to change the measured value to the true value (see Figure 5.17). On the other hand, removing the error of a thermometer is simpler. Just remove the old calibration and then carefully calibrate the thermometer again.

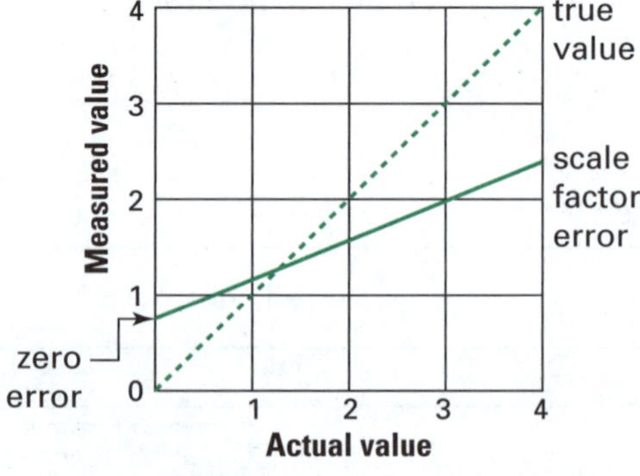

Figure 5.17 If the scale is out, the measured reading could be more or less than the actual value giving a systematic error.

Reliability in first- and second-hand investigations

Reliability is a measure of how much you can trust the conclusion drawn from your experimental results. If the results of some investigation can be repeated, then your conclusion may be considered reliable. Scientists constantly carry out similar and, sometimes even identical, investigations to check the reliability of someone else's experiments and see whether they can replicate the results. They aim to maximise the inherent repeatability or consistency of an experiment.

One way to check reliability is to compare your results with those of others. Reliability can be improved by carrying out repeat measurements; this is why you are often asked to obtain at least five measurements from your experiments.

If you repeat the experiment several times with different samples, and one generates completely different results from the others, then there may be something wrong with the sample you are using. If it is too extreme, you may want to remove it from the experiment and replace it with another. If you still keep getting wildly different results, then perhaps there is something wrong with the design of the experiment itself.

For many experiments, results will not be identical and there is always a chance that your sample group produces results clustering around the true value. Some very few may even be lying at one of the extremes. Using multiple sample groups will allow you to smooth out these extremes and generate a more accurate spread of results.

A first-hand investigation is one that you conduct yourself. A second-hand investigation is when you report on experimental results carried out by others. Researching on the internet or in the library is an example of a second-hand investigation. Even here you must check your sources to see that they are reliable.

For example, in 1989 a couple of research scientists announced that they had managed to generate heat from cold fusion at normal temperatures. These reports raised hopes of a cheap and abundant source of energy, and the announcement was widely reported in the media. Many researchers around the world tried to replicate the experiment, without success. Soon there were a number of reasons as to why it was unlikely to occur, experimental flaws were discovered in the original experiment and there were sources of experimental errors. Whether the researchers lied, or genuinely made a mistake, is uncertain but their results were clearly unreliable. So it is important to check your results with other, independent sources to see that your gathered information is correct.

The following points should be noted when gathering information from secondary sources.

- Use a wide range of resources. Do not just rely on the internet. Not all the information on the internet is accurate or correct. Much of it is not peer reviewed (checked by other experts). CD-ROMs, periodicals, library books and even newspapers are valuable sources of information.
- Not all the gathered data will be reliable. You need to check reliability by comparing it with other sources. Science articles that appear in respected journals and that have been peer reviewed are more likely to be reliable than an article that is self-published. Articles written by creation scientists are not reliable as they do not follow the scientific method. Some companies may be unethical in that they may publish data that is favourable to them.

Reliability simply describes the repeatability and consistency of an experiment or investigation. Accuracy refers to how far the measurements are from the true answer. Validity refers to whether or not a chosen procedure measures what is intended to be measured. Figure 5.18 provides a way of distinguishing between the terms accuracy and reliability. Each dot on the diagram represents an experiment result.

5.8 Presentation of experimental data and analysis of trends in experimental results

Scientists and other researchers use a variety of methods to present experimental data and analyse trends in experimental results.

Constructing models

Figure 5.18 shows a model used to compare accuracy and reliability. Generally, a model is anything used

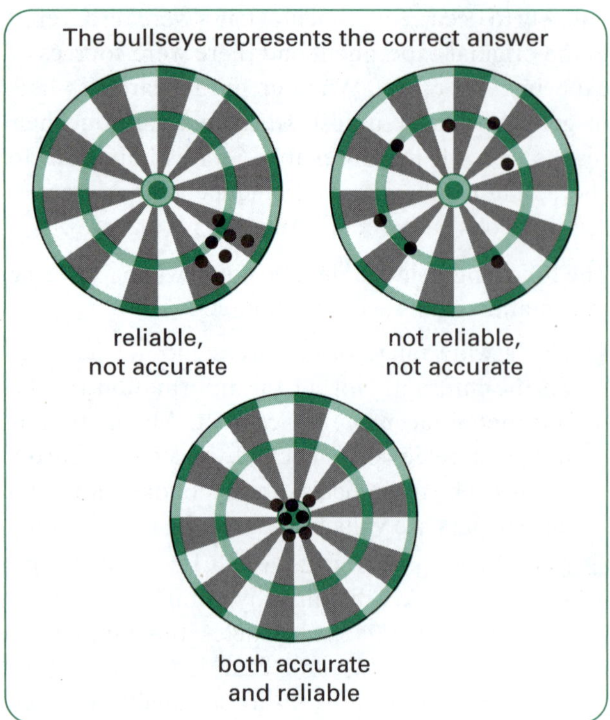

Figure 5.18 Comparing accuracy and reliability

in some way to represent something else. Some models are physical objects, such as a scale model of a building or fighter plane that can be assembled, and can even be made to work like the object it represents. Scientists use models as a way to extend human thought processes. All models are simplified views of reality but, even though they are not entirely correct, are nevertheless extremely useful. They are used to help us know and understand the subject matter they represent. Models are often used when it is either impractical or impossible to create appropriate experimental conditions where scientists can directly measure outcomes.

The following are some characteristics of a good model.

- It is based on a given set of observations.
- It must be able to explain as many characteristics of these observations as possible.
- It should be as simple as possible.
- It should be able to explain phenomena other than the ones used to develop the model in the first place.
- It should enable predictions to be made, which can then be tested.

Centuries ago there was strong debate over whether the Earth was flat or round. This led to differences in predictions, depending on the model used. A spherical-Earth model, for example, would predict you could not fall off the end of the Earth. Like most educated people of the time, Christopher Columbus understood that the Earth was spherical and this allowed him to complete four voyages across the Atlantic Ocean.

The spherical-Earth model also explained the observation that a boat appears to 'sink' as it goes over the horizon. Over large distances the curvature of the Earth gives the illusion of a sinking boat as seen in Figure 5.19. The last part of a departing boat seen is the top of the mast. This happens regardless of the direction the boat is moving, and is downwards in all directions.

Figure 5.19 The sailboat in the distance appears to have dipped below the horizon.

But this model can also explain phenomena which appear to be different from the ones used to develop the model in the first place. For instance, a lunar eclipse occurs when the Earth passes between the Sun and the Moon. The spherical-Earth model predicts that the shadow of Earth will be round as it passes across the face of Moon; it is.

No scientific model is ever totally complete. Sometimes new observations conflict with the model's predictions, so something must give as either the data or the model is incorrect. While Columbus picked the right model, he underestimated distances and how long it would take him to reach India. No ship in the 15th century could carry enough food and fresh water for such a long voyage, and the dangers in navigating through uncharted waters were formidable. Later, the model was refined with a better estimate of the Earth's radius, allowing for improved predictions about distances to locations on its surface.

All models have limitations; there is no model that can possibly explain every detail of scientific phenomena.

For example, while we can use a globe (a model) to estimate the distance from one side of Australia to the other, this would not be exact. A globe is smooth (spherical) but crossing the Australian terrain is not. While these features could be added to make it more realistic, it would complicate the model.

Using diagrams to present data

Mathematical data can be presented graphically. A variety of different graphs are used to organise and represent information, allowing trends and patterns to be identified. Sometimes patterns are not immediately obvious when looking at a table of numbers until a graph is drawn.

A line graph is a very common method of plotting experimental data. This is generally used where the information presented is continuous. The following are rules you should remember when graphing (see also Figure 5.20).

1. Use grid paper to draw your graph. Alternatively, you can draw it on a computer using appropriate software, but include a feint grid in the background.
2. Your graph should occupy at least 80% of the available grid space.
3. The variable that you control (i.e. the independent variable) is placed on the horizontal axis. Number the grid lines along this axis. Label this axis and its units.
4. The variable being observed (i.e. the dependent variable) is placed on the vertical axis. Number the grid lines along this axis. Label this axis and its units.
5. Choose a suitable scale for each axis to ensure that the graph fills most of the grid space.
6. Plot the data points as small crosses (×) or dots (•) using a pencil.
7. Draw the 'line of best fit' through the data points. This line will not necessarily pass through all the points but should be as close as possible to them.
8. Write a title above the graph.
9. Once a line graph is drawn, new data can be extracted from the graph in two ways.
   a) Extrapolation. The line can be extended beyond the plotted points and new values of the dependent variable predicted from values of the independent variable that were not actually measured experimentally. This process assumes that the line shape does not change.
   b) Interpolation. New values of the dependent variable can be extracted from the graph for values of the independent variable between the measured data. This process is more accurate than extrapolation as it is within the range of measurements.

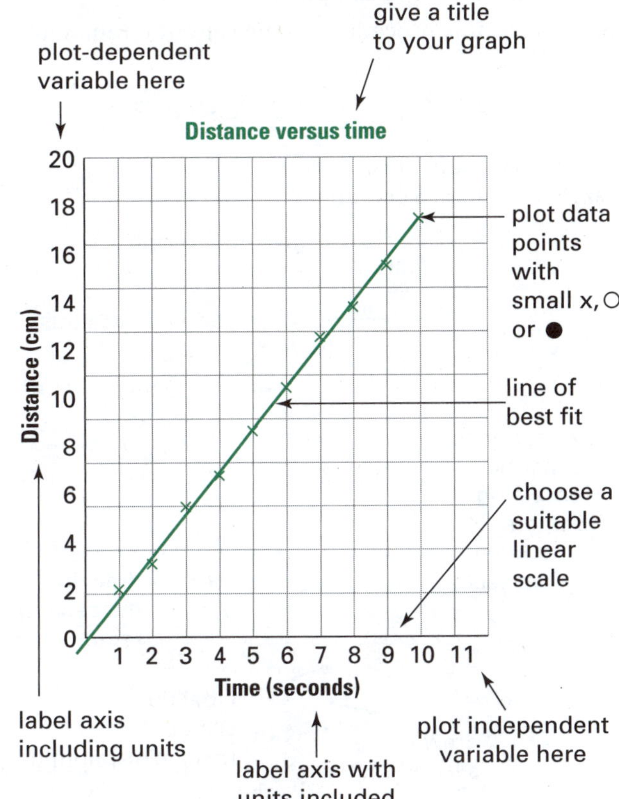

Figure 5.20 Drawing a line graph

When the data collected is discrete (i.e. not continuous), you should not represent the data as a line graph. Continuous data can take any of an infinite number of values between whole numbers. Discrete data is best represented using a column graph, pie chart or bar graphs.

Mathematical data may also be presented as a pie graph. For example, the percentage composition of a mixture can be presented this way. The following rules should be used to draw a pie (sector) graph (see also Figure 5.21).

1. Use a compass to draw a suitably sized circle that will fit on your page. A circle of radius 5 cm may be a good option.
2. Convert the percentage data to degrees. Thus, if one component is 30% then $\frac{30}{100} \times 360 = 108°$.

(See the next section for more detail.) Repeat this with the remaining percentage data.

3. Draw a radius on the circle using a ruler and pencil.
4. Use your protractor to measure the first angle (e.g. 108°) and draw a second radius to produce the first sector of the pie graph.
5. Repeat this procedure to measure the remaining sectors.
6. Label each sector and give the graph a title.

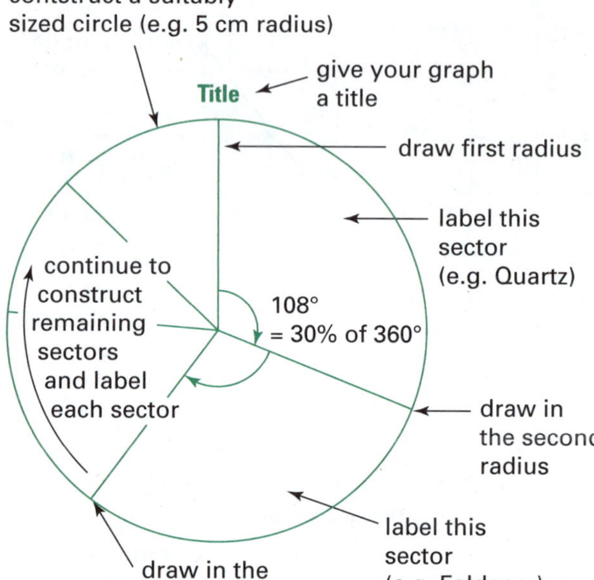

Figure 5.21 Drawing a pie graph

Diagrams and maps often have different scales. The scale should be stated on the map or diagram as shown in Figure 5.22. The scale should be easy to use.

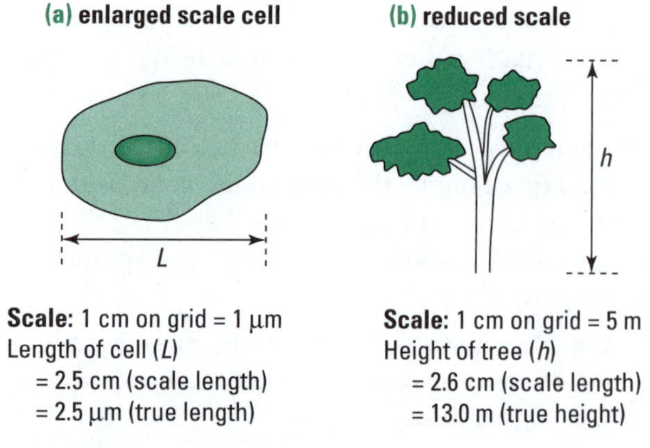

Figure 5.22 Using scales on maps and diagrams

Computer software can be used to generate graphs, tables, flow charts and diagrams. Graphing software often allows users to produce a variety of different graphs of the same data, such as those shown in Figure 5.23.

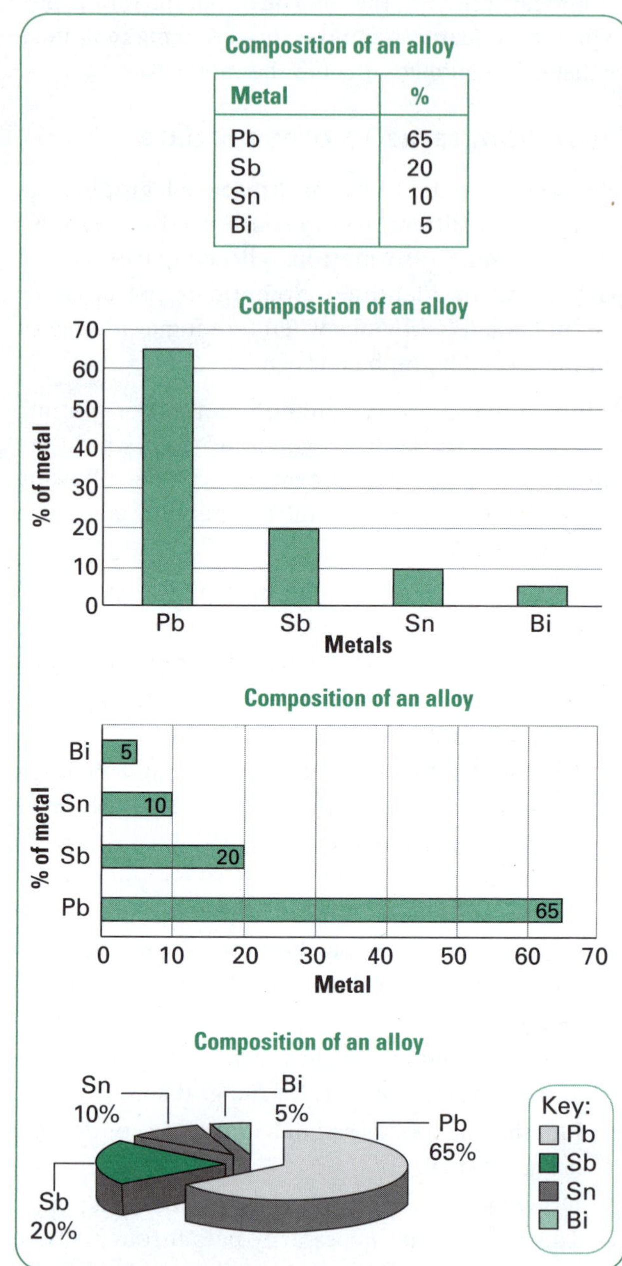

Composition of an alloy

Metal	%
Pb	65
Sb	20
Sn	10
Bi	5

Figure 5.23 Computer graphing

Mathematical analysis of data

In many experiments you are asked to make several measurements, and then take an average. The reason for this was discussed earlier. For example, suppose you weighed four samples and their masses were 4.56 g, 4.40 g, 4.44 g and 4.51 g.

The range of this data is *highest value – lowest value* = 4.56 – 4.40 = 0.16 g.

The mean (average) value is

$$\frac{4.56+4.40+4.44+4.51}{4}=\frac{17.91}{4}=4.4775 \text{ g}.$$

However, notice that each measurement was made correct to two decimal places, so your average value cannot be more accurate than that. You should report your mean as 4.48 g. That is, round off your answer (in this case) correct to two decimal places.

Suppose your measurements were 4.56 g, 4.40 g, 4.44 g and 7.51 g. Can you see that the last measurement is not close to the other values? An outlier is an observation that is numerically distant (unusually large or small and out of place) from the rest of the data. If you were now to take an average:

$$\frac{4.56+4.40+4.44+7.51}{4}=\frac{20.91}{4}=5.2275=5.2.$$

By including this outlier, you get an average that doesn't represent (is not somewhere near the middle) of the other three values.

Outliers can occur by chance; some data points will be further away from the sample mean than what is considered reasonable. This can be due to incidental systematic error. The best thing you can do is to take another measurement and ignore the outlier. So, for example, 4.56 g, 4.40 g, 4.44 g, ~~7.51 g~~ (deleted measurement) and 4.52 g (the new measurement) would lead you to ignore the outlier and work with the other four values. If you make careful measurements, you should not often come across outliers.

Graphical analysis using spreadsheet applications

Many of the graphs shown in this chapter can be drawn using spreadsheet applications such as Microsoft Excel. This can make your report look more appealing while still clearly presenting your information. There are other graphical programs, many of which are free and can be downloaded from the internet. Computer programs can present a variety of graphs and tables in many different and interesting orientations. This can save much time from manually drawing the graph, although you should still understand how to create one.

1. Open Microsoft Excel.
2. Enter the values you wish to draw (see Figure 5.24).
3. Click on one of the chart icons at the top, and then follow instructions to create your graph.

You may need to practice to obtain a graph that contains all the features you want.

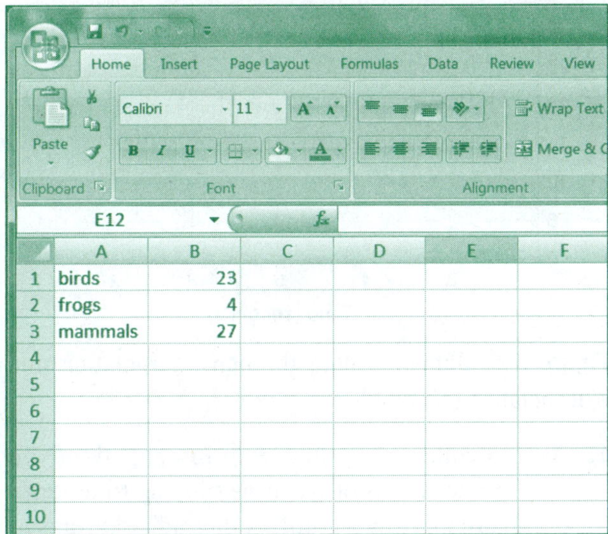

Figure 5.24 Using Microsoft Excel

Analysing trends

The following points should be noted when processing information that has been gathered from various first-hand and second-hand sources.

- Where applicable, use mathematical procedures to analyse the data. For example, if you have collected data on the velocity of a ball falling from rest as a function of time, then the formula $v = u + at$ can be used to determine the acceleration of the ball. (There is no need for you to remember this formula; the important idea is for you to follow the procedure used to obtain the acceleration of the ball.) When v is plotted as a function of t, then your data should fall on a straight line that passes through the origin. The slope of this line will be equal to the acceleration a.
- Data can also be determined from a graph by interpolation or extrapolation as shown in Figure 5.25. When extrapolating a graph, it is important to determine whether the extrapolation is valid.
- Reliability is improved by making repeated measurements (at least five). The collected measurements should then be averaged.
- When conducting a fair test, your data will either support or discount the hypothesis. If the hypothesis is supported, it does not mean that it is necessarily correct, as further experiments may be contradictory and not support it.

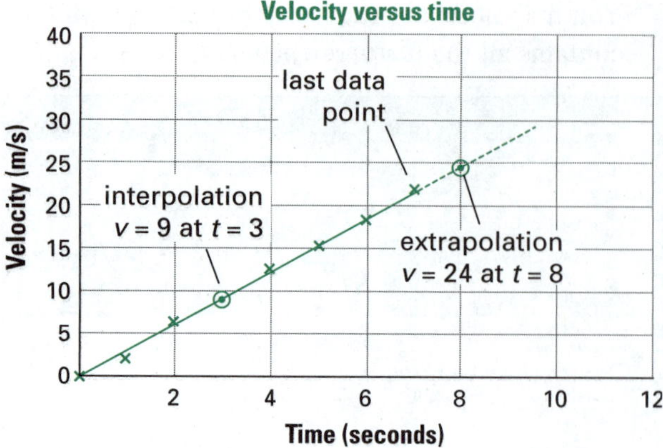

Figure 5.25 Obtaining data through interpolation and extrapolation of a graph

- As you collect data during an investigation, you may notice trends or patterns starting to emerge. If you are plotting a graph of the radii of atoms as a function of their atomic weight, you will begin to notice that there is a repeating pattern where the radius drops from Group I to Group VIII across any one period. This is shown in Figure 5.26.

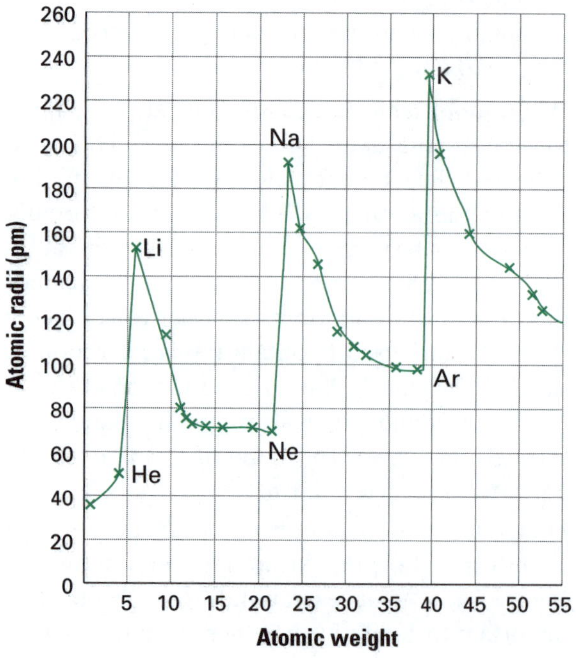

Figure 5.26 Patterns in data emerging when data is processed

- Once you have processed your results, you will need to draw valid conclusions. The conclusions must relate directly to the experimental results. You cannot make a conclusion about information that you have not gathered or measured.
- As part of the scientific method, you will then provide plausible explanations of the phenomena being investigated. These possible explanations will be written in the discussion section of the report. There is also an opportunity at this point to justify any inferences you have made about the collected data. In the discussion, you may make some predictions that will lead to further experimentation.
- As more experiments are completed, you may be able to make some generalisations about your area of investigation.

5.9 Explanations and conclusions

On 17 July 1998 what was thought to be a 7.1-magnitude earthquake occurred in the sea 25 km off the coast of Papua New Guinea. Soon, a tsunami 10 to 15 m high pounded the shore killing over 2200 people and displacing around 10 000 people more.

Local people had different ideas on what caused it, as shown in Figure 5.27.

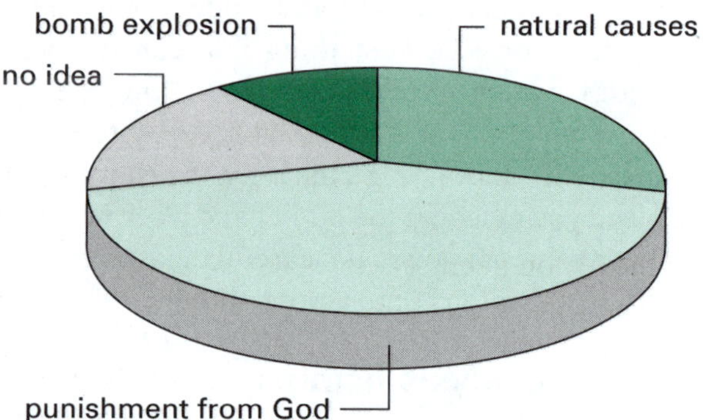

Figure 5.27 How local people interpreted the events

But there were inconsistencies.

- A 7.1 earthquake is not typically strong enough to create such a large tsunami.
- The tsunami took longer than usual (around 20 minutes) to reach the shore.
- A loud rumbling sound lasting over a minute heard by locals is not typical of earthquakes, or its after-shocks.

This led scientists to do more research and look for an alternate explanation. Besides interviewing people and looking at surface damage, they conducted surveys of the ocean floor around where this earthquake occurred. A bowl-shaped depression in the sea floor slope and recent movements were observed. This confirmed their theory of an undersea landslide.

Case study 1: Animal extinctions

There are 23 birds, 4 frogs and 27 mammal species which are believed to have become extinct since European settlement occurred in Australia. This information can be presented as a pie chart (sector graph) or divided bar graph. Both of these are useful for showing parts of a whole. Figure 5.28 is a pie chart of this data.

Vertebrate extinctions in Australia since 1788

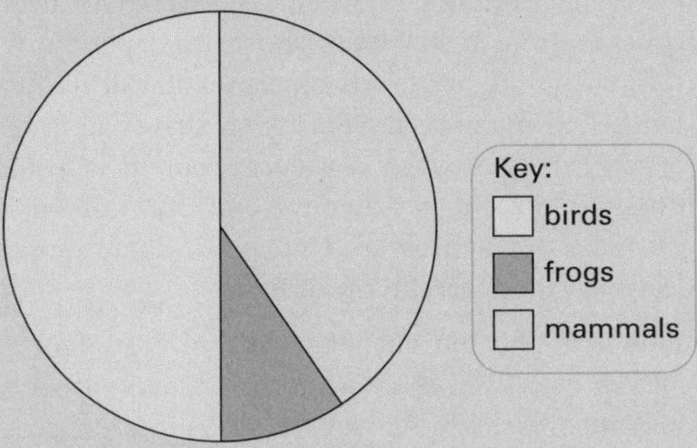

Figure 5.28 Pie chart of vertebrate extinctions

In order to calculate the central angle (the angle you measure at the centre of the circle) for each of these, you need to find:

$$\frac{\text{number in the category}}{\text{total number}} \times 360°$$

For example, 23 birds became extinct out of (23 + 4 + 27 =) 54 vertebrates. So the central angle for birds is $\frac{23}{54} \times 360° = 153°$ (to the nearest degree).

Do the same for the other categories.

This information can also be shown as a divided bar graph, like those in Figure 5.29.

(a) Vertebrate extinctions in Australia since 1788

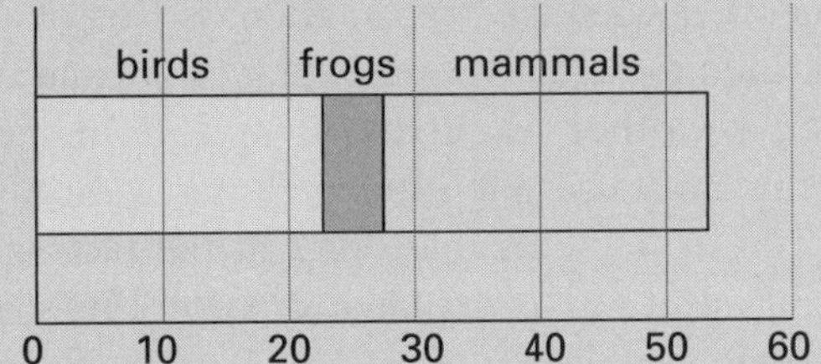

(b) Vertebrate extinctions in Australia since 1788

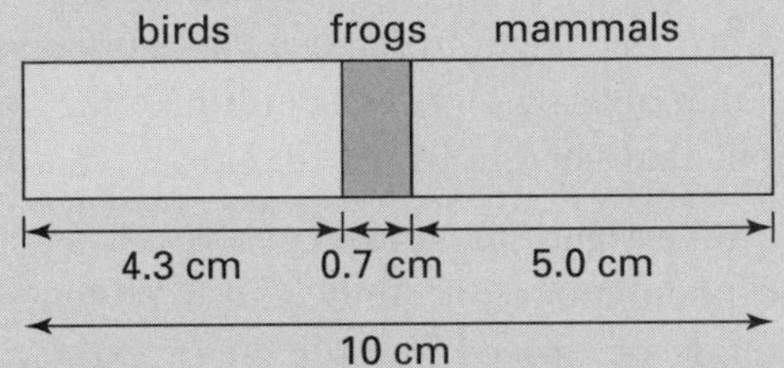

Figure 5.29 Two ways to show this information using divided bar graphs

In a divided bar graph, a scale can be drawn and each of the bars representing the category placed next to each other. For example, 4 frogs became extinct, so its bar begins at 23 (where the bird bar ends) and ends at 27.

Another way is to draw a bar, say, 10 cm long and determine what length each category occupies. This is similar to the sector graph:

$$\frac{\text{number in the category}}{\text{total number}} \times \text{length of whole bar}$$

For the birds this is $\frac{23}{54} \times 10 = 4.26 = 4.3$ cm.

Do the same for the other categories.

It is now generally accepted that the tsunami was caused by an undersea landslide and not by an earthquake. While the tsunami was originally thought to have been created by a 2-m vertical drop in the Pacific Plate, it is now known that the earthquake caused a huge underwater landslide. It was this that generated the tsunami.

And the implications? This event has changed scientists' ideas about small earthquakes producing undersea landslides, making them realise that landslides can be more hazardous. Undersea earthquakes near the shore are normally felt by people on land, and detected by seismic stations. But undersea landslides give no warning and can generate 'surprise' tsunamis.

Alternative explanations

This example shows that scientists may interpret the same experimental data differently. Science advances through legitimate scepticism. Asking questions and querying other scientists' explanations is part of any scientific inquiry. Scientists evaluate the explanations proposed by other scientists by examining and comparing evidence, identifying faulty reasoning, pointing out statements that don't fit the facts, and suggesting alternative explanations for the same observations.

Different kinds of questions suggest different kinds of scientific investigations. So depending on what scientists are investigating, and what emphasis is placed on that investigation, will determine the kinds of questions that need to be asked and answered.

Scientific investigations sometimes result in new ideas and phenomena for study, and may generate new methods or procedures for an investigation. Applying new technologies to existing data can lead to new conclusions.

Cause and effect relationships

Cause and effect refers to an action or event that produces a certain response to the action in the form of another event. In other words, a cause is something that makes something else happen. Out of these two events, the cause happens first and the effect occurs afterwards.

Connecting words are often used to link the cause and effect. These include 'because', 'the reason for', 'therefore', 'consequently', 'since', 'as a result', 'due to' and 'thus'. Here are some examples.

- *Because* of the moon's gravitational attraction, there are ocean tides.
- The moon's gravitational attraction *affects* the ocean tides.
- *Since* the moon has a gravitational attraction, so it follows that there are ocean tides.

There are several factors that cause and effect have in common.

- Cause and effect must be linked in space and time.
- The cause must be prior to the effect.
- There must be some constant relationship between the cause and effect.
- The same cause always produces the same effect.
- Where several different objects produce the same effect, there must be something common among them.

In science, especially physics and chemistry, it is reasonably easy to establish causality. A good experimental design neutralises any potentially confounding variables. The only difference between a control group and an experimental group is one variable with the aim being to determine whether that one variable produces different results.

But cause is not always obvious. For example, there is a high correlation between depression and alcohol consumption. It has been shown that people who suffer from higher levels of depression drink more. Does this mean that depression drives people to drink? Since alcohol is a depressant it is equally possible that heavy alcohol consumption makes people more depressed. That is, the same research leads to two different interpretations.

This is a 'chicken and egg' type of argument and makes establishing causality a difficult aspect of scientific research. As well, it allows interest groups to manipulate the results. For instance, alcoholic drink producers and sellers could claim that alcohol is not a factor in depression, and that depression is a problem that should be tackled by society. On the other hand, anti-drinking groups could claim that alcohol is harmful as it leads to depression and make a case for harsher drinking laws.

Critical thinking strategies

When you evaluate your work and draw conclusions, you will use critical thinking strategies. Here are some important points about these strategies.

- Evaluate which of a number of possible strategies is the best approach to solve a problem. For example, if you plan to study the ecology of a local swamp, will you use photographic records, physical and chemical measurements, statistical methods or recounts from personal observations? Some strategies may be more useful than others.
- Be flexible and willing to change your view when confronted with contradictory evidence. Your hypothesis is not necessarily correct and you need to be prepared to change it if collected evidence discounts it.
- Think creatively when solving a problem.
- Distinguish between facts and opinions. Facts are pieces of information that can be tested to

check that they are correct. Opinions are personal attitudes and feelings that individuals have about an issue. In science, opinions carry no validity.

- Use logical reasoning when interpreting collected data. Poor reasoning is responsible for incorrect conclusions.

5.10 Report presentation

Collected and processed data must be presented to an audience in an appropriate way. The presentation of an experimental report is different to the way you would present an exposition. It is important to select the appropriate text type to present information. Note that the report on your investigation may be made up of several text types.

Different types of text

The following are some different types of text.

- Procedure. This type of text is used to describe how an experiment that has not yet been done is to be performed. Present tense is used. The steps of the procedure are usually numbered or presented as sequential bullet points.
- Procedural recount (experimental record). This type of text is used to record the procedure of the experiment that has already been performed. It is also used for the results and conclusion. Past tense is used in a recount.
- Report. This type of text is used to present information on a particular topic (e.g. the different types of frogs in Australia) that has been investigated using first-hand and/or second-hand data. Factual and descriptive information is presented.
- Discussion. This type of text identifies issues and presents points for and against. The report that you write for an investigation of a problem will have a discussion at the end. The discussion will often include explanations where causes and effects are discussed.

Including an abstract in a report

An abstract is an abbreviated version of a final report. Think of an abstract as a condensed summary of the entire paper. It is often limited to around 250 words, but there is no rule on this. An abstract appears at the beginning of the report. It lets people quickly decide if they want to read your entire report. In other words, it is like an advertisement for what you've done.

An effective abstract:

- uses one or several well-developed paragraphs that are well written, concise and able to stand alone
- contains an introduction, body and conclusion
- follows the time line of the report
- has a logical connection between the material
- simply summarises the report and adds no new information
- can be understood by a wide audience.

The following are some tips for writing an effective report abstract.

- Read the report, keeping the idea of abstracting in mind.
- Write a rough draft. Do not just copy key sentences from your report.
- Revise your rough draft. This is an opportunity to correct any weaknesses, remove irrelevant information, insert important information originally left out, cut down the number of words and correct grammatical and other errors.
- Carefully proofread the final copy.

When writing abstracts, it is important to be brief and state only what is relevant. A successful abstract is compact, accurate and self-contained. Writing a good abstract is hard work, but will repay you as it is easier to read and will entice other people to read your efforts.

Correct methods of referencing

Referencing is a standardised method of acknowledging the information sources you have used in your assignments or written work. Its two main purposes are to:

- acknowledge the source of the information if you are using the ideas of others
- allow the reader to trace the source, to verify your work and/or to obtain more information.

Referencing:

- shows that you are maintaining standards of writing
- acknowledges the work of other writers and gives them due respect
- shows that you have read and considered the relevant sources of information
- makes your work more credible.

At the end of your assignment, place a list of the references you have mentioned, used for information or quoted in the text. Arrange this in alphabetical order based on the author's surname.

There are no absolute rules for referencing. However, the essential information that needs to be included is the author(s), the title and details concerning the imprint (date, publisher, date of publication).

The following are some examples of referencing.

- Book, 1 author:
  Stamell, J, 2008, *Excel HSC Chemistry*, Pascal Press, Sydney.
- Book, 2 or more authors:
  Thickett, G, Stamell J & Thickett, L, 2000, *Science Tracks 9*, MacMillan Education, Sydney.
- Journal article:
  Stamell, J, 2000, 'Dependent vs independent variables', *Reflections*, 25(2), p. 26.
- Newspaper article:
  Lennon, T, 2011, 'Early communication', *The Telegraph*, 16 August, pp. 28, 45.
- Web document:
  Helmenstine, A M, 2011, *What is distillation?*, http://chemistry.about.com/cs/5/f/bldistillation.htm.

Often you don't need to use the whole book, only several pages from it. So it may be necessary to cite page numbers. Do this by writing the numbers as the final item of your referencing. For example, p. 34 for one page, pp. 34–55 for more than one page. If the author is unknown, start with the title of the work and place the year immediately after it.

Presentations using digital technologies

Digital technology is everywhere. What was once thought to be the realm of science fiction has now become science fact. Mobile phones, MP3 players and digital cameras are part of this digital revolution.

Digital technology allows for a variety of information, including images and sounds, to be easily recorded and transmitted electronically. The technology is continually developing at a rapid rate. This is an improvement on the static presentation of paper alone.

As well, digital technology allows for a simple and seamless method of manipulating those digitised images and materials making it easy to rearrange, add to or subtract from them, and to organise them far more easily and efficiently. This provides you with a variety of ways to present information easily, but also makes it easier to plagiarise other people's works.

A digital presentation can take many forms. Microsoft PowerPoint, for example, has been a popular tool used for over two decades to help presenters provide image-rich and visually appealing multimedia presentations. Another form of presentation is using video clips from YouTube to demonstrate a process. In recent years, technology tools have become available online to make presentations even more engaging, interactive or able to reach a diverse audience.

Presenting information using these technologies makes your presentation more appealing than just handing out sheets of paper. Consider the appropriate use of digital technologies when putting together your presentation.

Experiment 3

Stretching a rubber band

Aim

- To measure the tensile strength of samples of different types of rubber bands

Background

Tensile strength measures the force required to pull something such as rope, wire, rubber or other material to the point where it breaks. A material's tensile strength is the maximum amount of force that it can take before it fails (breaks). In this experiment, the largest amount that a rubber band can be stretched before breaking can be used as a measure of its tensile strength. It could also be determined by recording the total mass added to the carrier just before the rubber band snaps.

Method

1. Obtain a new packet of flat rubber bands of the type that are about 0.5 cm wide.
2. Take several samples of rubber bands from the packet and cut them into 3- or 4-cm lengths.
3. Clamp individual lengths as shown in Figure 5.30, so that initially exactly 2 cm of rubber is exposed between the clamps.

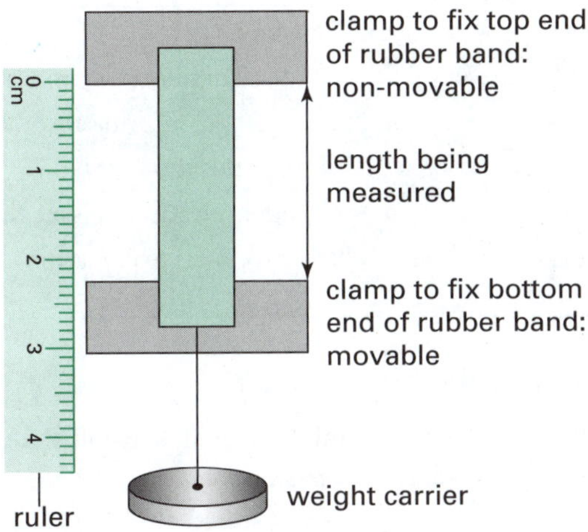

Figure 5.30 Experiment to measure the stretch of a rubber band

4. On the floor just below the set up, place a cushion, a bucket of sand, foam rubber or other soft material to lessen the impact when the rubber band fails.
5. Hang a mass from the bottom of each sample and record the mass and new length.
6. Increase the mass added in regular increments and continue to measure the length until the rubber snaps. You will need to keep track of the expansion as, when the rubber band snaps, it will be very quick.
7. Repeat the experiment with your other pieces of rubber band, making at least five measurements. For example, you might like to leave some of the same type of rubber bands in a glass jar exposed to the Sun for several days and see whether the breaking point remains the same.
8. Repeat the experiment with other types of rubber band.

Analysis

Typical results are shown in Table 5.6.

Table 5.6 Typical results of the rubber band experiment

	Sample 1	Sample 2	Sample 3	Sample 4	Sample 5
Breaking point (cm)	12.2	11.5	14.6	15.3	13.7

Range of values: 15.3 – 11.5 = 3.8 cm

Average breaking point:

$$\frac{12.2+11.5+14.6+15.3+13.7}{5} = 13.5 \text{ cm}$$

Questions

1. Why is it important that the rubber bands come from the same packet?
2. Does it matter if older rubber bands are used, along with newer ones?
3. What factor is being investigated? What is the outcome?
4. Why are several samples used and not just one?
5. List some of the factors that must be controlled in this experiment.
6. What might result if other factors are not controlled?
7. Account for the variation in stretching length in this experiment.
8. Suggest another experiment that can be done in the home to determine the tensile strength of a common material.

Go to p. 252 to check your answers.

Conclusion

Write a suitable conclusion.

Go to p. 253 to check your answer.

Test yourself 2

Part A: Knowledge

1. One major problem of using information obtained from the internet is that *(1 mark)*
   A there is usually little relevant information on science topics.
   B not all the information is reliable as it is not always peer reviewed.
   C the author's name is not recorded on the articles.
   D the information in many articles is untrue and unreliable.
2. A student gathers second-hand data to write a report on the marsupials found outside Australia. Which of the following article titles would *not* be useful for the student? *(1 mark)*
   A Kangaroo fossils in Queensland
   B Types of South American mammals
   C Evolution of mammals in New Guinea
   D Fossils of mammals around the world

3. A suitable type of text to use when recording the experimental method used in a series of investigations would be *(1 mark)*
   A narrative.
   B discussion.
   C procedural recount.
   D exposition.

4. Which of the following is *not* an example of a critical thinking strategy? *(1 mark)*
   A logical reasoning
   B willingness to change one's point of view when new evidence is presented
   C evaluating possible strategies
   D selecting experimental data that supports the hypothesis

5. Which of the following is *not* a satisfactory method of citing a science reference? *(1 mark)*
   A Waterhouse, G. R. (1846–48) *A Natural History of the Mammalia*, 2 vols, Bailliere, (SU) London.
   B Russell, T & Watt, D, 1990, *Evaporation and Condensation*, Science Processes and Concept Exploration (SPACE) Research Report, University of Liverpool Press, Liverpool.
   C Mason, Stephen, New York. F. *A History of the Sciences* New. Collier.
   D Fermi, Laura (1954) *Atoms in the Family*. University of Chicago Press. Chicago.

6. Complete the following restricted-response questions using the appropriate word. *(1 mark for each part)*
   a) The weight of the boy was measured and found to be 459 .... (units).
   b) Once the results of an investigation are processed, the scientist will need to draw .... conclusions.
   c) Tables and scaffolds are ways in which .... can be organised.
   d) Once information has been gathered from secondary sources, it needs to be .... and summarised.
   e) The limit of reading of a measuring instrument is normally .... a scale division.

7. Use the code letters to match the terms or phrases in each column. *(1 mark for each part)*

Column 1	Column 2
A hypothesis	F factual information
B database	G testable information
C report text type	H organisation of data
D facts	I identification of issues
E discussion text type	J proposed explanation

Part B: Skills

8. Figure 5.31 shows a number of scales. Read the value on each scale. *(3 marks)*

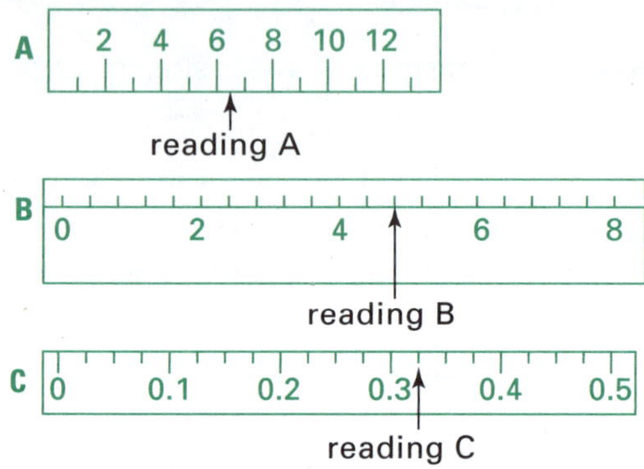

Figure 5.31 Reading scales

9. Table 5.7 shows the percentage of different components of a sandy loam soil. Use this information to construct a pie graph. *(3 marks)*

Table 5.7 Components of sandy loam soil sample

Soil type	% sand	% silt	% clay
Sandy loam	60	30	10

10. The data in Table 5.8 was collected from an experiment involving light refraction in a plastic prism.

Table 5.8 Angles of light refraction in a plastic prism

Angle of incidence (*i*) (degrees)	Angle of refraction (*r*) (degrees)
20	7
30	
40	13
50	16
60	19
70	22

   a) Plot a line graph of the incidence angle (horizontal axis) versus the refracted angle. Draw the line of best fit. *(3 marks)*

b) Use your graph to interpolate the refracted angle when the incidence angle is 30°. *(1 mark)*

c) Extrapolate your graph to determine the incidence angle when the refracted angle is 25°. *(1 mark)*

11. Use the scale on Figure 5.32 to determine the true length of the insect. *(1 marks)*

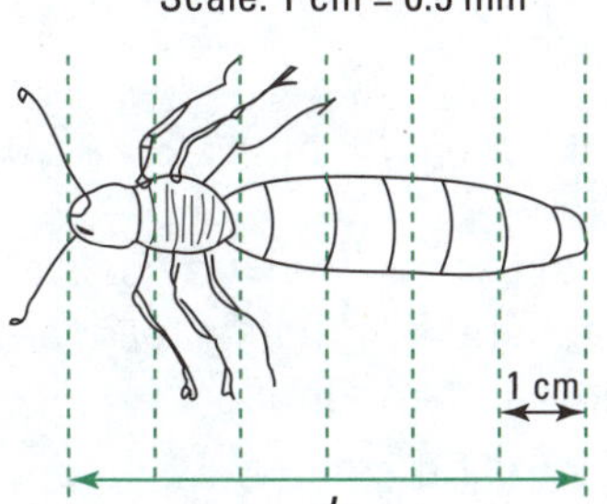

Figure 5.32 Insect drawn to an enlarged scale

12. Figure 5.33 shows the graph of a uniformly accelerated body.

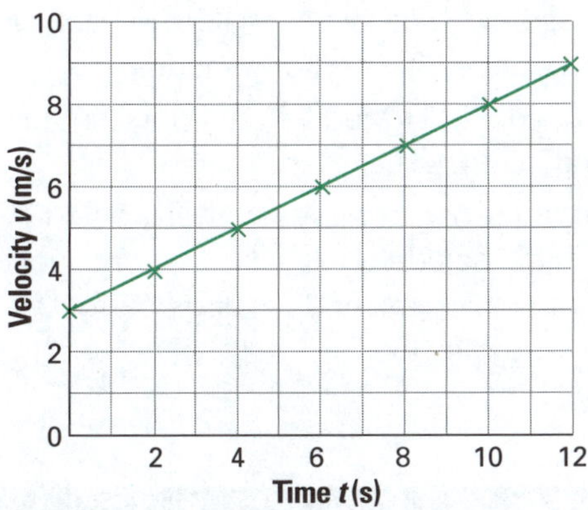

Figure 5.33 A uniformly accelerated body

a) Show that the graphical data satisfies the equation $v = u + at$, where u = initial velocity, v = final velocity, a = acceleration and t = time. *(2 marks)*

b) What is the initial velocity of the body? *(1 mark)*

c) What is its acceleration? *(1 mark)*

13. Mark uses a measuring cylinder that has an incorrectly marked scale. He draws the graph shown in Figure 5.34 to convert from his measured reading to the true volume.

a) If Mark measures 12 mL using his measuring cylinder, what is the true volume? *(1 mark)*

b) Rachel reports a true value of 12 mL. What reading should Mark's cylinder read? *(1 mark)*

c) Write a simple equation linking true value to the measured value. *(1 mark)*

d) Is this an example of a random error or a systematic error? *(1 mark)*

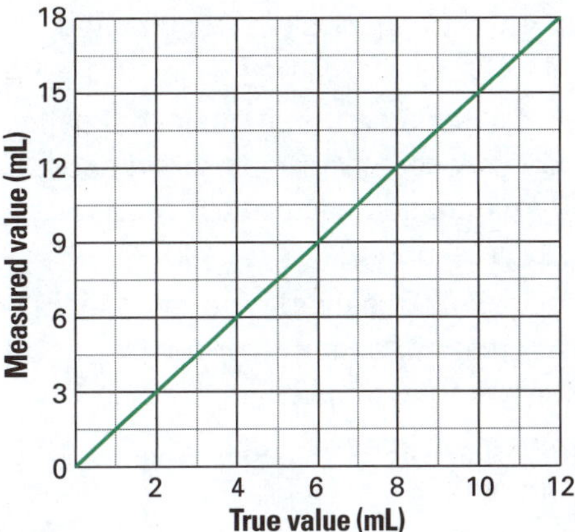

Figure 5.34 Converting true and measured volumes

14. Consider the following extract, with some minor changes, from *Life on the Mississippi* by Mark Twain, written in 1883. (The Mississippi River is the largest river system in North America, flowing entirely in the United States. It originates in western Minnesota and meanders southwards for 2320 miles, or 3730 km, to the Mississippi River delta at the Gulf of Mexico; 1 mile = 1.6 km.)

The Mississippi between Cairo and New Orleans was 1215 miles long, 176 years ago. It was 1180 miles after the cut-off of 1722. It was 1040 miles after the American Bend cut-off. It has lost 67 miles since. Consequently its length is only 973 miles at present. Please observe. In the space of 176 years the Lower Mississippi has shortened itself 242 miles. This is an average of a trifle over one mile and a third per year. Therefore, any calm person, who is not blind or idiotic, can see that ... just a million years ago next November, the Lower Mississippi River was upward of 1 300 000 miles long, and stuck out over the Gulf of Mexico. And by the same token any person can see that 742 years from now the Lower Mississippi will be only a mile and three-quarters long.

Comment on this extract using your knowledge of science. *(4 marks)*

Go to pp. 254–255 to check your answers.

Summary

1. The steps in a first-hand investigation are observation, inference, prediction, hypothesis, experimentation, analysis of results and making a conclusion.
2. Before conducting a first-hand investigation, background research is undertaken and secondary data sources are checked for validity and reliability.
3. There are similarities and differences in the ways that experiments are performed individually and in a team.
4. Appropriate equipment must be selected for an experiment.
5. Fair testing involves controlling variables and allowing only one variable to change at a time.
6. Blind and double-blind controls are used to avoid bias in experiments involving human subjects.
7. A knowledge of laboratory safety rules is essential before first-hand investigations are conducted.
8. Accuracy of measurements depends on the quality of the measuring apparatus and how carefully the measurement is taken.
9. Reliability refers to the consistency of a measure. A test is considered reliable if the same result is obtained repeatedly.
10. Random errors are inherently unpredictable and are usually scattered about the true value.
11. Systematic errors are predictable and typically constant or proportional to the true value.
12. Scientists use a variety of methods to present experimental data and analyse trends in experimental results.
13. A model is anything used in some way to represent something else.
14. A variety of different graphs can be used to organise and represent information, allowing trends and patterns to be identified.
15. Mean, median and range are some of the calculations that can be made to analyse data.
16. Many graphs can be drawn using spreadsheet applications.
17. Scientists may interpret the same experimental data differently.
18. The cause explains why something happens. The effect describes what happens.
19. Critical thinking strategies are important in applying science.
20. Presenting reports involves different text types, may include an abstract, often involves digital technologies, and requires correct referencing methods.

Syllabus checklist

Are you able to answer every syllabus question in this chapter? Tick each question as you go through the list if you are able to answer it. If you cannot answer it, turn to the appropriate page in the guide as is listed in the column to find the answer.

	For a complete understanding of this topic	Page no.	✓
1	Can I remember the steps in a first-hand investigation?	172	
2	Can I explain why before conducting a first-hand investigation, background research is undertaken and secondary data sources are checked for validity and reliability?	174	
3	Can I recall the similarities and differences in the ways that experiments are performed individually and in a team?	175	
4	Can I explain why appropriate equipment must be selected for an experiment?	175–176	
5	Can I explain why fair testing involves controlling variables and allowing only one variable to change at a time?	176	
6	Can I explain why experiments involving human subjects use blind and double-blind controls?	177	

	For a complete understanding of this topic	Page no.	✓
7	Can I explain why knowledge of laboratory safety rules is essential before first-hand investigations are conducted?	178–179	
8	Can I explain accuracy and carry out an investigation accurately?	185–186	
9	Can I describe and give examples of random errors?	186	
10	Can I describe and give examples of systematic errors?	186	
11	Can I explain reliability and carry out an investigation reliably?	187	
12	Can I use a variety of methods to present experimental data and analyse trends in experimental results?	187–192	
13	Can I describe a model?	187–189	

	For a complete understanding of this topic	Page no.	✓
14	Can I use a variety of different graphs to organise and represent information, allowing trends and patterns to be identified?	188–190	
15	Can I make calculations when analysing data?	190–191	
16	Can I use spreadsheet applications to draw graphs?	191	
17	Can I explain why scientists may interpret the same experimental data differently?	194	
18	Can I explain cause and effect?	194	
19	Do I know why critical thinking strategies are important in applying science?	194–195	
20	Can I use different text types, use an abstract and digital technologies, and use correct referencing methods when presenting reports?	195–196	

Chapter test

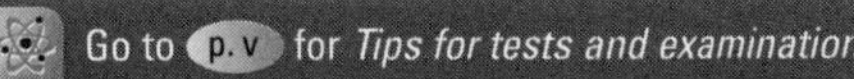
Go to p. v for *Tips for tests and examinations* 80 MIN

Part A: Multiple-choice questions

(1 mark for each)

1. In experiments involving the testing of new drugs on humans, a placebo is a
   - **A** tablet containing the new drug to be tested.
   - **B** person who volunteers to take the drug.
   - **C** blind person who cannot see what is being done.
   - **D** control that is used for comparison.

2. In double-blind experiments involving testing drugs on humans
   - **A** the control group receives the drug.
   - **B** only the person preparing the drug samples knows the code and that person does not communicate with the scientist.
   - **C** the scientist is masked so he cannot see the volunteers.
   - **D** two scientists work together to prepare and administer the drug.

3. Repeating an experimental procedure five or more times increases the
   - **A** accuracy.
   - **B** validity.
   - **C** reliability.
   - **D** safety.

4. Which piece of equipment would be appropriate for measuring 210 mL of water?
   - **A** 500 mL beaker
   - **B** 250 mL beaker
   - **C** 250 mL measuring cylinder
   - **D** 1 L measuring cylinder

5. Identify which second-hand source of scientific information is the most reliable.
   - **A** daily newspaper
   - **B** *Australian Journal of Science*
   - **C** science blogs on the internet
   - **D** creation science journals

6. Which of the following is *not* true about cause and effect?
   - **A** Cause and effect must be linked in space and time.
   - **B** The cause must come after the effect.
   - **C** There must be some constant relationship between the cause and effect.

D The same cause always produces the same effect.

7. Figure 5.35 shows the relationship between two variables.

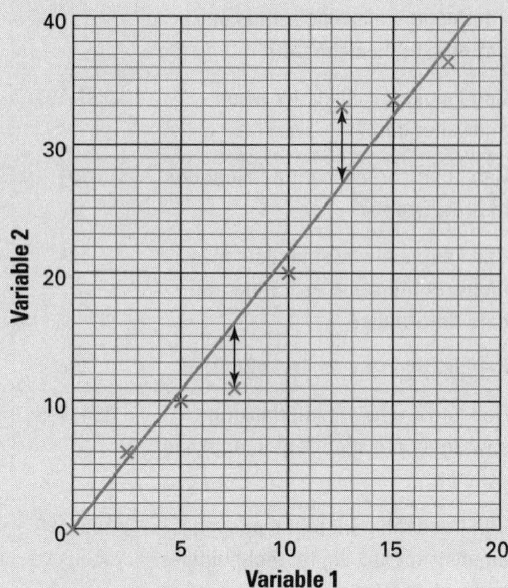

Figure 5.35 Variable 2 versus variable 1

Two values are shown a reasonable distance from the line of best fit. This shows an example of

A random error.
B systematic error.
C validity.
D reliability.

The next two questions refer to Figure 5.36.

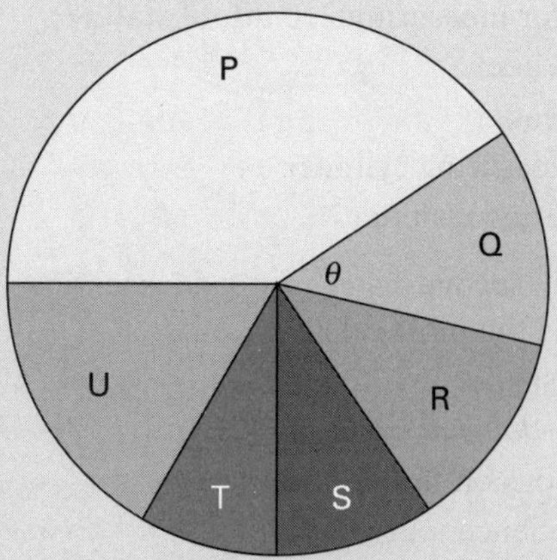

Figure 5.36 A sector graph (pie graph)

8. A sector graph is used to show

A continuous data over a range.
B the relationship between two variables.
C parts of a whole.
D data that repeats in a cycle.

9. In order to find the size of angle θ, calculate

A $\frac{\text{total size}}{\text{size of } Q} \times 360°$.

B $\frac{\text{total size}}{360°} \times \text{size of } Q$.

C $\frac{\text{size of } Q}{360°} \times \text{total size}$.

D $\frac{\text{size of } Q}{\text{total size}} \times 360°$.

10. Consider these two statements:

P. Reliability refers to the extent to which measurements are consistent.

Q. Validity refers to the accuracy of a measurement—whether or not it measures what it is supposed to measure.

Which of the following is true?

A Both P and Q are correct.
B P is correct, but Q is incorrect.
C P is incorrect, but Q is correct.
D Both P and Q are incorrect.

Part B: Short-answer questions

11. Bernadette and John are watching their friend Kim playing football and are horrified when he is hit in a tackle and breaks his collarbone. Bernadette and John wonder which bones are most commonly broken by football players.

a) Where would they go to gather information about broken bones? *(1 mark)*

b) They decide to survey the school football teams and ask 34 players about their broken bones. Do you think these results will be a fair representation of the whole school population of 903? Explain. *(2 marks)*

12. Leah notices that certain grasses disperse their seeds by sticking onto passing animals (especially ones wearing school socks). She decides to investigate this further. Leah notices the seeds have tough hairs at one end that act like velcro. She decides she is going to test how well these seeds grip to different sock fabrics—cotton, wool and nylon.

a) Write an experimental method that would allow Leah to measure the ability of seeds to grip the fabrics. *(7 marks)*

b) Which variables should she control to create a fair test? *(4 marks)*

13. Soft water is water that allows soap to lather. The softer the water, the more bubbles that form. Hard water is water that does not allow soap to lather. This is caused by the presence of certain mineral salts in the water. Bath crystals and washing soda are known to affect the hardness of water. Given this information, design an experiment to test the hypothesis that bath crystals and washing soda soften hard water. Include:
   a) the aim of the experiment *(1 mark)*
   b) six variables that need to be controlled *(6 marks)*
   c) the method of the experiment *(7 marks)*
   d) the type of data you would collect that would support the hypothesis. *(2 marks)*

14. Lynden and Milber are sitting on their grandfather's wharf fishing (see Figure 5.37) when they are surprised by a loud noise behind them.
   Lynden: It seemed to come from the mangrove tree.
   Milber: Something fell into the water! Look at the ripples.
   Lynden: It could have been a mangrove seed. There are a lot of them floating in the water and there is a wind blowing.
   Milber: Maybe one of the cormorants dived in after a fish. There were two of them sitting on the railing before.
   Lynden: The wind could have caught the door and hit it against the handrail.
   Milber: No! It came from the other direction. Maybe it was a water rat; Grandpa said there is a family living in the roof of the boatshed.

Figure 5.37 Milber and Lynden fishing on the wharf

   a) Which statements are inferences and which statements are about actual observations? *(10 marks)*
   b) What will Milber and Lynden have to do in order to work out which of their inferences (if any) are correct? *(2 marks)*

15. A drug company wants to test a new anti-asthma drug, AsthmoZ, which it believes prevents asthma attacks. It selects 100 young asthma sufferers and organises 50 to get AsthmoZ in an orange juice each morning and 50 to get just orange juice. After 6 months of the trial, the company releases the results shown in Table 5.9.

Table 5.9 AsthmoZ trial results

	Total number of asthma attacks	
	In the 6 months before treatment	**In the 6 months of the drug trial**
orange juice only (50 subjects)	15	17
orange juice and AsthmoZ (50 subjects)	17	7

   a) On the basis of this preliminary test of AsthmoZ, should the company scientists continue with their test of this substance? Explain. *(1 mark)*
   b) What are some of the variables that the scientist would find difficult to keep controlled? *(2 marks)*
   c) Before the drug was trialled on humans, what other trials would the scientists have conducted on AsthmoZ? *(3 marks)*

16. Jai is impressed with an ad on television that claims that a new sole plate for an iron lets you get the toughest wrinkles out of any fabric because the iron can be set on the highest setting without scorching the fabric. Jai buys one of these special sole plates and tests its claim for himself. He decides to test denim, polyester and silk fabrics. He cuts his fabric into six identical pieces. One piece of each he leaves unwashed and unironed; the other five pieces are washed and rolled into identical cylinders and allowed to dry. When they are dry, one piece of each fabric is ironed according to the normal temperature settings on the iron. Jai then sets up the equipment as shown in Figure 5.38, fits the special sole plate and sets the iron on its hottest setting. Jai then has each piece of fabric ironed, using the set-up in

the figure, and compares each strip with the normally ironed strip.

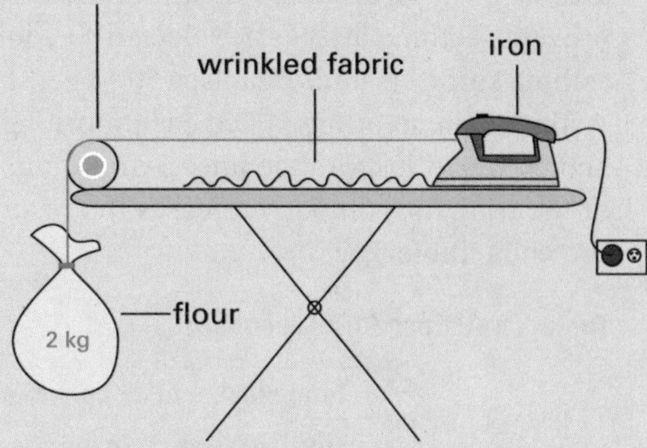

Figure 5.38 Ironing board set-up

Jai designs a rating scale where ✗ = minimum wrinkles and ✗✗✗✗✗ = maximum wrinkles, and records his results in Table 5.10.

Table 5.10 Ironing board experiment

	Denim	Polyester	Silk
control piece	✗	✗	✗
sample 1	✗✗✗✗	✗✗	✗
sample 2	✗✗✗✗✗	✗✗✗	✗✗
sample 3	✗✗✗✗	✗✗	✗
sample 4	✗✗✗✗	✗✗	✗

a) Which variables is Jai controlling? *(3 marks)*

b) Is a ratings scale that relies on personal judgement suitable for this experiment? *(1 mark)*

c) What conclusion can Jai draw from his experiment? *(1 mark)*

d) How could a simple change in equipment, such as a weight on top of the handle of the iron, affect the outcome of this experiment? *(1 mark)*

17. What is wrong with the following statement? *(1 mark)*

Once a scientist has proposed a hypothesis or theory, his or her next task is to prove it through experimentation or observation.

18. The flow chart in Figure 5.39 shows the steps of the scientific method. Five labels (A to E) are missing. Use the following list to match these labels: report; hypothesis developed; observations; experiment redesigned; experiment performed. *(5 marks)*

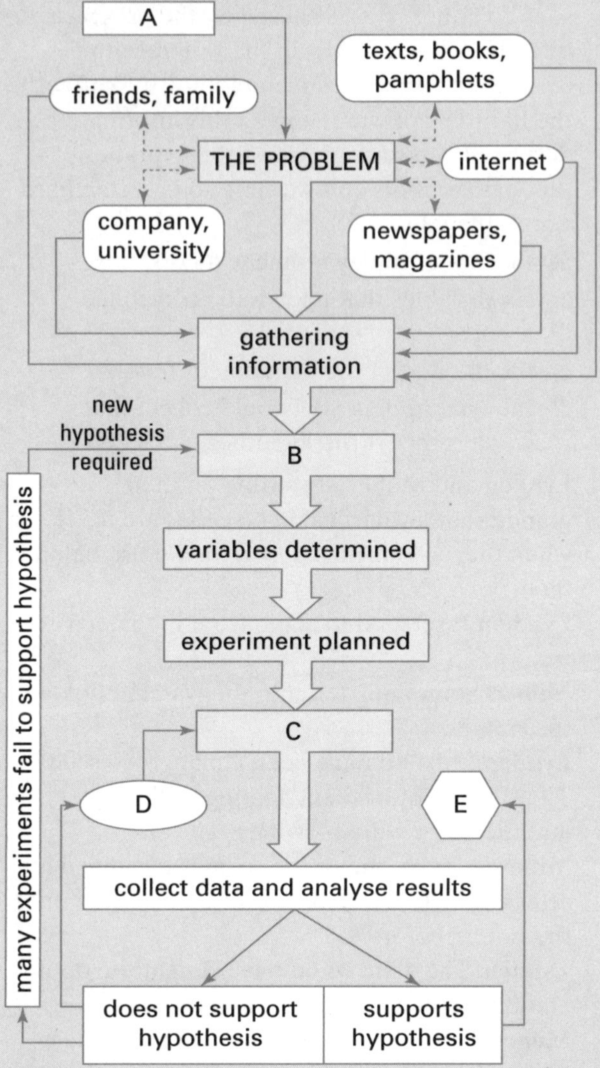

Figure 5.39 Scientific method

19. Three words used in science are 'hypothesis', 'theory' and 'law'. Each of the following statements refers to one of these. Choose the correct word to go with the right statement. *(3 marks)*

A. This is basically an untested conjecture explaining an observation. One or more of these are proposed to explain that observation. They can undergo extensive scrutiny to find the one that survives the tests.

B. This is a generalisation about the behaviour of nature from which there has been no known deviation after numerous observations or experiments.

C. This is basically a tested conjecture. In order to become a tested conjecture, quite a bit of testing has occurred and it seems to be correct. If new data proves it wrong, it must give way to improved versions. Scientists can only explain observations to the best of their ability with the existing set of data at the time. They are also scrutinised by other experts in the same field to test it.

20. Explain why the following statements cannot be shown to be false and hence are not scientific statements.
    a) The moon is populated by little yellow men who hide from the prying eyes of humans and flee into space whenever anyone comes near. *(1 mark)*
    b) In a fluorescent light tube are small, invisible and undetectable microbes that glow whenever the light switch is turned on. *(1 mark)*
    c) God keeps the planets moving in a path around the Sun. *(1 mark)*
21. Explain why the following can be called scientific theories.
    a) There are no little yellow men living on the Moon. *(1 mark)*
    b) The Loch Ness monster does not exist. *(1 mark)*
    c) There is no such creature as an abominable snow man. *(1 mark)*
    d) The Earth is flat. *(1 mark)*
    e) All the planets in our solar system revolve around the Sun. *(1 mark)*
22. Until the time of Isaac Newton, the modern scientific method was not used in investigations. People would observe things around them and only guess as to how they happened. They did not experiment, and relied on the authority of ancient scholars. How is this different to the way scientists operate today? *(2 marks)*
23. Analyse the data in Table 5.11 concerning the shoot and root growth of pine seedlings and draw appropriate conclusions. *(4 marks)*

Table 5.11 Shoot and root growth of pine seedlings

Night temperature (°C)	Day temperature (°C)	Shoot growth (cm)	Root growth (cm)
15	21	3	37
21	28	16	13

24. Assess the advantages and disadvantages of using models in chemistry. *(7 marks)*
25. Suggest a logical inference that could be made for each of the following observations.
    a) Jenna noticed that *Agapanthus* plants only flowered on the northern and western sides of her house. *(1 mark)*
    b) Ten boys from the same class all turned up at school with runny noses and a cough. *(1 mark)*
    c) The label on the mayonnaise said to store it in the refrigerator after opening. *(1 mark)*
    d) Water floats on mercury. *(1 mark)*
    e) All six children of a family have olive skin and brown eyes. *(1 mark)*
26. a) What are outliers? *(1 mark)*
    b) Suggest some causes for outliers. *(5 marks)*
27. a) What are random errors? *(1 mark)*
    b) Give three examples of random error. *(3 marks)*
    c) Can they be detected and compensated for? Why/why not? *(2 marks)*
28. It is often useful to distinguish between the cause and the effect. In each pair of statements, identify which is the cause and which is the effect.
    a) There is no life on the Moon. There is no atmosphere on the Moon. *(1 mark)*
    b) The element copper melts. Copper is heated to above 1085 °C. *(1 mark)*
    c) Solar batteries need frequent recharging. Solar batteries are impractical for powering vehicles on long trips. *(1 mark)*
    d) Cakes and other sweets are high in carbohydrates. Too many cakes and sweets can cause a person to gain weight. *(1 mark)*
29. From its discovery in 1930, and until 2006, Pluto was considered a planet. Now there are officially eight planets in the solar system. Pluto has been demoted to a 'dwarf planet', along with dozens of other dwarf planets. A planet is now defined as an object that:
    - orbits the Sun
    - is large enough to have become rounded due to the force of its own gravity
    - dominates the neighbourhood around its orbit.

Pluto does not dominate its neighbourhood. Its large moon, Charon, is only about half the size of Pluto, while all the official planets are much larger than their moons. As well, Pluto's orbital characteristics are very different from those of the planets, which follow nearly circular orbits around the Sun. In contrast, Pluto's orbit is highly inclined relative to the plane of the ecliptic (over 17°) and very elliptical.

a) Comment on the fact that for generations Pluto was considered a planet, and that now it no longer is. Does this mean that scientists got it wrong for almost 80 years? *(3 marks)*

b) Will all astronomers be happy with the decision? *(2 mark)*

30. Figure 5.40 shows an example of systematic error.

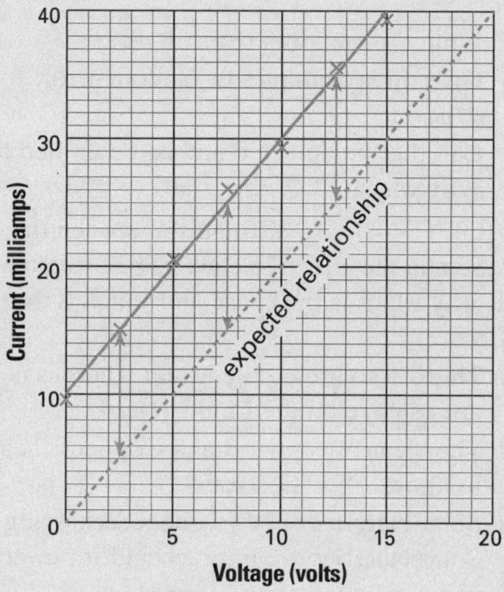

Figure 5.40 Current versus voltage

a) Describe what the graph shows. *(3 marks)*

b) Give an instance of how a systematic error occurs. *(1 mark)*

c) How can the systematic error shown in this diagram be corrected? *(1 mark)*

31. Consider the graph shown in Figure 5.41.

a) Describe the relationship shown. Is it linear or not linear? *(2 marks)*

b) There is an outlier in the plotted data. When does it occur, and what is its value? *(2 marks)*

c) Do the points exactly follow a curve? Why or why not? *(2 marks)*

d) Estimate what the atmospheric concentration will be in 2020. *(1 mark)*

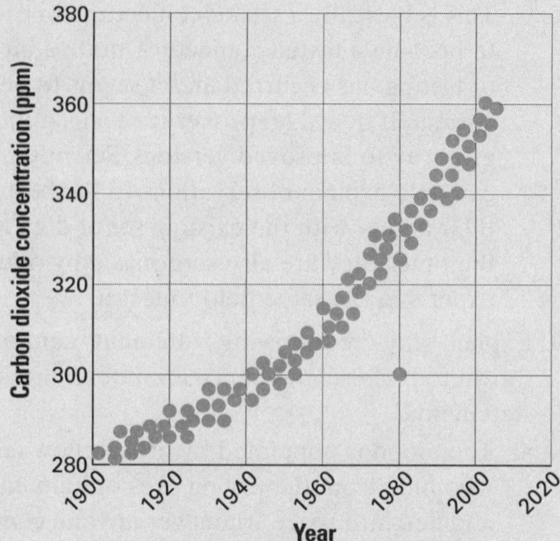

Figure 5.41 Atmospheric CO_2 concentration

e) What *process* are you undertaking when estimating the value in Question d)? *(1 mark)*

f) What *assumption* are you making when estimating the value in Question d)? *(1 mark)*

32. Use one of the words in this list to match each of the following statements: predicting; observing; modelling; investigating; measuring; inferring; communicating; classifying.

a) making observations using standardised units *(1 mark)*

b) making a representation of an object or an idea *(1 mark)*

c) grouping objects based on their characteristics *(1 mark)*

d) stating what is expected to happen *(1 mark)*

e) testing a hypothesis with an experiment and recording the results *(1 mark)*

f) sharing information through charts, graphs and laboratory reports *(1 mark)*

g) using one or several senses to take in information *(1 mark)*

h) making statements based on observations made *(1 mark)*

33. Australia's greenhouse gas emissions for a particular year are shown in Figure 5.42.

a) What percentage is represented by 'other emissions'? *(1 mark)*

b) Calculate the sector angle represented by agriculture. *(1 mark)*

c) Does this sector graph show the amount of greenhouse gases emitted? What does it show? *(2 marks)*

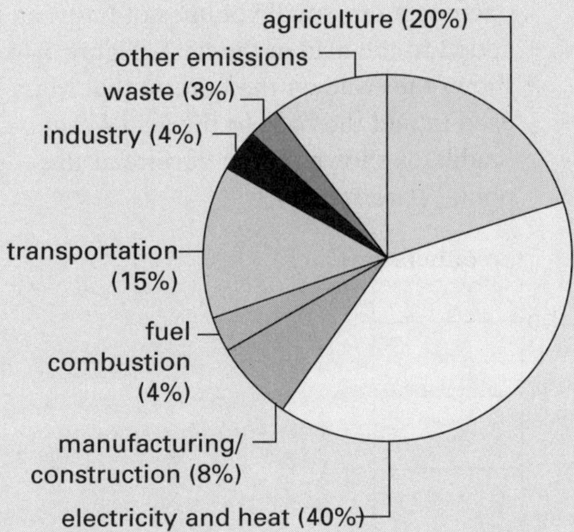

Figure 5.42 Australia's greenhouse gas emissions

d) In this particular year 517 million tonnes of CO_2 (or its equivalent) were emitted. How much of this came from transportation? *(1 mark)*

e) By 2010 this emission reached 580 million tonnes of CO_2. Calculate the percentage increase in this period. *(2 marks)*

f) Would the size of the sectors, as shown in Figure 5.42, be the same for 2010? Why or why not? *(3 marks)*

34. Figure 5.43 shows the gas composition of the atmosphere.

a) What name is given to this type of graph? *(1 mark)*

b) What percentage of the atmosphere is represented by argon? *(1 mark)*

c) There has been much discussion of CO_2 pollution and global warming, but this graph does not show any CO_2. What does this indicate about carbon dioxide in the atmosphere? *(1 mark)*

d) If the length of the whole bar is 15 cm, how long is the nitrogen rectangle? *(1 mark)*

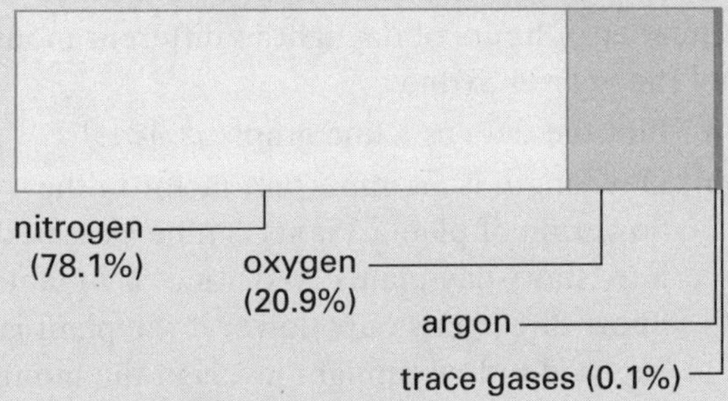

Figure 5.43 Composition of Earth's atmosphere

35. A company that manufactures alternative medicines has made some claims about its latest product, Oxideze. Which of the following claims are opinions that cannot be easily tested scientifically? *(3 marks)*

A. Oxideze prevents the build-up of damaging free-radicals in the blood.

B. Oxideze is made only from natural ingredients.

C. Oxideze is the best alternative medical product on the market.

36. Bertram investigated the effect of different gases (X and Y) on the growth of cress seedlings. He placed five wheat seeds on damp cotton wool in each of two glass vessels as shown in Figure 5.44.

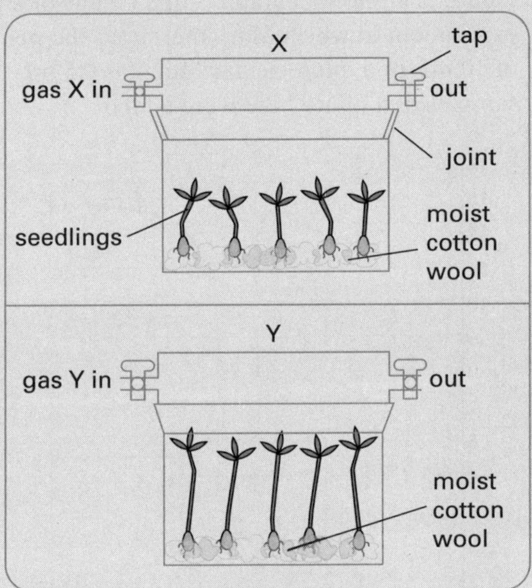

Figure 5.44 Bertram's experiment

Bertram waited until they germinated before introducing the selected gas into containers through the tubes and taps. Plants 1 to 5 were exposed to gas X and plants 6 to 10 were exposed to gas Y. After 6 days he measured the length of the shoot for each plant. His results are shown in Table 5.12.

Table 5.12 Results of Bertram's experiment

Plant number	Length of shoot (cm)
1	5.3
2	4.9
3	no growth
4	5.5
5	4.8

Plant number	Length of shoot (cm)
6	6.6
7	6.8
8	7.0
9	6.8
10	7.2

a) Calculate the average length of the wheat shoots in each experimental condition. *(4 marks)*

b) In order to conduct a fair experiment, Bertram tried to control his variables. Name the variables that needed to be controlled. *(3 marks)*

c) What other experiment did Bertram need to do to determine whether these gases (X and Y) had an effect on the growth of the wheat seeds? *(1 mark)*

d) Bertram concluded in his report that gas Y promoted the growth of the seedlings more than gas X. Is this a valid conclusion? Explain. *(1 mark)*

37. Figure 5.45 shows a graph of the results of an experiment in which Mary measured the pH of 10 mL of a dilute acid solution as 0.5-mL samples of a dilute base were added.

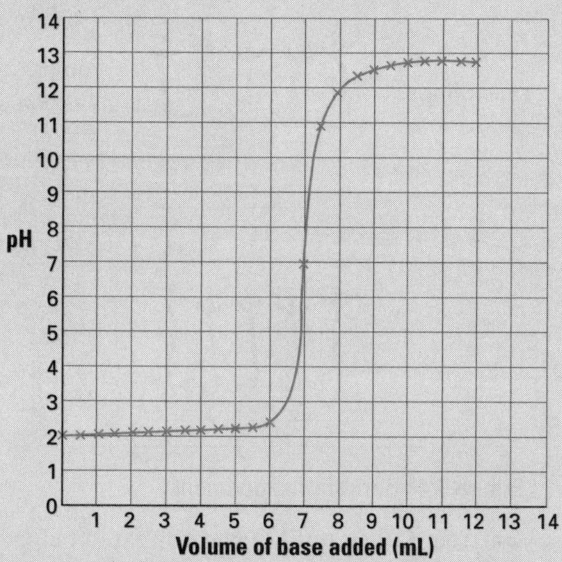

Figure 5.45 pH graph

a) What volume of base solution did Mary add over the whole experiment? *(1 mark)*

b) Describe how the pH of the mixture changed throughout the experiment. *(2 marks)*

c) At what volume of added base did the solution become neutral? *(1 mark)*

d) Mary did not record the pH of the mixture when 6.5 mL of base had been added. What would the pH of the mixture have been at this point? *(1 mark)*

e) The base solution was placed in an accurate piece of glassware known as a burette. By opening a tap, small volumes of base can be added to the acid in the flask. Figure 5.46 shows the scale on the burette that Mary used to add the base to the acid. What reading is shown on the burette at this point? *(1 mark)*

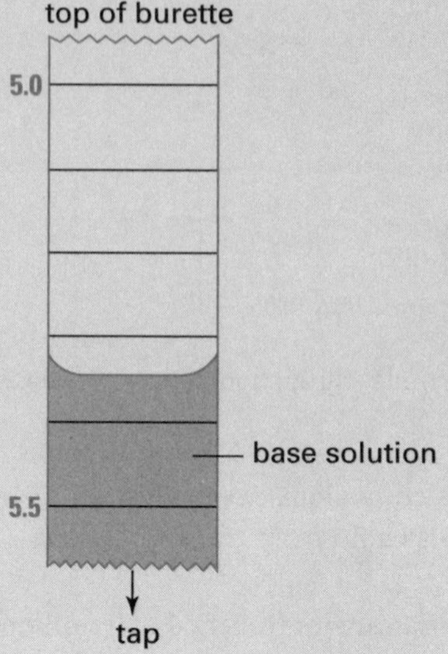

Figure 5.46 Close-up of a burette scale

38. A rock sample is analysed for its mineral content. Table 5.13 records the percentage by mass of each mineral.

Table 5.13 Mineral composition of a rock sample

Mineral	% (by mass)
quartz	47
feldspar	40
ferromagnesian	
mica	4

a) Calculate the percentage of ferromagnesian minerals in the rock. *(1 mark)*

b) Present this composition data in the form of a divided bar graph. *(2 marks)*

39. Table 5.14 provides information about the number of hours of daylight in different months of the year in Sydney.

a) Plot the data as a line graph. *(3 marks)*

b) Day length is an important factor in the flowering of plants. Plants can be classified into ‘short-day plants’ and ‘long-day plants’. Short-day plants only flower if the plant is exposed to low sunlight levels in the month prior to flowering. Long-day plants only flower if the plant is exposed to high sunlight levels in the month prior to flowering.

i) Plant X is a short-day plant that requires exposure to less than 10 hours of sunlight in the month before flowering. In what season(s) is this plant most likely to flower? *(1 mark)*

ii) Plant Y is a long-day plant that requires exposure to a minimum of 13 hours of sunlight in the month prior to flowering. In what months of the year is this plant able to flower? *(1 mark)*

Table 5.14 Hours of daylight per day in Sydney

Month	Hours of daylight/day
January	17.0
February	15.0
March	13.0
April	11.5
May	10.0
June	8.5
July	8.5
August	10.0
September	11.5
October	13.0
November	15.0
December	17.0

40. Solutions of cobalt chloride are pink. The higher the concentration of cobalt ions, the more pink is the solution. Clement made up four solutions of cobalt chloride by dissolving different masses of the cobalt chloride crystals in different volumes of water (see Table 5.15).

Table 5.15 Hours of daylight per day in Sydney

Solution	Mass of solute (g)	Volume of water (mL)
A	3.6	150
B	4.4	200
C	2.8	120
D	3.3	130

a) Calculate the concentration of cobalt chloride in solution A in units (mg/L). *(2 marks)*

b) Rank the solutions in decreasing order of pink colour. *(1 mark)*

Go to pp. 255–259 to check your answers.

Course test 1

Go to p. v for *Tips for tests and examinations*

This test covers material from all the chapters in this book. It consists of the following.

- Section A: 25 multiple-choice questions *(1 mark for each)*
- Section B: 15 restricted-response questions *(1 mark for each)*
- Section C: 11 knowledge, skill and processing data questions *(total 60 marks)*

Total marks for this test: 100

Time: 2 hours

Part A: Multiple-choice questions

(25 marks—1 mark for each question)

1. What is the reading on the thermometer scale shown in Figure T1.1?

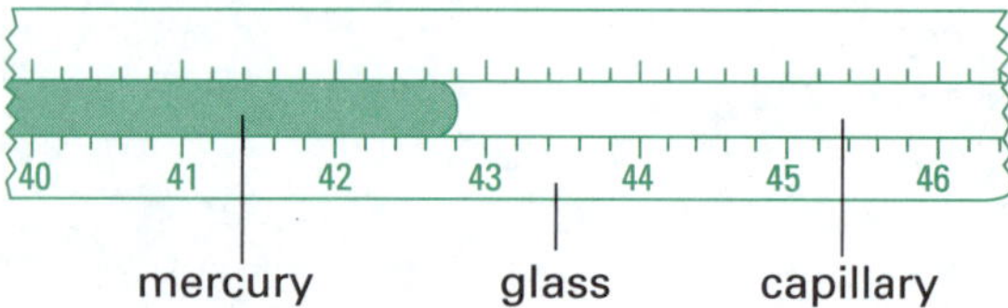

Figure T1.1 Thermometer scale

A 42.8 °C
B 42.4 °C
C 40.8 °C
D 43.2 °C

2. Use the divided bar graph in Figure T1.2 to determine the percentage of zinc in the alloy.

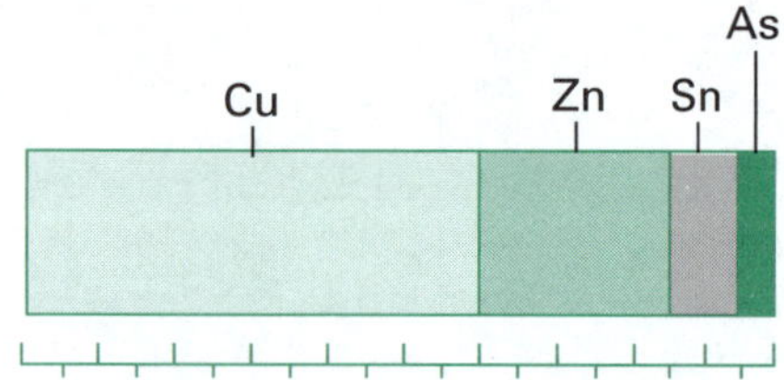

Figure T1.2 Divided bar graph of alloy composition

A 5%
B 10%
C 25%
D 60%

3. Figure T1.3 shows the structure of a magnesium atom.

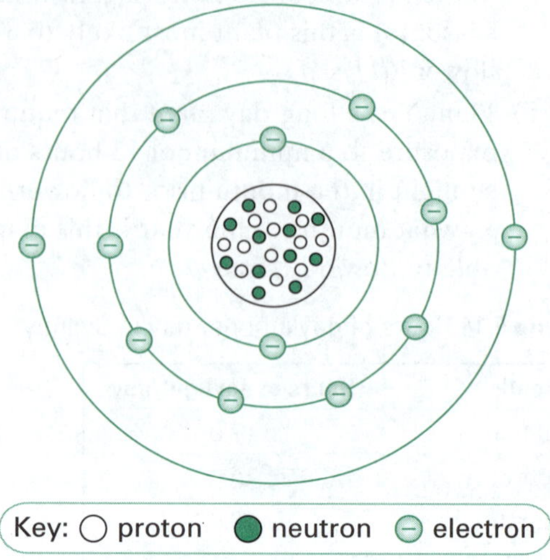

Figure T1.3 Structure of a magnesium atom

Select the correct response about magnesium.

A Magnesium has two outer-shell electrons.
B The magnesium nucleus contains 24 protons.
C There are more protons than neutrons in the nucleus.
D The charge on the nucleus is –12.

4. A useful chemical that is used to determine the pH of the water in a swimming pool is

A chlorine.
B universal indicator.
C sodium chloride.
D sodium hydroxide solution.

5. Classify the wave shown in Figure T1.4 and determine its wavelength.

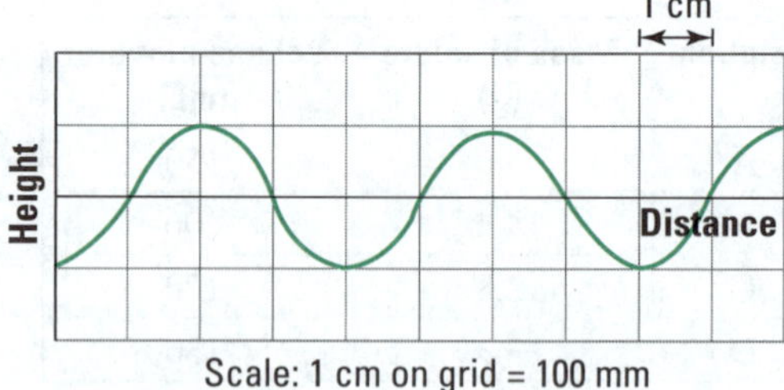

Figure T1.4 Waves

A transverse wave; wavelength = 4.0 m
B longitudinal wave; wavelength = 4.0 m
C transverse wave; wavelength = 400 mm
D compression wave; wavelength = 400 mm

6. Select the response that correctly identifies a component wave of the electromagnetic spectrum and its use.

	wave	use
A	sound	sonar
B	seismic	location of mineral deposits
C	gamma	sterilisation of instruments
D	infra-red	detection of bone fractures

7. Select the response in Figure T1.5 that correctly identifies the path of a light ray as it travels through the glass.

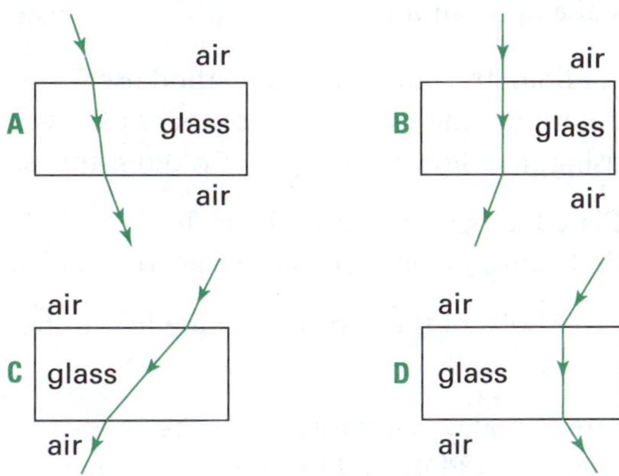

Figure T1.5 Light refraction in glass

8. Which of the following systems are components of the coordination systems of the body?
   A digestive system and respiratory system
   B endocrine system and circulatory system
   C nervous system and endocrine system
   D endocrine system and skeletal system

9. Which response contains both a biotic and an abiotic feature of a rainforest ecosystem?
   A rainfall; wind speed
   B competition for mating; light intensity
   C competition for food; predation
   D light intensity; water quality

10. A common location for earthquakes and volcanoes is
   A the Pacific rim.
   B the south Pacific Ocean.
   C Antarctica.
   D England.

11. On 21 September 2003 the *Galileo* spacecraft, which was launched in 1989, was intentionally allowed to crash into the planet Jupiter. Gravitational forces and frictional melting tore the craft apart. During its operational life *Galileo* relayed over 14000 images of Jupiter and its moons back to Earth by
   A compression waves.
   B ultraviolet waves.
   C visible light waves.
   D radio waves.

12. Select the statement that is true of mumps.
   A The disease is caused by a bacterium.
   B Mumps is not an infectious disease.
   C Mumps is caused by a protozoan that enters the body via a mosquito bite.
   D Mumps is a viral disease.

13. The Black Death (1348–1350) was a pandemic caused by an outbreak of bubonic plague. The microbe responsible was a
   A virus.
   B fungus.
   C bacterium.
   D protozoan.

14. In 1867 Joseph Lister introduced the use of antiseptics in surgery and the survival rate from operations increased greatly. The first antiseptic used was
   A carbolic acid.
   B sulphonamide.
   C penicillin.
   D hydrochloric acid.

15. Silvery magnesium is heated in a Bunsen burner flame. The magnesium burns with a dazzling bright white light and a crumbly white powder is formed. Select the true statement.
   A The reaction is exothermic.
   B The reaction is endothermic.
   C The white powder is a carbonate of magnesium.
   D The magnesium can be recovered from the white powder by filtration.

16. Select the true statement about mid-ocean ridges.
   A The rocks on either side of the ridge get older as the distance from the ridge increases.
   B A mid-ocean ridge is a destructive plate boundary.
   C Old crust undergoes subduction at the mid-ocean ridge.
   D The mid-ocean ridge in the Atlantic is the only one present in the Earth's crust.

17. Which of the following molecules is tetra-atomic?
   A carbon dioxide
   B methane
   C ammonia
   D oxygen

18. The chemical formula for nitric acid is
   A HCl
   B HNO_3
   C H_2SO_4
   D H_2CO_3

19. A salt called barium sulfate is formed in a neutralisation reaction. Identify the acid and the base that could produce this salt.
   A barium chloride and sulfuric acid
   B barium hydroxide and sulfuric acid
   C calcium hydroxide and hydrochloric acid
   D barium oxide and nitric acid

20. The products formed from the respiration of glucose in a cell are
   A carbon monoxide and carbon dioxide.
   B carbon dioxide and water.
   C carbon dioxide and oxygen.
   D water and carbon monoxide.

21. Select the most active metal in the following list.
   A zinc
   B copper
   C iron
   D potassium

22. A subduction zone exists between the
   A African Plate and Eurasian Plate.
   B South American Plate and Nazca Plate.
   C Antarctic Plate and the Indo-Australian Plate.
   D Arabian Plate and the Eurasian Plate.

23. Of the electromagnetic waves listed, the one with the greatest frequency is
   A radio.
   B microwave.
   C ultraviolet.
   D visible.

24. The bending of light rays when they travel into different media is called
   A reflection.
   B scattering.
   C translucency.
   D refraction.

25. Select the statement that is true of sound waves.
   A They travel faster in gases than solids.
   B They travel faster in cold air than warm air.
   C They travel at speeds close to the speed of light.
   D They are examples of compression waves.

Part B: Restricted-response questions

(15 marks—1 mark for each question)

Complete the following restricted-response questions using the appropriate word.

26. Radiometric dating is the method used to determine the .... of rocks or fossils using the known half-lives of radioisotopes.

27. Plate tectonics is the study of the .... that cause the movement of the crustal plates.

28. Over 80% of the Earth's volume lies in the ....

29. The Himalayas are formed as the Indo-Australian Plate and the .... plate collide.

30. Hormones are chemical messenger molecules that are produced by ....

31. The small intestine is the place in which digestion is completed and the nutrients pass through the intestinal wall into the ....

32. High-dose X-rays can be used in radiation therapy as .... cells can be killed by X-ray treatment.

33. The term PET is an acronym for .... emission tomography.

34. Some vaccines, such as the measles vaccine, give life-long ....; others, such as polio vaccine, are given to children in a series of vaccinations over 15 years.

35. Isotopes are atoms with the same .... number but different mass numbers.

36. When acids are added to water, the hydrogen atoms break free and form hydrogen ....

37. Soluble salts can be recovered from a solution by .... the water.

38. Carbon dioxide fire extinguishers can smother a fire by cutting off the .... supply.

39. A pollutant is any substance that does not naturally occur in a particular environment or one that has a .... outside the normal limits.

40. Mechanical waves require a .... (such as solids, liquids or gases) in which to move.

Part C: Knowledge, skill and processing data questions

(60 marks)

41. Use the scientific method to comment on each of these statements made by commercial companies.
   a) Nine out of 10 dentists recommend Dento toothpaste. *(1 mark)*
   b) University tests show that Wizzo powder washes 22% whiter. *(1 mark)*
   c) Our products are available from all leading retailers. *(1 mark)*

42. Read each of the following statements and decide whether they are examples of observations, inferences, predictions or generalisations.
   a) A virus caused the cough. *(1 mark)*
   b) Most compounds of copper are blue. *(1 mark)*
   c) The minerals in the rock were a lighter colour to those in the rock from the bottom of the cliff. *(1 mark)*
   d) If a more concentrated acid is added to zinc, the zinc will dissolve faster. *(1 mark)*

43. Read the following statements. Each one describes a student doing something wrong in a school laboratory Write down what is wrong, and the correct way of performing the task.
   a) Anita uses her hand to scoop out a small amount of solid chemical from a jar. *(1 mark)*
   b) Gianni looks down the barrel of a test tube being heated to see what is going on. *(1 mark)*
   c) Charles leaves the beaker boiling on the tripod to go and talk to his friends. *(1 mark)*
   d) Frances sticks her nose directly over a bottle to smell the chemical. *(1 mark)*
   e) James sets up some glassware close to the edge of the bench. *(1 mark)*
   f) Noha puts her thumb over the end of the test tube and shakes it up and down to mix the contents. *(1 mark)*
   g) After boiling some liquid, Andrew picks up the beaker with his hand to pour the contents down the sink. *(1 mark)*
   h) Alyssa pours some liquid and solid wastes into the sink. *(1 mark)*

44. Burglars breaking into houses may leave impressions of the soles of their shoes in garden soil under windows.

   Figure T1.6 shows the partial sole prints of five shoes. Four of the prints are from suspects' shoes and the last is from the garden.
   a) Examine each of the patterns carefully and determine which shoe pattern, if any, matches the impression made in the garden. *(1 mark)*
   b) If the suspect's shoe is a common brand, what features of the print will help to bring this evidence into court? *(1 mark)*

45. The amount of oxygen in water depends on the temperature of the water and the level of pollution.

Figure T1.6 Shoe prints

a) Table T1.1 shows the maximum concentration of oxygen in fresh water as a function of temperature.

Table T1.1 Maximum concentration of oxygen in fresh water at different temperatures

Temperature (°C)	Dissolved oxygen (ppm)
5	12.4
10	10.9
15	9.8
20	8.8
25	
30	7.5
35	7.0

i) Plot a line graph of this data and use the line of best fit to interpolate the concentration of dissolved oxygen at 25 °C. *(5 marks)*

ii) Extrapolate your graph to 40 °C and predict the oxygen concentration in fresh water at this temperature. *(1 mark)*

b) All fish die if the oxygen concentration is lower than 1 ppm. Carp and perch will die if the oxygen level is lower than 3 ppm and trout will not survive at levels lower than 5 ppm. At zero ppm, only anaerobic organisms can survive. Use this information to construct a column graph. *(4 marks)*

c) At what temperature is the dissolved oxygen concentration in unpolluted fresh water 11.5 ppm? *(1 mark)*

46. After examining the evidence collected at a crime scene, a forensic scientist developed a list of the most likely characteristics of any suspect.

P. The suspect is a male of short stature.
Q. He has black hair.
R. He is probably of northern Asian origin.
S. He has scarring on his right fingers.
T. He possibly walks with a limp.

Match these five statements to the following types of evidence she would be basing her hypotheses on. *(2 marks)*

A. Footsteps with a non-uniform pace or dragging footprints
B. Black hairs at the crime scene.
C. Fingerprints with cuts across the prints.
D. DNA analysis of hair/microscopic examination of hair.
E. Small foot size from footprints.

47. Figure T1.7 shows UV radiation recorded in Adelaide during a morning in November.

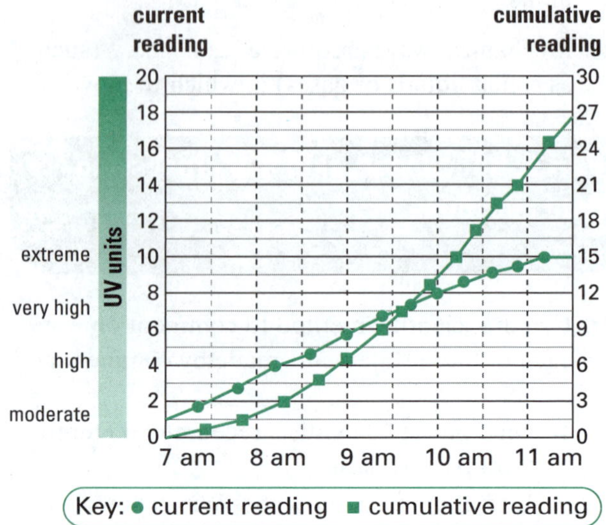

Figure T1.7 UV readings in Adelaide

a) What UV reading is regarded as extreme? *(1 mark)*

b) The units of UV radiation on the vertical axis are related to energy. One unit of UV = 9 millijoules/centimetre. If 10 000 square centimetres (1 m^2) of ground is exposed to 540 joules of UV radiation, what would be the UV reading? *(2 marks)*

c) What is the UV reading at 9 am? What is it at 11 am? Why is it increasing? *(3 marks)*

d) Exposure to UV rays is additive. That is, the longer you are exposed, the more damage caused. Use the second curve on the graph (the accumulative reading) to work out the UV accumulative reading for a person who has been exposed to sunlight from 7 am until 11 am. *(1 mark)*

48. The graph in Figure T1.8 compares the death rate from lung cancer for people aged between 45 and 64.

a) Does this graph support the hypothesis that smoking increases the risk of lung cancer? Explain. *(2 marks)*

b) Calculate the ratio of death from lung cancer (in the 45 to 64 age group) between all smokers and all non-smokers (all smokers : all non-smokers). *(1 mark)*

c) In a population of 8 000 000 people between the ages of 45 and 64, calculate the total number of deaths that would be expected from lung cancer. (Assume 50% of the population is female.) *(2 marks)*

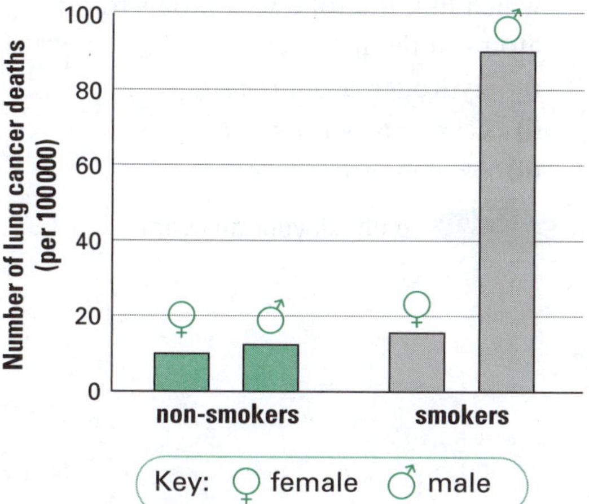

Figure T1.8 Cancer deaths in smokers and non-smokers

49. Bushfires are quite common in many parts of Australia. Examine Figure T1.9 and answer the questions that follow.

a) What type of electromagnetic radiation is mainly radiated from the approaching fire? *(1 mark)*

b) Explain why cooler air containing oxygen moves in towards the burning trees. *(1 mark)*

c) Eucalyptus trees contain oils and waxes. Why are fires in eucalyptus forests so intense? *(1 mark)*

d) Fragments of burning plant material are carried up to heights of 4 km. Why is this a serious problem? *(1 mark)*

50. Lauren constructed a model of the human eye as shown in Figure T1.10.

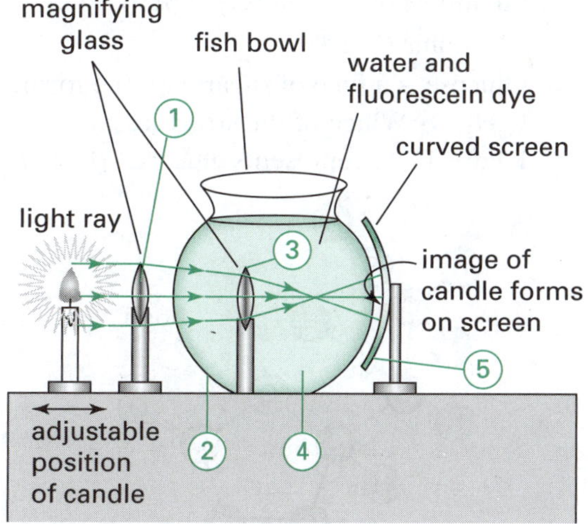

Figure T1.10 Eye model

a) Why did Lauren colour the water with fluorescein dye rather than just use clear water? *(1 mark)*

b) How is the image different from the object? *(1 mark)*

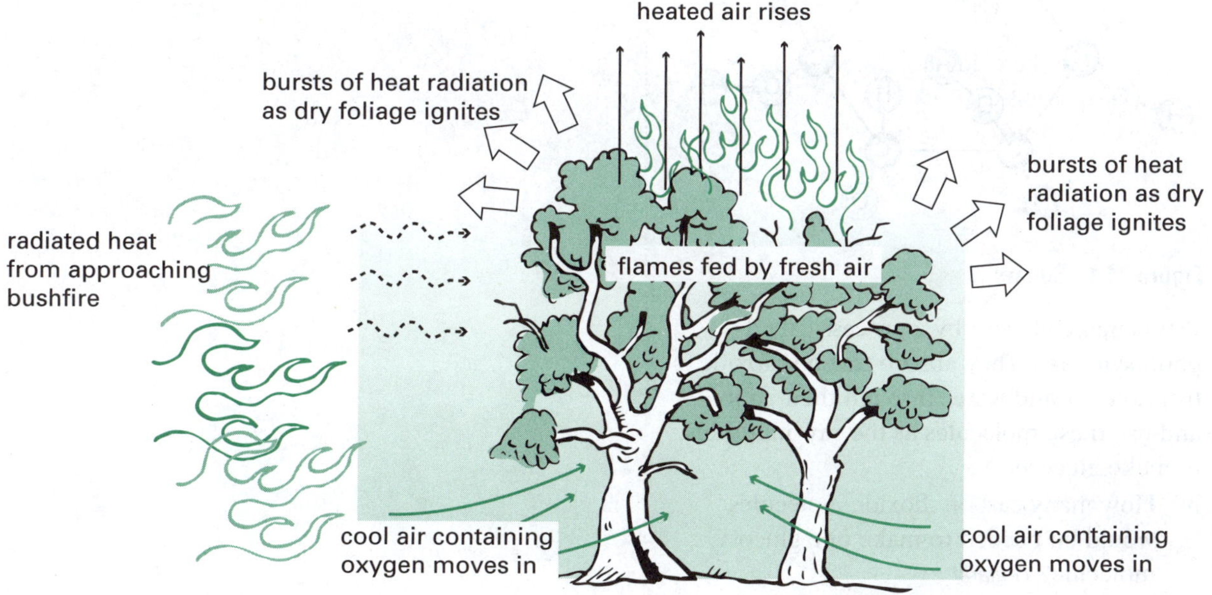

Figure T1.9 Bushfire

c) Which parts of the eye correspond to the labels 1 to 5 in Lauren's model eye? *(2 marks)*

51. The sugar molecules you put on your morning cereal have the formula $C_{12}H_{22}O_{11}$.

a) In one molecule of sugar, how many of the following are there?

i) atoms of carbon (C) *(1 mark)*

ii) atoms of hydrogen (H) *(1 mark)*

iii) atoms of oxygen (O) *(1 mark)*

iv) atoms *(1 mark)*

b) Glucose is a type of sugar with the formula $C_6H_{12}O_6$. Which of the structures in Figure T1.11 represents glucose? *(1 mark)*

A

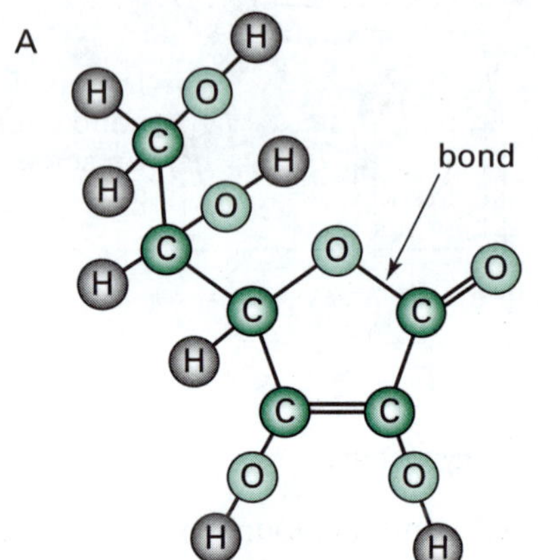

B

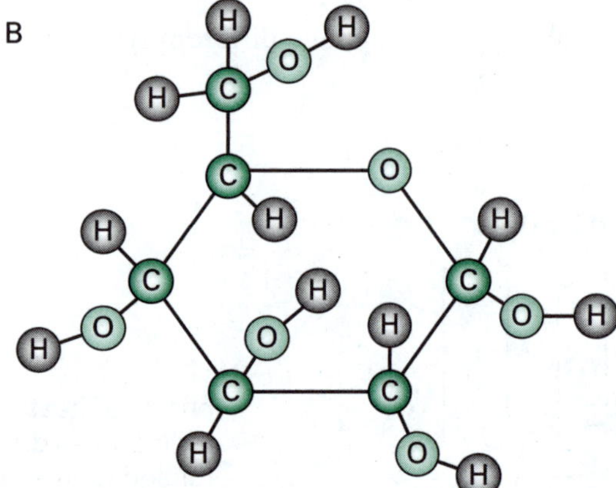

Figure T1.11 Sugars

c) Plants make glucose by a process called photosynthesis. They absorb carbon dioxide from the air and water through their roots and use these molecules as the raw materials to make glucose.

i) How many carbon dioxide molecules would be needed to make one glucose molecule? *(1 mark)*

ii) How many water molecules would be needed to make one glucose molecule? *(1 mark)*

d) Examine the structure of glucose again. What is the maximum number of bonds which link the following atoms to other atoms in the molecule?

i) hydrogen atom *(1 mark)*

ii) carbon atom *(1 mark)*

iii) oxygen atom *(1 mark)*

Go to pp. 259–262 to check your answers.

Course test 2

Go to p.v for *Tips for tests and examinations*

This test covers material from all the chapters in this book. It consists of the following.

- Section A: 25 multiple-choice questions *(1 mark for each)*
- Section B: 15 restricted-response questions *(1 mark for each)*
- Section C: 11 knowledge, skill and processing data questions *(total 60 marks)*

Total marks for this test: 100

Time: 2 hours

Part A: Multiple-choice questions

(25 marks—1 mark for each question)

1. The chamber of the heart which pumps blood out and into the aorta is the

A left atrium.
B left ventricle.
C right atrium.
D right ventricle.

2. An element is determined by its number of

A protons.
B neutrons.
C electrons.
D atoms.

3. Which of the following is *not* a hereditary disease?

A Down's syndrome
B cystic fibrosis
C sickle cell anaemia
D emphysema

4. Which part in Figure T2.1 best fits the following description?

This part of the eye possesses about 130 million light-sensitive photoreceptor cells (rods and cones) that convert light into electro-chemical impulses, which then travel to the brain.

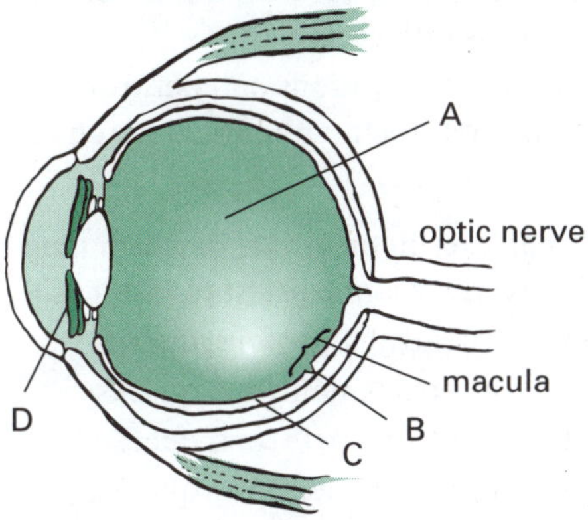

Figure T2.1 A cross-section of the eye

5. Which scientist came up with a model of the atom that looked like Figure T2.2?

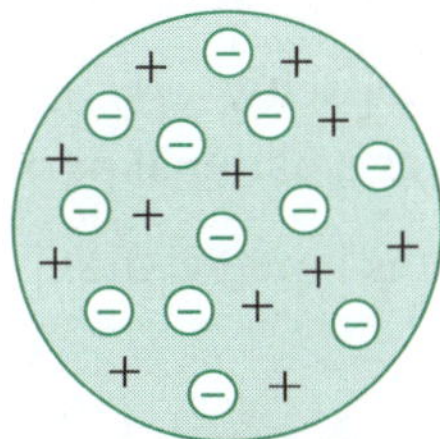

Figure T2.2 An early atomic model

A John Dalton
B JJ Thomson
C Ernest Rutherford
D Neils Bohr

6. An appropriate balance is needed between predator and prey to avoid

A running out of food resources.
B one species over-populating.
C extreme changes in numbers of organisms.
D all of the above.

7. Nutrients move from the small intestine and into the blood stream through the walls of the

A villi.
B alveoli.
C skin.
D stomach.

8. Which one of these statements is true about an isotope of an element?

A The number of neutrons remains the same, but the number of protons is different.

B The number of protons remains the same, but the number of neutrons is different.

C The number of protons and electrons remain the same, but the number of neutrons is different.

D The number of protons and neutrons remain the same, but the number of electrons is different.

9. Atomic particles in the order in which they were discovered, from earliest to latest, is

A electrons, neutrons, protons, quarks.

B protons, electrons, neutrons, quarks.

C electrons, protons, neutrons, quarks.

D protons, neutrons, electrons, quarks.

10. A major reasons that wildlife populations become threatened is

A a loss of habitat.

B a decrease in landslides, forest fires, floods and weather extremes.

C an increase in carrying capacity.

D the declaration of land as a national park.

11. The layer of the Earth that is liquid is the

A crust.

B mantle.

C outer core.

D inner core.

12. Suppose a large earthquake occurs 100 km away from where you are. You would feel the P-waves about 17 seconds after the earthquake occurred, then the S-waves arrive 13 seconds later. Finally, the L-waves arrive causing the ground to move more than the other two types of waves. This is because

A S-waves move less rapidly and so do less damage than P-waves.

B P-waves do the most damage as they move along the surface like ocean waves and arrive first.

C L-waves do the most damage as they move through the body of the Earth by compressing and stretching.

D L-waves do the most damage as they move along the surface like ocean waves.

13. Which of the following is an anomaly as to where volcanoes and earthquakes are normally located as the result of tectonic plate movement?

A the mid-oceanic ridge

B collision zones

C Hawaii

D the Andes mountains

14. The part of the brain that controls memory, intelligence, behaviour, emotions and speech is the

A cerebrum.

B hypothalamus.

C cerebellum.

D brain stem.

15. Prey numbers in a population will decline when

A climatic conditions allow prey to breed.

B prey have access to more resources than usual.

C predator numbers increase.

D predator numbers decrease.

16. Rutherford's alpha (α) scattering experiment was important because it showed that

A atoms consist of protons in a bed or matrix of electrons.

B atoms can be broken up by alpha particles.

C the nucleus is relatively large, compared to the atom, and positively charged.

D the nucleus is very small, compared to the atom, and positively charged.

17. A mid-oceanic ridge is an example of

A sea floor spreading.

B transform faulting.

C subduction.

D fold mountain building.

18. A CT (computerised tomography) scanner uses a computer to make pictures of the inside of a person's body. Which of the following statements is *not* true about CT scans?

A It is used for diagnosis to show detail of parts inside the body.

B Internal examinations can be made instead of using surgery.

C CT scans do not use radiation.

D CT scans are painless, accurate and fast.

19. Which of the following represents the energy changes that occur in a flashlight?

A chemical $\rightarrow$ electrical $\rightarrow$ light + heat

B electrical $\rightarrow$ light + heat

C light + heat $\rightarrow$ electrical

D electrical $\rightarrow$ chemical $\rightarrow$ light + heat

20. Which continent has no active volcanoes?

A Australia
B Antarctica
C Europe
D Asia

21. Figure T2.3 shows an example of

A conduction.
B convection.
C radiation.
D vaporisation.

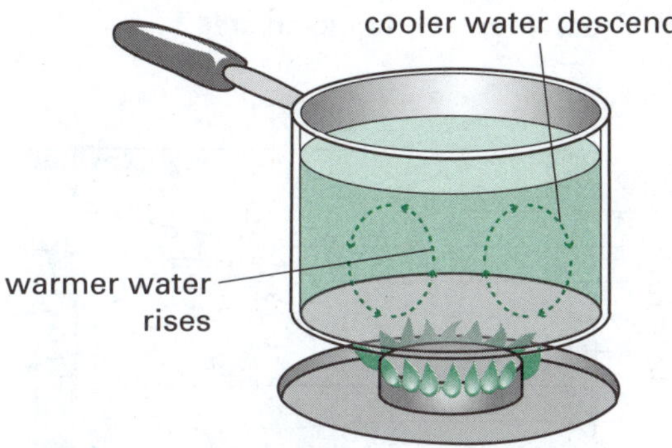

Figure T2.3 Transferring heat

22. The theory of continental drift is supported by

A matching mountain ranges and rocks on different continents.
B similar fossils matching on several continents.
C continental shapes appearing to fit together like jigsaw puzzles.
D all of the above.

23. In the electromagnetic spectrum shown in Figure T2.4, the form of radiation marked ? in this diagram is

A radio waves.
B X-rays.
C sound waves.
D microwaves.

24. Which in the following list is *not* shown in Figure T2.5?

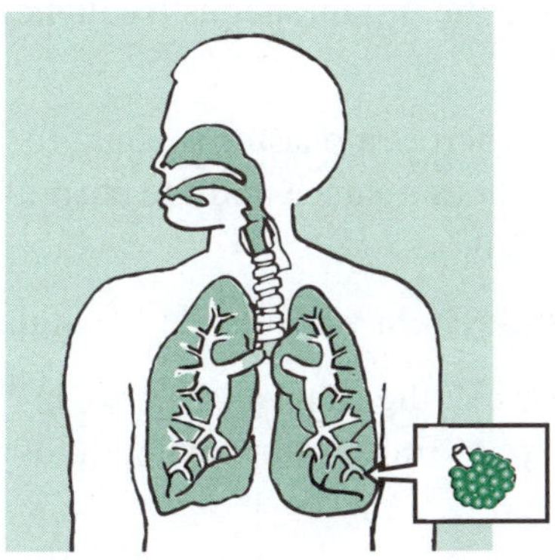

Figure T2.5 Internal image of human

A alveoli
B trachea
C bronchi
D oesophagus

25. A communication satellite is powered by

A transmitted energy from Earth.
B fuel cells.
C solar cells.
D on-board fuel.

Part B: Restricted-response questions

(15 marks—1 mark for each question)

Complete the following restricted-response questions using the appropriate word.

26. The .... system includes arteries and veins.

27. The muscle that contracts and relaxes so that you can breathe is the ....

28. A proton has a .... electrical charge.

29. The function of the 35 mm-long .... tube is to equalise pressure on either side of the ear drum.

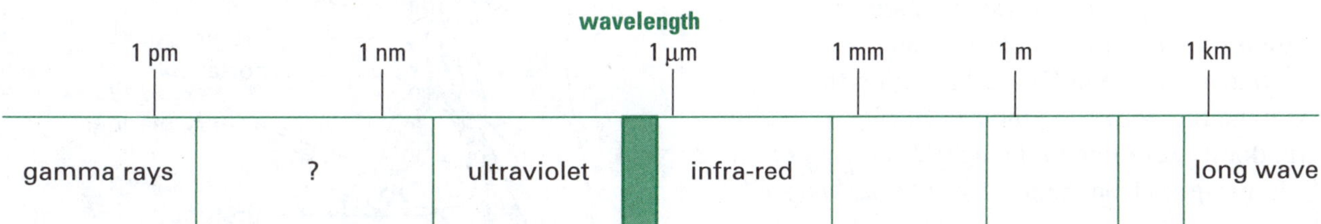

Figure T2.4 Electromagnetic spectrum

30. Frequency .... (FM) is a method of impressing data onto an alternating-current wave by varying the instantaneous frequency of the wave.

31. The .... across a resistor is equal to the product of the resistance and the current flowing through it.

32. The .... system controls our breathing.

33. Many scientists call the slow .... in the temperature of Earth's atmosphere global warming.

34. A volcanic .... is a body of magma that cuts through (and across) adjacent rock, but does not strike through to the surface. The magma hardens to form rock that is always younger than the rocks which surround it.

35. The .... is a region of highly viscous, mechanically weak and easily deforming region of the upper mantle of the Earth on which the tectonic plates float.

36. .... is the molten rock that erupts from volcanoes.

37. The quantity marked X on Figure T2.6 is ....

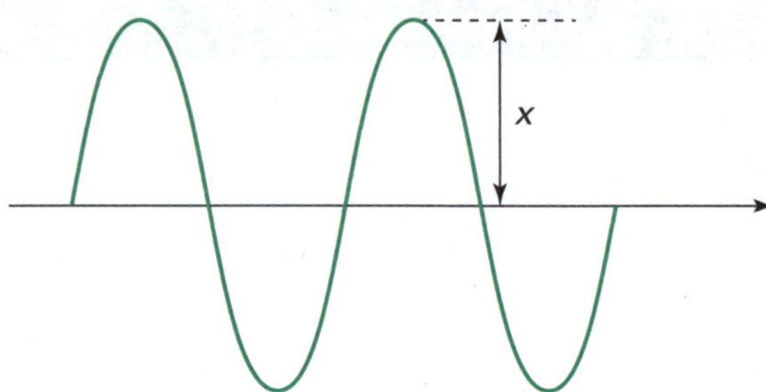

Figure T2.6 Oscillating wave

38. Electrical .... is a measure of how much an object opposes the passage of electrons.

39. The super-continent that existed about 225 million years ago was known as ....

40. A type of window glass can be made from liquid crystals whose molecules align with voltage as shown in Figure T2.7. With the voltage turned off, the molecules align randomly, resulting in the light dispersing so the window goes from being clear to not clear or ....

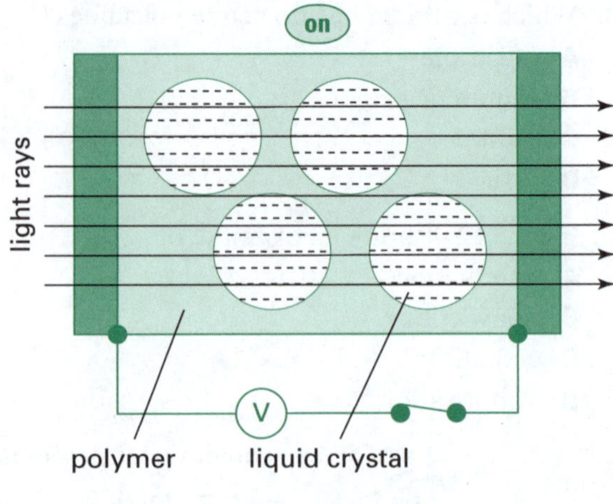

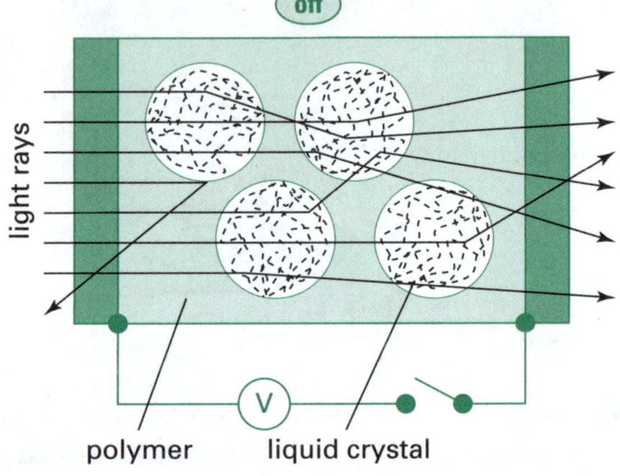

Figure T2.7 Electronic light-controlled glass

Part C: Knowledge, skill and processing data questions

(60 marks)

41. a) Describe the function of each of the components labelled in the diagram of the human urinary system shown in Figure T2.8. *(6 marks)*

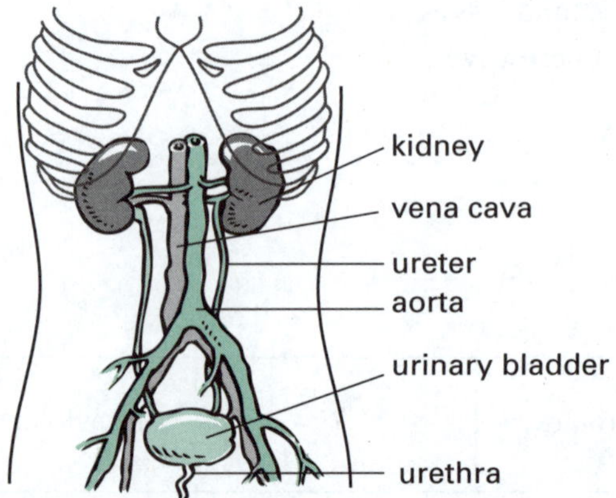

Figure T2.8 Components of the urinary system

b) Where do the wastes in the bloodstream come from? *(2 marks)*

42. Creatinine is a waste product in the blood produced by the normal breakdown of muscle cells during activity. This substance is removed by the kidneys. Normal blood creatinine levels range between 0.6 to 1.2 mg/dL.

a) Explain the units mg/dL. *(1 mark)*

b) A certain person has a creatinine level of 2.0 mg/dL. What could this indicate? *(1 mark)*

c) Figure T2.9 shows a molecule of creatinine.

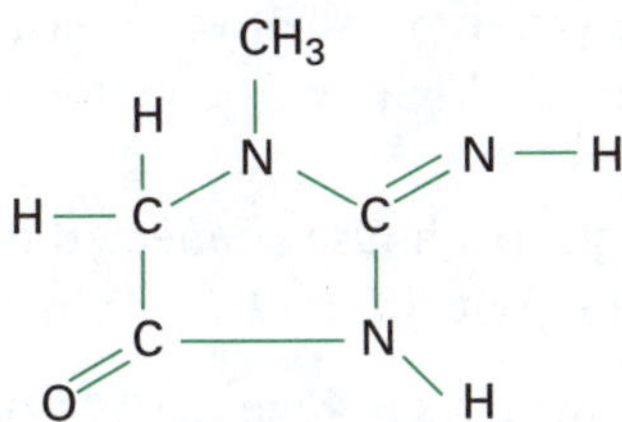

Figure T2.9 Creatinine

i) A student wrote the formula for creatinine as $C_wH_xN_yO_z$. What numbers should replace the letters *w, x, y* and *z*? *(2 marks)*

ii) Each line (—) between atoms represents a pair of electrons shared between those atoms. How many electrons can an atom of carbon share? *(1 mark)*

iii) What is indicated by = between C=N? *(1 mark)*

43. Figure T2.10 shows the energy levels in a hydrogen atom. The numbered orbits refer to the allowed energy levels in the atom.

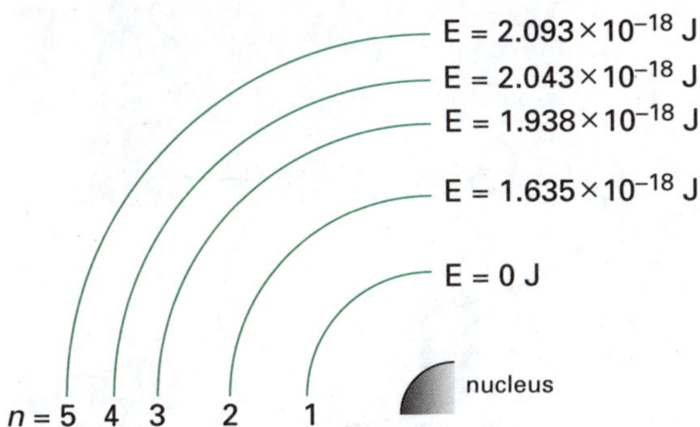

Figure T2.10 Energy levels in a hydrogen atom

a) How much energy, in joules, would be released as an electron moved from the $n = 3$ to the $n = 2$ energy level? *(1 mark)*

b) How much energy does it take to move an electron from the first energy level to the fourth energy level? *(1 mark)*

c) True or false: It takes the same amount of energy to move an electron from $n = 3$ to $n = 4$ as it does to move an electron from $n = 4$ to $n = 5$. *(1 mark)*

d) True or false: The energy difference between energy levels becomes less the further the electron is from the nucleus. *(1 mark)*

44. Figure T2.11 shows the numbers of rabbits in a particular region over a period of time.

Figure T2.11 Rabbit population

a) What time period (how many years) does the graph cover? *(1 mark)*

b) What does the *y*-axis represent? *(1 mark)*

c) Hunters were given permission to kill foxes in the region. About when was this permission granted? *(1 mark)*

d) Describe what happened to the rabbit population in this region when foxes were removed. *(2 marks)*

45. a) What is a reflex arc? *(2 marks)*

b) Figure T2.12 shows an example of a reflex arc. Describe the action shown. *(4 marks)*

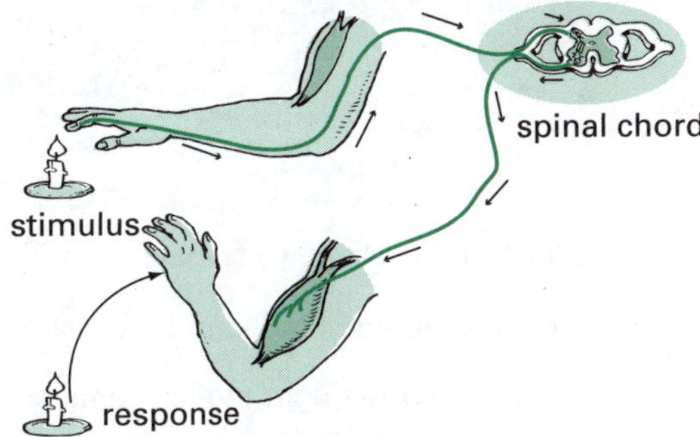

Figure T2.12 An example of a reflex arc

c) Why is a reflex arc important to the safe functioning of an organism? *(1 mark)*

46. There are four types of microbes (microorganisms) that cause disease: viruses, bacteria, fungi and protozoa.

a) Identify the type of microbe from the description. *(4 marks)*

i) Their DNA is contained inside nuclei-like plant and animal cells. Many consist of thread-like structures called hyphae that branch as they grow and expand to penetrate whatever they are growing on.

ii) These are single-celled organisms, only visible with a good light microscope. They are larger than viruses but smaller than ordinary human cells. While they contain DNA, it is not inside a structure like a nucleus.

iii) They are very small, with most being only seen using an electron microscope. They just contain a core of nucleic acid (DNA or RNA) surrounded by a covering layer of protein.

iv) Generally, these are single-celled organisms having true nuclei and a cell membrane, and are very similar to our body cells.

b) Name one disease cause by each of these four types of microbes. *(4 marks)*

47. a) What is a tsunami? *(2 marks)*

b) Describe how the event shown below in Figure T2.13 can cause a tsunami. *(2 marks)*

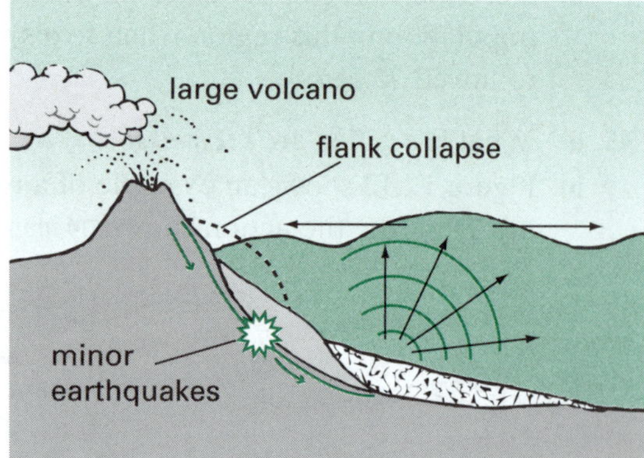

Figure T2.13 Initiating a tsunami

c) How else might a tsunami be caused? *(1 mark)*

48. Figure T2.14 shows light globes connected in series and in parallel.

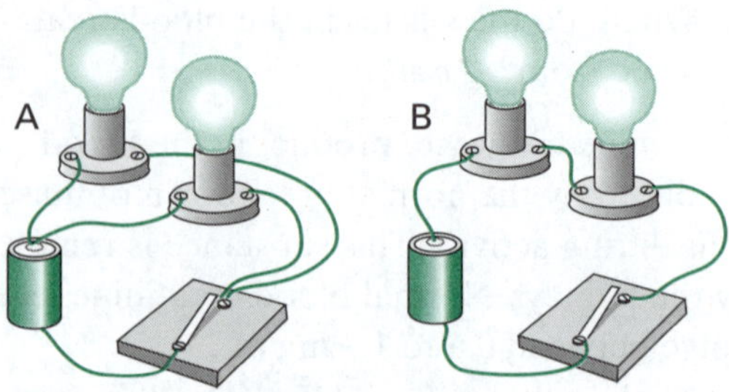

Figure T2.14 Series and parallel circuits

a) Which circuit shows the globes connected in series? *(1 mark)*

b) Explain the differences between series and parallel circuits. What are the features of each? *(4 marks)*

c) Are the lights in a house connected in series or in parallel? Why? *(2 marks)*

49. a) Which two switches in Figure T2.15 need to be closed so that all globes light up? *(1 mark)*

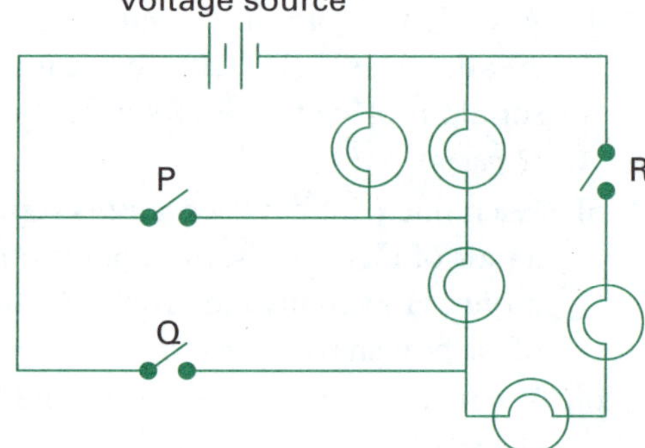

Figure T2.15 Light globes connected in a circuit

b) Four identical globes are connected, as shown in Figure T2.16.

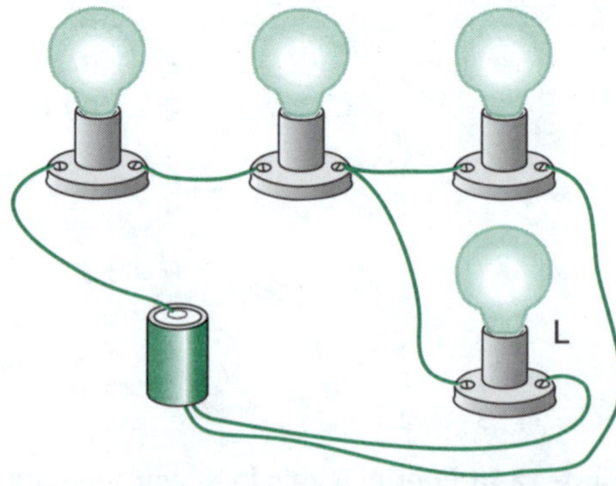

Figure T2.16 Light globes in a series/parallel circuit

Suppose the light globe marked L burns out; comment on the brightness of the remaining three globes. *(1 mark)*

50. Look at Figure T2.17 and answer the questions that follow.

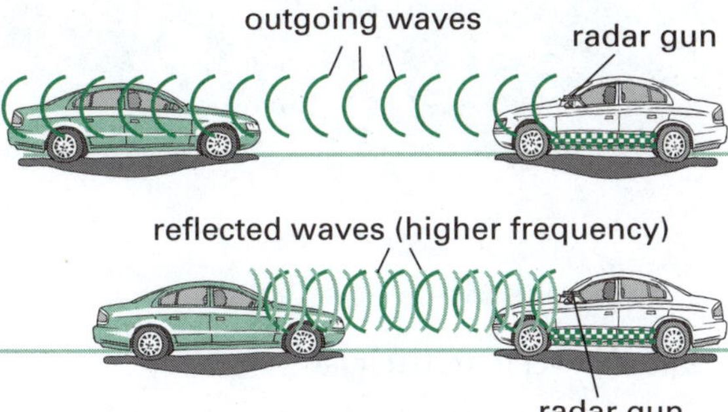

Figure T2.17 A police radar at work

a) Explain how radar can be used to locate an object such as an aeroplane or moving car. *(1 mark)*

b) Explain how police radar is able to determine the speed of a moving vehicle. *(1 mark)*

51. Catalytic converters have been standard on cars since the mid-1970s.

a) What is the function of a catalytic converter? *(1 mark)*

b) Palladium (Pd) and rhodium (Rh) metals in the converter change the nitrogen dioxide and carbon monoxide to nitrogen and carbon dioxide. Are the Pd and Rh changed in the process? Why or why not? *(1 mark)*

c) Palladium (Pd) and platinum (Pt) metals in very small amounts convert hydrocarbons in the unburned petrol and carbon monoxide to carbon dioxide and water.

i) What is a hydrocarbon? *(1 mark)*

ii) What other element is involved in this process? *(1 mark)*

d) Rhodium has an electronic configuration of 2, 8, 18, 16, 1.

i) How many protons does an atom of rhodium have in its nucleus? *(1 mark)*

ii) The most common isotope of rhodium is Rh-103. Determine the number of neutrons in this isotope. *(1 mark)*

Go to pp. 262–266 to check your answers.

Chapter 1 answers

Experiment 1

Analysis:

- Region A is both salty and sweet.
- Region B does not detect anything.
- Region C is bitter.
- Regions D and E are sour. The front regions of D and E (closer to A) also detect salty substances.

Conclusion: Different regions of the tongue contain sensory cells that detect different food tastes.

Experiment 2

Analysis: At all dilutions the effectiveness of the antiseptics in killing bacteria was (in decreasing effectiveness): Ozo; Septro; Chloro.

Conclusion: The more concentrated the antiseptic solution, the fewer bacterial colonies that grow. Ozo brand antiseptic was more effective in killing bacteria, even at low concentrations.

Experiment 3

The results you obtain will depend on how many objects you originally chose and the size of your sample in the recapture. Obviously, the larger the number in the capture and then recapture, the better your results will agree with the actual number.

Results:

2. estimated population $= \dfrac{220 \times 50}{35}$
$= 314$ rabbits.

5. These are just estimates, and estimates will vary but should be around the true figure.

Conclusion: The estimate of the population of objects should be reasonably close to the true number of objects.

Test yourself 1

Part A: Knowledge

1. **D** ✓ A is wrong as this is the site where air enters the body. B is wrong as the diaphragm is a muscle that contracts and relaxes to change the volume of the chest. C is wrong as this is a tube that coveys air to the lungs.
2. **B** ✓ The pulmonary artery conveys deoxygenated blood away from the heart to the lungs where is becomes oxygenated. Thus A and C are incorrect. D is wrong as the oesophagus is part of the digestive system.
3. **C** ✓ A is wrong as connector neurons do this. B is wrong as sensory neurons are not stimulated by hormones. D is wrong as this is the role of the motor neurons.
4. **D** ✓ A is wrong as the endocrine system regulates many organs. B is wrong as the pituitary gland sends out hormones and not electrical stimuli. C is wrong as thyroxine is produced in the thyroid gland.
5. **A** ✓ B is wrong as HIV is a virus. C is wrong as viruses do not respond to antibiotics. D is wrong as there is no red rash over the whole body.
6. **a)** blood ✓; **b)** water ✓; **c)** oxygenated ✓; **d)** air ✓; **e)** brain ✓
7. A/G ✓; B/I ✓; C/F ✓; D/J ✓; E/H ✓

Part B: Skills

8. **a)** X ✓ (the pupil is very small to exclude too much light)
   b) involuntary ✓
   c) Light sensory neurons are stimulated by the bright light and an electrical message is sent to the spinal cord where a connector neuron passes the message to a motor neuron. ✓ The electrical signal moves along the motor neuron to the effector iris muscle that constricts and reduces the size of the pupil. ✓

9. A = Z ✓; B = X ✓; C = Y ✓

10. C, B ✓, E, D, A ✓

11. 1A Organism has long branching structures emerging from one end .... go to 2
1B Organism has no long branching structures .... go to 3
2A Body is multicellular..................... *Penicillium* ✓
2B Body is unicellular......................... *Heliobacter* ✓
3A Organism is spherical with small projections on its surface .... Measles ✓
3B Organism is elongated with many fine hairs on its surface .... *Paramecium* ✓

12. **a)** The broth was boiled to kill any microbes inside the flask and the glass/plastic tubing. ✓
b) This temperature is an optimal temperature for the growth of microbes. It is the temperature of the human body. ✓
c) The control is an open flask of pre-boiled broth with no tubing attached. ✓ This will test whether microbes enter via the air.
d) The test broth stayed clear as air-borne microbes could not travel through the bent plastic tube whereas they could enter the open control flask. ✓

13. Antibiotic A is very effective in killing all three types of bacteria. ✓ Antibiotic B is only effective in killing bacteria X. ✓ Antibiotic C is effective in killing bacteria Y and Z but not X. ✓

14. A/G ✓; B/H ✓; C/F ✓; D/J ✓; E/I ✓

Test yourself 2

Part A: Knowledge

1. **D** ✓ Producers are able to make their own food using CO_2 and sunlight. The types of organisms in answers A, B and C all derive their energy by consuming (eating) other living or dead organisms. A (decomposers) is wrong as they obtain their energy by breaking down dead remains. B (scavengers) is wrong as they eat the remains of animals attacked by carnivores. C (consumers) is wrong as they obtain their energy by eating other organisms.

2. **D** ✓ An omnivore is a living (biotic) component of an ecosystem. Answers A, B and C are examples of abiotic (non-living) components. A is wrong as temperature is not a living thing. B is wrong as rainfall is independent of living things. C is wrong as the sun's energy is not living.

3. **C** ✓ Energy derives from the Sun and is eventually lost from an ecosystem as heat. The other components are recycled. A is wrong as minerals move between parts of the living and non-living world. B is wrong as water is recycled between the living and non-living world. D is wrong as nitrogen atoms cycle between the living and non-living world in a variety of compounds.

4. **A** ✓ A habitat is the natural home or environment of an animal, plant or other organism. B is wrong as a community refers to the living things in an ecosystem. C is wrong as it refers to the total number of a species in a community. D is wrong as it refers to the relational position of a species in an ecosystem.

5. **B** ✓ In order to estimate the number of organisms in an area, you often need to take a sample of the total population. A is wrong as scientific modelling refers to establishing a model that predicts the behaviour of a system under study. C is wrong as tagging refers to the process of attaching a tag to a captured animal before release. D is wrong as culling refers to killing a group of animals and reducing their population numbers (selective slaughtering).

6. **a)** recognise ✓; **b)** food ✓; **c)** population ✓; **d)** energy ✓; **e)** ecology ✓

7. A/J ✓; B/I ✓; C/H ✓; D/F ✓; E/G ✓

Part B: Skills

8. They produce urushiol to protect the plant from herbivores. ✓

9. **a)** As the orchid is pollinated at night, it is pointless flowering during the day. It flowers when the moths are most active. ✓ (Plants flower to attract birds or insects. They don't do it so we can cut their flowers and stick them in a vase.)
b) The scent is to attract the moth. ✓ As there is not enough light at night, a showy and colourful flower would not be of much use. ✓ Moths have a good sense of smell and can fly upwind looking for the source.
c) It should have a long and slender proboscis (tongue) to reach the nectar at the end of the long tube. ✓ (In 1903 such a moth,

Xanthopan morgani was discovered in Madagascar.)

10. Energy has a 'one-way ticket' through an ecosystem ✓ while matter can continually be recycled through an ecosystem ✓.

11. a) Habitat destruction is the process where a natural habitat is made functionally unable to support the species present. ✓

    b) The organisms which previously used the site are displaced ✓ or destroyed ✓, and this reduces biodiversity.

    c) Humans have caused habitat destruction mainly to harvest natural resources for industrial use ✓ and urbanisation ✓. Clearing habitats for agriculture ✓ is the principal cause of habitat destruction. Other important reasons include mining, logging, trawling and urban sprawl.

12. a) Chemical: Animals release chemicals, such as CO_2 and body odours. ✓ Mosquitoes can sense these odours and fly upstream to their source. ✓

    Visual: Moving organisms tell mosquitoes 'I'm alive'. ✓ Also certain colours (of clothing) might be more visually attractive to them than other colours. ✓

    Heat: Warm-blooded animals (birds and mammals) release heat ✓ and this can direct the mosquito to them ✓.

    b) Yes, because it relies on its host for its existence without making a suitable or adequate contribution in return. ✓

    c) The blood provides proteins that the females need to lay eggs. ✓

    d) The purpose is to prevent the blood from the host from clotting. ✓ If the host's blood clotted too soon, the mosquito would not be able to obtain its fill.

13. a) number of fish $= 232 \times 329 \div 16$

    $= 4770.5$

    $= 4771$ (approx.) ✓✓

    b) There should be no immigration, emigration, births or deaths between the release and recapture times. ✓ The chance of being caught should be the same for all individuals in the population. ✓ There should be no possibility that tags become lost between capture and recapture. ✓ (These points are ideal; the first two might be hard to achieve.)

14. a) The snake population would die out ✓, as the only food shown for the snakes is frogs. The insect population should increase ✓, and this should in turn create a greater food source for the spiders.

    b) The rabbit numbers would increase. ✓ This would then have an effect on the number of owls which, in turn, could affect the numbers of seed-eating birds. ✓

Chapter test

Part A: Multiple-choice questions

1. **B** ✓ A is wrong as this is the site where water is removed from the waste that leaves the small intestine. C is wrong as the stomach is where digestion of proteins is started. D is wrong as the liver provides bile to emulsify fats in the small intestine.
2. **A** ✓ The kidney filters the blood of waste materials. Thus B, C and D are all incorrect.
3. **A** ✓ B and C are wrong as the cerebrum controls these functions. D is wrong as this is the function of the brain stem.
4. **A** ✓ You do not have voluntary control of the heart muscle. Thus B, C and D are wrong as you have voluntary control of these functions.
5. **D** ✓ A is wrong as this is the site of sensory nerves. B is wrong as connector neurons are found here. C is wrong as the brain coordinates these functions.
6. **D** ✓ Abiotic refers to non-living things. Answers A, B and C refer to living organisms.
7. **B** ✓ A parasite lives on or in a host and derives its energy from it. The host does not obtain any benefit, so A is incorrect. A mutualist is an organism which lives in close association with another organism and both derive benefit, so C is incorrect. A cannibal eats others of its own species, so D is incorrect.
8. **C** ✓ Predators keep the numbers of prey under control. Without them, prey numbers will increase, so A and B are wrong. With increase in prey, especially if they are plant-eaters, the numbers of plants will decrease, so D is incorrect.
9. **B** ✓ Producers generally are green plants, while a consumer is an animal. So an organism that eats both must be an omnivore. Carnivores are meat-eaters, so A is incorrect; herbivores

are plant eaters, so C is incorrect; prey is the organism being eaten, not the eater, so D is incorrect.

10. **C** ✓ Having no natural predators to keep it in check could wreak havoc in an environment. Think of the Queensland cane toad. Dying off or being killed off are not risks, just a failure of the organism to thrive, so A and B are incorrect. D is wrong as this is the outcome required, so it is not a risk.

Part B: Short-answer questions

11. **a)** Yes ✓ The reflex operates via the spinal cord but the person does not feel the reflex test.
b) No ✓ The connector neurons in the spinal cord cannot transmit the signals to the brain.
c) The patient has no voluntary control of their leg muscles. The brain cannot send signals to the legs to allow the muscles to contract. ✓

12. A/H ✓; B/E ✓; C/F ✓; D/G ✓

13. **a)** A and B ✓
b) A ✓
c) brainstem ✓
d) sweating ✓; hairs lie flat on skin ✓; increased blood flow to skin ✓

14. stimulus = C ✓; receptor = E ✓; control centre = A ✓; effector = D ✓; response = B ✓

15. **a)** **i)** false ✓; **ii)** true ✓; **iii)** true ✓; **iv)** true ✓
b) The cortisol stimulates metabolic processes to provide energy during the time of stress. Glycogen will be converted to glucose to provide this energy. ✓

16. **a)** food ✓; **b)** enzymes ✓; **c)** fat ✓; **d)** bile ✓

17. $L = 100\ \mu m$ ✓

18. **a)** acidosis ✓
b) brainstem (medulla oblongata) ✓
c) through the nervous system ✓
d) pH rises ✓
e) hydrogen carbonate ions ✓
f) It is faster to remove acidic carbon dioxide from the blood by increased breathing rates which are controlled by the nervous system. ✓ The kidneys are stimulated by hormones flowing in the blood. This process is much slower. ✓

19. **a)** See Figure A.1.
b) The populations in various countries in the world where smallpox had been endemic had received sufficient vaccinations ✓ to raise immunity and prevent new cases ✓.

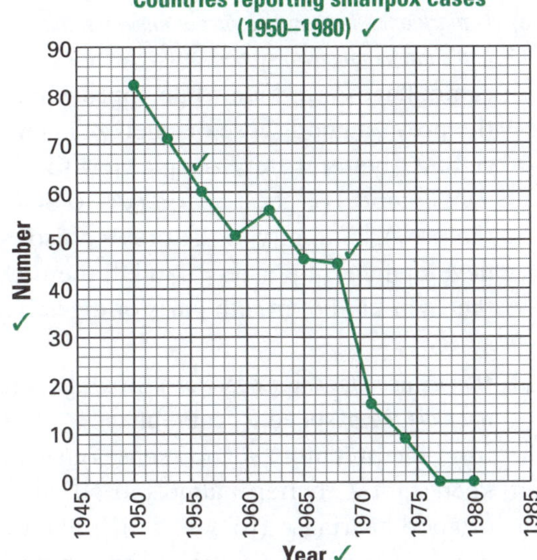

Figure A.1 Smallpox cases

20. **a)** *Heliobacter pylori* ✓
b) antibiotics ✓
c) Scientific findings have to be verified by other research workers. This process takes some time. ✓

21. Koch isolated the tuberculosis bacteria from infected animals. ✓ He then transferred these bacteria to uninfected animals and showed that these animals now exhibited the symptoms of TB. ✓

22. **a)** burning fossil fuels ✓
b) Changes in humidity, temperature and rainfall have produced a milder, warmer climate in Europe. ✓ Disease-carrying mosquitoes have moved from Africa to Europe and have found places to breed and feed. ✓
c) **i)** malaria ✓; **ii)** dengue fever ✓

23. **a)** bacterium ✓;
b) eating contaminated food ✓;
c) malaria ✓; **d)** virus ✓; **e)** measles ✓;
f) fungus ✓

24. Our bodies are healthy when they are in a state of balance. If you are too hot then the body responds by removing the heat. This is achieved by sweating and blood vessels coming close to the skin surface. ✓ If the body is too cold then we start to shiver to increase the amount of heat and thus warm the body. ✓

25. a) virus ✓; **b)** contaminated ✓; **c)** dog ✓; **d)** transmit ✓; **e)** inflamed ✓; **f)** fungus ✓; **g)** cows ✓; **h)** anaemia ✓

26. a) *(1 mark each for any six correct answers; this list is not exhaustive)* speed ✓, stealth ✓, camouflage ✓ (to hide while approaching the prey), a good sense of smell ✓, keen sight ✓ (better if stereoscopic view to gauge depth) or good hearing ✓ (to find the prey), immunity to the prey's poison ✓, development of poison ✓ (to kill the prey), the right kind of mouth parts or digestive system ✓

b) *(1 mark each for any six correct answers; this list is not exhaustive)* speed ✓, camouflage ✓ (to hide from the predator), a good sense of smell ✓, keen sight ✓ (better if almost a 360° view) or good hearing ✓ (to detect the predator), thorns ✓ or prickles (such as the echidna) or poison ✓ (to spray when approached or bitten)

27. a) Its function is to deter animals (particularly insects) from eating it, especially its fruit. ✓

b) It can act as a repellent, keeping insects away by its smell. ✓ The chemical can also act as an insect growth regulator (causing deformities in the insects' young) or as an insect poison. ✓ It can make plants unpalatable to insects or suppress the insect's appetite. ✓

c) It can act as an insect repellent for other plants. ✓ (For example, it can help control pests like leafminers. These feed within leaves and are not normally affected by sprays that only cover the outsides of the plant.)

28. a) *(1 mark each for any two answers)* cutting them down ✓, setting fire to them ✓, poisoning ✓

b) In South America prickly pear was not a problem. Therefore, this would be a good place to go searching for a predator that keeps it in check. ✓

c) Cactoblastis needed to be specific to prickly pear. That is, it should only eat prickly pear. ✓ A problem would be created if, once it had reduced prickly pear numbers, it turned to some Australian native or other plants. (The cactoblastis caterpillar is a black and yellow striped 'grub' that tunnels into and devours the insides of prickly pear.)

d) With plenty of prickly pear (food) around, cactoblastis numbers increased. ✓ Once there was a shortage of prickly pear, then moth numbers decreased through starvation. ✓

e) Once the cactus becomes sufficiently rare, the moths also become rare. They are unable to find and eliminate every last plant. There will always be some prickly pear somewhere that was missed by the moth. ✓ So if it starts to multiply, the moth will eventually find it. ✓ This is perhaps the only factor that keeps the cactus moth from completely exterminating its principal food source.

29. a) This is to attract the animal pollinators. ✓ Using sight (colours) and smell (odours) allows insects to better find them. ✓

b) Wind picks up pollen from one plant and blows it onto another. These wind-pollinated plants would often have long stamens and pistils ✓, to get them into wind currents. As they do not need to attract animal pollinators, they can be of dull colours ✓, unscented ✓ and possibly lacking any petals ✓ since there is no need for insects to land on them.

30. a) shrimp ✓ and rock cod ✓

b) Producer: microscopic plants or seaweed. ✓ Herbivore: mussel or chiton. ✓ (Notice that the shrimp is an omnivore, eating both producer and consumer.)

c) seaweed → chiton → crab → seagull ✓✓

d) Going via the path described in c), it is a third-order consumer. ✓ Via the path microscopic plants → mussel → seagull, it becomes a second-order consumer. ✓

e) The starfish would die out. ✓ Sea anemones would now be wholly dependent on shrimp ✓ in their diets, so shrimp numbers should in turn decrease. ✓

f) Rock cod increasing: no shrimp, more seaweed and hence more chiton, and so more food for rock cod. ✓✓ Rock cod decreasing: sea anemone would be totally dependent on rock cod in their diet, so rock cod numbers would decrease. ✓✓ (What actually happens is more complex, and lies somewhere between these two scenarios.)

31. a) *P. aurelia* grows the fastest ✓, reaching a population size of around 150 units in 5 days, compared to *P. caudatum* which reaches about 75 units in 8 days ✓.

b) The culture needs to be 'seeded' initially ($t = 0$) ✓ with an identical number of each species, hence the identical short distance along the vertical axis ✓.

c) Their respective environments become full of paramecium ✓, putting pressures on the populations to stop increasing.

d) *P. aurelia* is more successful in competing with *P. caudatum* ✓ and so denies nutrients and other resources to it. The population of *P. aurelia* will continue to increase ✓ at the expense of *P. caudatum*, which falls off to zero after several days ✓.

32. a) population size ✓

b) Q ✓

c) Predator size lags slightly behind prey size. ✓ When there is an increase in prey, soon afterwards there is a corresponding increase in predators. ✓ When prey size decreases, similarly the predator size would decrease. ✓ This zig-zag of the curves continues in the predator–prey relationship.

33. availability of food and water ✓; availability of living space (geographical habitat) ✓; nesting materials and shelter ✓; average temperature and temperature variations ✓; natural disasters (hurricanes, earthquakes, floods, droughts) ✓; and, of course, the impact of humans

34. a) protect horse food from being contaminated by the fluids of flying foxes ✓; isolate sick horses ✓; follow standard hygiene and cleaning practices

b) wear protective clothing and avoid physical skin contact with the sick horse ✓; if any fluids do contact unprotected skin, wash off with soap and water urgently ✓

35. a) In sharing the same environment, barred owls are generally more aggressive ✓ and out-compete spotted owls ✓, leading to decreased populations of the native owls. (Barred owls are bigger, more versatile and able to hatch more offspring.)

b) Spotted owls nest in these trees and hunt animals that live in these forests. Destroying trees limits nesting sites ✓ and refuge, and lessens the numbers of organisms ✓ they feed on. (Spotted owls nest in the cavities of or on platforms in large trees and will use abandoned nests of other species. They also need a large amount of land for hunting and nesting.)

c) i) limit the amount of timber that can be removed from the area ✓ (timber companies are required to leave at least 40% of the old-growth forests intact within a 2-km radius of any spotted owl nest or activity site); culling barred owls to give spotted owls a chance of recovery ✓

ii) Local companies argue that limiting logging impacts employment ✓ and the survival of towns and industries that rely on it. ✓ Many animal rights and other activists would criticise controlling barred owl populations through culling. ✓ Also, culling would need to occur indefinitely, otherwise the populations of barred owl will increase again. ✓ And, unless done in expert hands, it may be easy to mistake a protected owl from the one to be culled.

d) See Figure A.2.

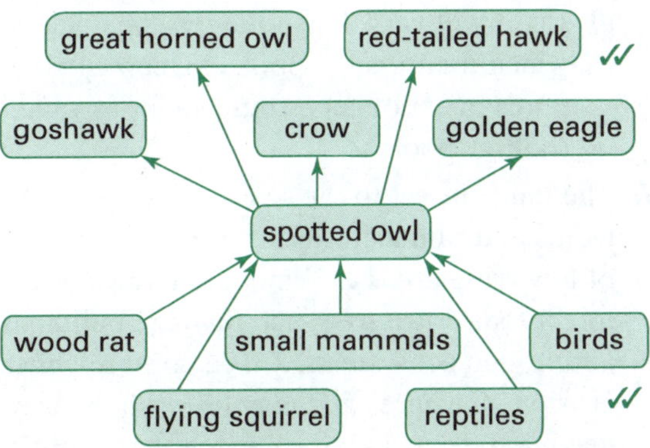

Figure A.2 Spotted owl

e) The decline of spotted owls is an indication of the demise of other species that inhabit these forests. ✓ (The spotted owl is considered an indicator species: a gauge of the health of the ecosystem that provides its habitat.) Society ought to preserve this species and its unique ecosystem because of their aesthetic value. ✓ The owl and its habitat provide immense scientific opportunities for inquiry and for increasing our understanding of this unique ecosystem. ✓ (To date, little research has been done on these forests.)

f) i) % decline = $4000 - \frac{2500}{4000} \times \frac{100}{1}$
$= 37.5\%$ ✓

ii) around 1600 ✓; assuming the decrease in owls continues at the same rate ✓

36. a) During the day, when the tide is out, rock pools are heated by the Sun and warmed up, but during the night they are cooled. ✓ There can be a substantial range in temperature. Also, since water is not moving, temperature changes at the top do not pass downwards easily ✓ so there is a temperature range from top to bottom.

b) On hot days, evaporating water leaves dissolved minerals behind in the remaining water. This makes the rock pool saltier. ✓ On the other hand, when it rains, fresh water makes the rock pool less salty. ✓

c) Plants and animals need oxygen to survive. At night, oxygen levels would fall ✓ as both plants and animals are consuming O_2. In the day, if there are enough plants in the rock pool O_2 levels may increase. ✓ If there are not enough plants, it could decrease.

d) Light will vary depending on the angle at which it enters the pool. ✓ Deep narrow pools may not get as much light as wide, shallow pools. ✓

37. The land closest to the sea is subjected to high winds and rain ✓, resulting in possible flooding of low-lying areas ✓. Wind along the coasts might tear down trees and damage buildings, and fragile ecosystems. ✓ The coast might suffer erosion ✓, where sand may be washed away leaving bare rocks. Storm surges can inundate low-lying coastal areas ✓. While coral reefs are often damaged, many can cope and recover. ✓ But if cyclones are too frequent then they don't have time to recover and hence die. At sea, large waves can cause problems for ships in the area. ✓

38. a) arid, dry areas ✓ (deserts)

b) When it is dry up top there is little food so the frog needs to conserve its energy ✓, and it can do this in a moist, cocooned environment. ✓

c) The cocoon it forms prevents any water from escaping ✓, and its bladder is able to hold large volumes of water. ✓

d) When rains do come (infrequently), puddles may only remain for a few weeks. ✓ So the frog must lay its eggs quickly, and the tadpoles that form need go through metamorphosis to adult frogs rapidly before the water dries up. ✓ (Even though they live in arid regions, *Cycloranas* still need water to complete their life cycle.)

39. a) During times of drought, rabbits would be trying to conserve their energies ✓ and not have many young. ✓ Drought would also limit the numbers of rabbits as there is a limited amount of food around. ✓ It would also be easier for farmers to kill rabbits as they can put out poison bait to which rabbits would be attracted. ✓ But when the rains come there is lots of green feed ✓ available, making it very difficult to get them to eat the bait. And now rabbit populations can again begin to surge, producing more young (kits). ✓

b) Suppose these control methods kill, say, 80% of rabbits. The 20% remaining are able to reproduce in a good year and replenish the ones lost. ✓✓ A level of at least 90 to 95% control is often regarded as having any long-term effect.

40. a) See Figure A.3.

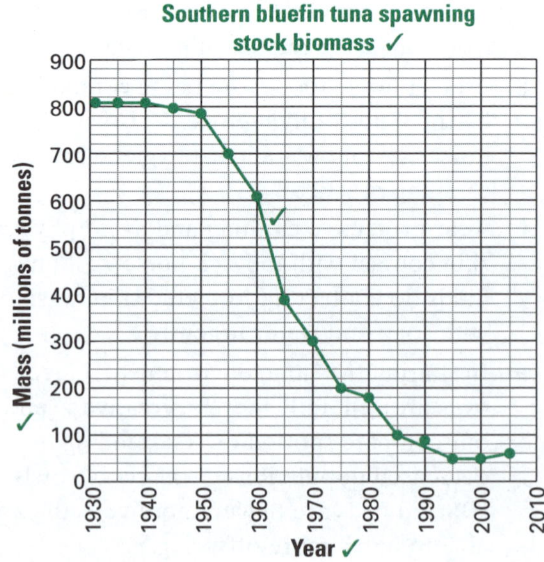

Figure A.3 Southern bluefin tuna spawning stock biomass

b) in the 1950s ✓; this is when the fish stocks began to decline ✓

c) Up until the late 1940s, the spawning stock remained fairly constant. ✓ Then it went through a steep decline ✓ until about 1995, where it has again flattened out. ✓

d) *(Several reasons may be possible, but only one is required for the mark.)* The fish stocks are so low that finding and fishing them becomes uneconomical. ✓ The remaining fish stocks are sparse so it takes a while for fishermen to find and catch them, and so they don't catch the numbers they did before. ✓ A halt or limit has been placed by countries to allow fish numbers to recover. ✓

e) Australia has control over its ocean waters, so it can put a moratorium on, or at least drastically limit, the number of southern bluefin tuna that can be caught. ✓ For fish in international waters, it can collaborate with other countries to set a worldwide quota. ✓

Chapter 2 answers

Experiment 1

Analysis: Atoms combine to form compounds in simple whole number ratios. Therefore, the weight of a compound is the sum of the atomic weights of each component element.

Prediction: Carbon and oxygen atoms combine in simple ratios such as 1:1 and 1:2.

Thus: Compound (1) = CO (i.e. 1 atom of C and 1 atom of O)

Compound (2) = CO_2 (i.e. 1 atom of C and 2 atoms of O)

The only difference between the two compounds is an 'O' atom. The difference in reacting weights is 44 – 28 = 16 g. Therefore, if the atomic weight of carbon is 12, then the atomic weight of oxygen is 16.

Conclusion: Carbon reacts with oxygen in a 1:1 atomic ratio and the atomic weight of oxygen is 16 u compared with carbon having a weight of 12 u.

Experiment 2

Analysis: The activity series for the five metals tested is determined by the rate of gas evolution. Calcium produced the greatest rate of effervescence and the magnesium was next. Following magnesium, the order was zinc, iron and finally copper which did not produce any gas.

Conclusion: Metals have different rates of reaction when placed in dilute hydrochloric acid. Of the five metals examined, calcium is the most active metal and copper is the least active. The full order is: calcium; magnesium; zinc; iron; copper.

Experiment 3

Analysis: The length of the candle has no effect on the flame and the candle diameter does produce some effect up to a certain diameter, after which no further increase in diameter has any effect. This is related to the pool of hot wax that can feed the wick.

Flames will not burn on wicks that are too long. The wick will burn down to establish a flame just above the pool of wax.

Conducting heat away will decrease the flame size and may cause the flame to go out.

Conclusion: Flames burn best when there is sufficient heat at the top of the wax candle to form a pool of melted wax that rapidly diffuses and vaporises on the wick to produce a combustible fuel/air mixture.

Test yourself 1

Part A: Knowledge

1. **B** ✓ The quote describes the plum-pudding model devised by Thomson. A, C and D are incorrect as these scientists had different concepts of the atom.
2. **C** ✓ Oxygen has a total of eight electrons in its orbits. Two of these electrons fill the first shell, leaving six for the second shell. So A, B and D are wrong.
3. **D** ✓ The first (innermost shell) is full as it contains two electrons, so is the second shell (8 electrons) and the third shell (18 electrons). The only shell which isn't full is the outermost.
4. **C** ✓ Rutherford made this comment after viewing the results of the gold-leaf experiment.
5. **B** ✓ The nucleus contains both protons and neutrons; electrons exist outside the nucleus. Electrons are much lighter than either protons or neutrons, which are *almost* the same in mass.
6. **a)** neutron ✓; **b)** Dalton ✓; **c)** two ✓; **d)** radioactive ✓; **e)** Democritus ✓
7. A/H ✓; B/I ✓; C/F ✓; D/J ✓; E/G ✓

Part B: Skills

8. a) Radioactive substances emit particles and/or radiation. ✓ This nucleus is radioactive. ✓
   b) gamma rays ✓; high energy ✓, no charge, highly penetrating
   c) particle = beta particle (single particle and not a composite particle like an alpha) ✓; *(any one of the following feature scores 1 mark)* small mass, negative charge, medium penetration
9. a) 2.5 ✓; b) 2.8.2 ✓; c) 2.8.8 ✓
10. Choose a radiation source that will pass through a small stack of paper. The thicker the stack, the less radiation that can penetrate and be picked up by the detector on the other side. ✓ The amount of radiation detected can be converted to a thickness measured in millimetres. ✓
11. See Table A.1. *(1 mark for each correct answer)*
12. a) protons = 15 ✓; neutrons = 31 – 15 = 16 ✓
    b) protons = 27 ✓; neutrons = 33 ✓
13. a) lead ✓; p = 82 ✓; n = 125 ✓; e = 82 ✓
    b) radon ✓; p = 86 ✓; n = 136 ✓; e = 86 ✓
    c) technetium ✓; p = 43 ✓; n = 56 ✓; e = 43 ✓
14. a) boron-11 ✓
    b) iii ✓
    c) That answer should be chosen because 20% of the atoms have a mass of 10, and 80% have a mass of 11. ✓ The average of the two numbers would be close to 11, but not quite. ✓ (While you don't need to be able to calculate the weighted average, it equals $\frac{20}{100} \times 10 + \frac{80}{100} \times 11 = 10.8$.)

Test yourself 2

Part A: Knowledge

1. **B** ✓ Acids attack carbonates to produce carbon dioxide. Thus, A, C and D cannot be correct.
2. **C** ✓ Propane is a hydrocarbon which will burn in air to produce carbon dioxide. A is incorrect as phosphorus oxide would form. In B, sulfur dioxide would form and in D water would form.
3. **D** ✓ Grapefruit is sour and contains citric acid. The other answers are incorrect because A and B are bases and C is neutral.
4. **A** ✓ If the bush ahead has already burned, then the advancing fire has no fuel. B is incorrect as there is always air available. C is incorrect as convection currents are already established as the fire advances. The advancing fire produces its own radiant heat and burning the scrub ahead has no effect on this, so D is incorrect.
5. **C** ✓ A salt is any ionic compound formed by neutralisation. A is wrong as water is only one of the products. B and D are wrong as the acid and base are not named and thus specific compounds cannot be determined.
6. a) hydrogen ✓; b) reactants ✓; c) water ✓;
   d) neutralised ✓; e) barium ✓
7. A/G ✓; B/J ✓; C/F ✓; D/I ✓; E/H ✓

Part B: Skills

8. a) green ✓ (an alkaline solution like sodium hydroxide turns the pigment green)
   b) control ✓
   c) It was acidic. ✓
9. a) A and B ✓ are combustion reactions in which oxygen is the oxidiser.
   b) A: $2H_2 + O_2 \rightarrow 2H_2O$ ✓
   B: $CH_4 + 2O_2 \rightarrow CO_2 + 2H_2O$ ✓
   C: $CO_2 + H_2 \rightarrow CO + H_2O$ ✓
10. a) 24.4 °C ✓
    b) chemical energy to heat energy ✓
    c) red to purple to blue ✓
    d) basic ✓ (an excess of sodium hydroxide is present)
    e) sodium sulfate ✓
    f) by evaporation ✓

Table A.1 Properties of some elements and subatomic particles

Element	Symbol	Atomic number	Mass number	Number of neutrons	Number of protons	Number of electrons
Oxygen	O	*8*	*16*	8	8	8
Carbon	*C*	6	*12*	*6*	6	6
Fluorine	*F*	9	19	*10*	9	*9*

11. A = one ✓; B = three ✓; C = three ✓
12. a) aluminium hydroxide + nitric acid → aluminium nitrate + water ✓
    b) copper hydroxide + sulfuric acid → copper sulfate + water ✓
    c) ammonium hydroxide + carbonic acid → ammonium carbonate + water ✓
13. a) neutralisation ✓
    b) to monitor the pH changes during the neutralisation ✓
    c) The reactions are exothermic since heat is released. ✓ The heat released during the reaction is taken up by the surroundings warming them up. This can then be registered by a temperature increase on a thermometer.
    d) Y is magnesium carbonate ✓ because the neutralisation leads to the formation of a gas. Acids on carbonates release carbon dioxide gas. ✓
    e) In X the final pH is high (> 10) ✓ whereas in Y the final pH is lower (between 8 and 9) ✓.
14. The oxygen has been partially used up and carbon dioxide has formed. ✓ The decreasing oxygen levels and the increasing carbon dioxide levels put out the flame. ✓ Carbon dioxide is used in fire extinguishers as it deprives the flame of oxygen.

Chapter test

Part A: Multiple-choice questions

1. **A** ✓ Neils Bohr first came up with a radical model of the atom which had electrons orbiting around a nucleus in special orbits. All other orbits just were not possible, but electrons could 'jump' between these given orbits. B, C and D are therefore incorrect.
2. **D** ✓ Democritus was one of the founders of ancient atomist theory. The atomists held that there are smallest indivisible bodies from which everything else is composed, and that these move about in an infinite empty space (void). Certainly Democritus was not the first to propose an atomic theory, but he was one of the earliest to popularise it. Therefore, A, B and C are incorrect.
3. **C** ✓ There are 40 – 19 = 21 neutrons. K-40 means there is a total of 40 protons and neutrons, so A and B are wrong. D is incorrect as K-39 and K-40 have the same number of protons, not neutrons.
4. **A** ✓ Protons and neutrons have very similar masses; neutrons are slightly heavier. Electrons are much, much lighter, so B and C are wrong. D does not answer the question about mass; rather, it refers to charge.
5. **B** ✓ 2 + 8 + 5 = 15 electrons = 15 protons in a neutral atom. The mass number is greater than 15, as neutrons have not been added, thus making C incorrect. From a periodic table, this element (Z = 15) is phosphorus, which is neither a metal nor unreactive making both A and D incorrect.
6. **B** ✓ An acid reacts with a metal to form hydrogen gas and a salt. A is wrong as the first reactant cannot be an acid or a base. C is incorrect as there are three products when an acid reacts with a carbonate. D is incorrect as none of the reactants can be oxygen, which is a diatomic molecule.
7. **B** ✓ A, C and D are wrong as these answers are not the definition of oxidation.
8. **A** ✓ B is wrong as respiration is not concerned with cleaning up pollution. C is wrong as catalytic converters are fitted to car exhaust systems. D is wrong as fine filters cannot trap gases.
9. **D** ✓ Sublimation requires an energy input whereas A, B an d C are all exothermic.
10. **C** ✓ Silver and copper are the least reactive metals in this list. Thus, A, B and D are incorrect.

Part B: Short-answer questions

11. Atoms are too small to be seen clearly using current techniques. Hence, making models of atoms allows people to picture (imagine) atomic structure. ✓ What scientists know about them has been obtained by observing the way a large number of samples of matter behave. Scientists find it useful to make diagrams or actual physical models to show what an atom 'looks' like. This can lead to predictions and future discoveries. ✓ Models also allow scientists to explain how things behave the way they do. Models, of course, are always subject to change with future discoveries improving on current models. ✓ It is very unlikely that

anyone will ever see atoms in the same sense that you see diagrams and drawings of atoms. ✓

12. a) Alpha particles are positively charged ✓, while beta particles are negatively charged ✓. So they would be affected differently by the electric field. ✓

b) Beta particles are just electrons that are much lighter ✓ than alpha particles (which consist of 2p + 2n) ✓. So it is easier to bend lighter particles than heavier ones. ✓

c) Gamma rays have no charge ✓, so are unaffected by electric fields ✓.

13. See Table A.2.

Table A.2 Properties of radiation

Radiation	Made up of	Charge	Distance travelled through air
alpha particle	2 protons + 2 neutrons ✓	+2 ✓	a few centimetres ✓
beta particle	electron ✓	–1 ✓	several metres ✓
gamma ray	electromagnetic radiation ✓	0 ✓	over hundreds of metres ✓

14. a) 6 units ✓ Since only 2 units of Al remain, the other 8 – 2 = 6 units must have been converted to Mg.

b) 1440000 years old ✓ After one half-life (720000 years), half of the original 8 units of Al remain. After a further 720000 years, half of that half (i.e. 2 units) remain. So its age is 720000 + 720000 = 1440000 years.

c) After 1 half-life ✓; the Al would have halved from 8 to 4 units, as the other 4 units became Mg.

d) In a further half-life, only half of the current amount remains. See Figure A.4.

15. a) See Table A.3.

b) X is an alpha particle. ✓ The daughter nucleus contains two less protons and two less neutrons. An alpha particle consists of 2p + 2n. Y is a beta particle. ✓ The daughter nucleus contains one more proton, but one less neutron. This means a neutron has decayed: $n \rightarrow p + e$

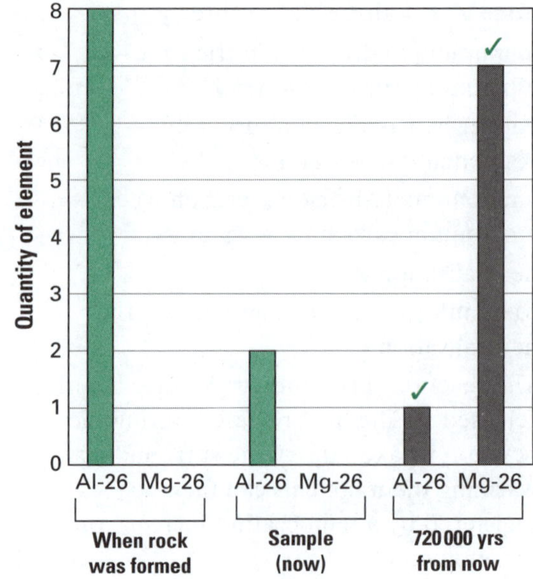

Figure A.4 Aluminium-26 decay

Table A.3 Properties of parent and daughter nuclei in nuclear reactions

	Symbol	Name	Proton number	Neutron number
parent nucleus	U-235	uranium	92 ✓	235 – 92 = 143 ✓
daughter nucleus	Th-231 ✓	thorium ✓	90 ✓	231 – 90 = 141 ✓
parent nucleus	K-40	potassium ✓	19 ✓	40 – 19 = 21 ✓
daughter nucleus	Ca-40 ✓	calcium	20 ✓	40 – 20 = 20 ✓

16. F, C, ✓ A, G, ✓ E, H, ✓ B, D ✓ *(½ mark for each correct answer)*

17. a) Gamma rays are more penetrating. ✓ Alpha particles would barely be able to pass through the skin ✓, as they are so large and charged.

b) High-energy gamma rays can be used to treat some types of cancer, since the rays kill cancer cells. ✓ They do this by damaging the DNA of the tumour cells, preventing them from multiplying. ✓ Radiation does most damage to fast-growing cells, such as cancer cells.

c) In gamma-knife surgery, multiple concentrated beams of gamma rays are directed on the growth in order to kill the cancerous cells. The beams are aimed from different angles to concentrate the radiation onto the growth ✓ while minimising damage to surrounding tissues ✓. Gamma rays need to pass through healthy tissue to get to the tumour. The less radiation that passes through a particular tissue, the less damage that is done. During treatment the patient is kept perfectly still and the gamma radiation source moves in a circle to limit the damage done to healthy cells.

18. In order to locate a leak in an underground pipe, a very small amount of radioactive material giving off gamma rays is introduced into the pipe. ✓ On the ground above the pipe, a detector follows the pipe's course. ✓ When there is a spike (sudden increase) in activity ✓ and little or no activity before and after that point ✓, the leak has been found. This is where a pool of material (including the radioactive tracer) is concentrated in the soil.

19. **a)** neutron ✓

b) Radioactive carbon reacts with oxygen in the atmosphere ✓, forming carbon dioxide.

c) Plants are able to convert CO_2 from the atmosphere ✓ into simple sugars during photosynthesis ✓.

d) No ✓, while radioactive C is incorporated into CO_2, there is plenty of CO_2 in the atmosphere which does not contain radioactive carbon ✓. Both types enter the food chain; plants don't discriminate.

e) While the tree is alive, it continues to take in CO_2. ✓ After the tree has been cut down, whatever CO_2 it has is it ✓; radioactive C in the carbon dioxide continues to decrease over time.

20. **a)** around 27 g ✓

b) The time until half (13.5 g) of Rn-220 remains is about 57 seconds. ✓

c) 136 seconds ✓

d) After 90 seconds, 9 g Rn-220 remains, so 27 – 9 = 18 g ✓ has been converted.

21. D. In order to date radioactive carbon, there must be carbon present in the sample. ✓ Cloth, charcoal and bone are all remains of carbon-based organisms. ✓ Clay pots don't contain carbon. ✓

22. E, B, ✓ H, G, ✓ J, A, ✓ F, D, ✓ C, I ✓ *(½ mark for each correct answer)*

23. **a)** See Figure A.5. *(Take 1 mark off for each incorrect feature. The shells have been named, but this is not a requirement for the marks.)*

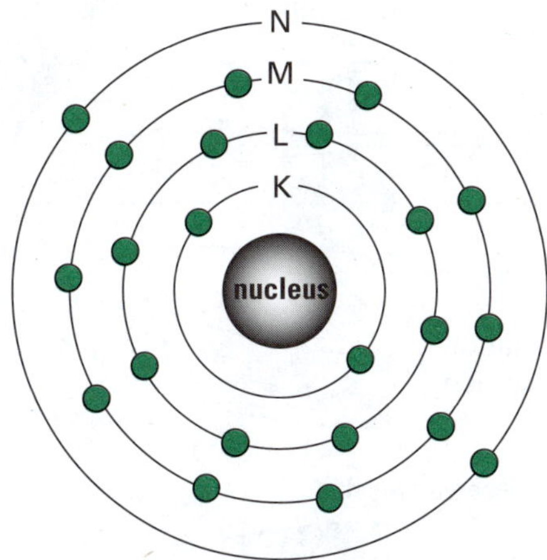

Figure A.5 Electron shells

b) 22 ✓ (2 + 8 + 10 + 2)

c) titanium ✓

d) 48 – 22 = 26 ✓

24. Up until 1911, Thomson's 'plum pudding' model of the atom described an atom where positive and negative charges were distributed regularly. In this model electrons were imbedded in a solid mass of positive charge. ✓ Rutherford devised the famous gold foil experiment and exposed the flaw in Thomson's model. ✓ In the experiment (which, by the way, was conducted by physicists Hans Geiger and Ernest Marsden) a sheet of gold foil was bombarded with alpha particles. According to the Thomson model, the scientists expected all the particles to pass through the foil barely altered in direction. The thinly spread positive charge would have a small, but consistent, effect on the very high positive charge of the alpha particles. ✓ Instead, a small percentage scattered, which showed that they had hit, or been bounced off, something very small and very dense. ✓ Clearly something having a strong positive charge was repelling some of the alpha particles. Rutherford concluded that the positive charge must be condensed into one place. And, since most of the beams passed through without any deflection, the rest of the atom had to be made up of empty space. This led Rutherford to the idea of a nucleus with a positive charge around which electrons orbited, somewhat like planets around a sun. ✓ See Figure A.6.

expected results using Thomson model

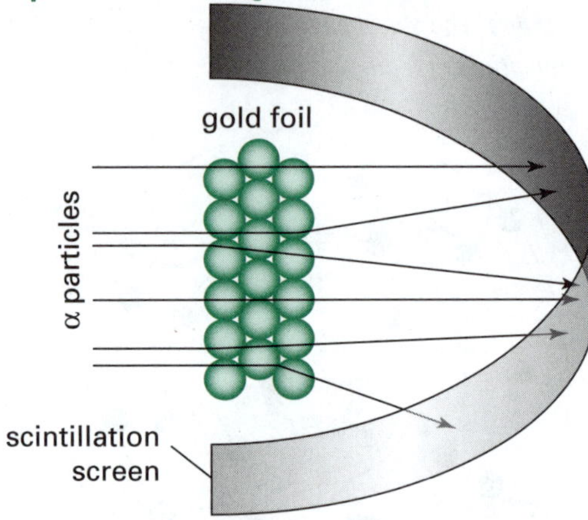

actual observed results

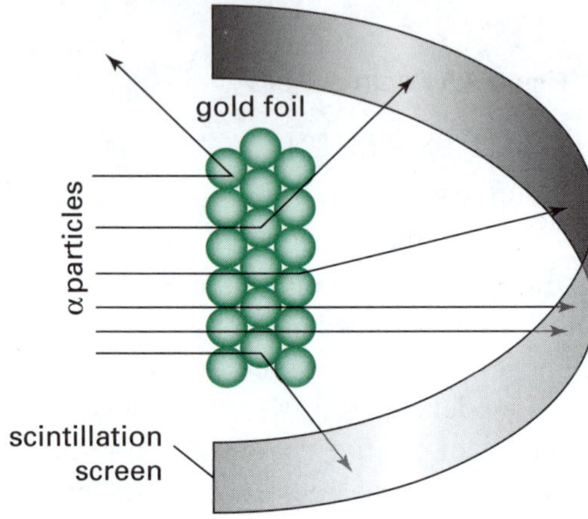

Figure A.6 Rutherford's experiment ✓

25. a) P shows gamma rays ✓; Q shows beta particles ✓

b) 1 cm ✓

c) around 17 cm ✓

d) After 8 cm, about 68% of radiation P is getting through ✓ whereas only 8% of radiation Q is getting through ✓.

26. a) Atoms and matter are conserved in a chemical change. ✓

b) Reactants are all diatomic molecules ✓; products are triatomic molecules ✓.

c) combustion reaction (H burns in O_2) ✓

27. a) sulfuric acid ✓; hydrochloric acid ✓

b) H_2SO_4 ✓; HCl ✓

c) hydrogen ✓

d) carbon dioxide ✓

28. a) 8 ✓

b) ethane + oxygen → carbon dioxide + water ✓

c) $2C_2H_6 + 7O_2 \rightarrow 4CO_2 + 6H_2O$ ✓

d) Pass the gas into limewater. ✓ If CO_2 is present in the gas, the clear liquid will turn a milky white colour. ✓

e) The heat released from the combustion keeps the combustion process going. ✓

29. a) heat released ✓; colour change ✓

b) sulfur dioxide ✓

c) sulfur + oxygen → sulfur dioxide ✓✓

d) $S + O_2 \rightarrow SO_2$ ✓✓

e) the gas is acidic, because the indicator turned red ✓

30. a) zinc sulfate ✓; **b)** calcium nitrate ✓;

c) sodium chloride ✓

31. a) pink to green (ammonia is a base) ✓

b) pink to red (lemonade is acidic) ✓

32. a) There is no ignition source (the temperature is too low for ignition to occur). ✓

b) Heat is liberated to the rest of the mixture and raises the temperature above the ignition point. ✓

c) Helium does not combust in air. ✓

33. a) excludes the oxygen required for the combustion ✓

b) lowers the temperature so that ignition cannot occur ✓

34. a/iii ✓; b/i ✓; c/ii ✓

35. a) carbon dioxide ✓, CO_2 ✓

b) Soot is formed when the oxygen concentration or the temperature of combustion is low. The soot can contaminate the air and be breathed in, causing respiratory difficulties. ✓ It blackens buildings and other parts of the environment. ✓ The soot may also contain carcinogenic compounds. ✓

36. a) Oxygen is required for both processes; energy is released. ✓

b) Respiration occurs at body temperature and combustion occurs at a high temperature. ✓

c) Limewater test: the limewater goes milky-white in the presence of carbon dioxide. ✓

37. a) water (vapour) ✓

b) $2H_2 + O_2 \rightarrow 2H_2O$ ✓

c) Water is produced; this product is natural and will not be a contaminant. ✓

d) It is dangerous as it can explode when mixed with air in the presence of a spark or flame. ✓

38. a) strength of acid (must be kept constant) ✓; size of zinc pieces (to be varied) ✓; weight of zinc pieces (controlled) ✓; volume of acid used (controlled) ✓; type of acid (controlled initial acid) ✓; temperature (controlled) ✓

b) i) the size of the pieces of zinc ✓

ii) No ✓; this would not be a fair test as there would be more than one variable. Leif should conduct one set of experiments with the same volume of acid and different sizes of zinc. In another set he can use a larger volume of acid to determine if that makes any difference. ✓

39. a) storage temperatures ✓; age of citrus juice ✓; type of citrus juice ✓; containers in which the juice is stored ✓

b) *(any 3 of the following)* ✓✓✓ sealable glass containers for the juice; storage cabinets in which the temperature can be controlled to any desired level; standard bases and indicators (e.g. sodium hydroxide and universal indicator) or a pH meter; flasks; thermometers; measuring cylinders

c) Citrus juices will have a lower pH (higher acidity) if they are stored at high temperatures for a long time. ✓ This will be shown by a pH that decreases as storage time and temperature increase. ✓

d) Method:

1. Select three different types of squeezed citrus juice (lemon, orange and grapefruit). ✓
2. Pour 100 mL of each juice into labelled glass containers and stopper them securely. Label the flasks. ✓
3. Store the flasks in temperature-controlled cabinets for known lengths of time. ✓
4. Remove each flask at the allotted time and test its acidity with a pH meter, or by adding a universal indicator and adding sodium hydroxide to neutralise the acid. Record the pH or volume of base used. ✓
5. Tabulate all results.

40. a) Photosynthesis is a process in which carbon dioxide and water combine together using solar energy collected by chlorophyll to produce glucose and oxygen. ✓ Respiration is the process in which glucose reacts with oxygen to form carbon dioxide and water. ✓

b) the grana in the chloroplasts ✓

c) mitochondrion ✓

d) 6 ✓

Chapter 3 answers

Experiment 1

Analysis:

1. They represent each side of the spreading oceanic plate.
2. The trench is formed where the ocean plate moves beneath the continental crust.
3. It represents the crust and mantle of the Earth.
4. Divergent boundaries are spreading zones. This occurs as the paper moves out of the mid-ocean ridge.
   Convergent boundaries include subduction zones. This occurs as the paper moves underneath and into the cardboard.
5. It models the process of sea floor spreading and subduction but the model does not show the movement of the continent away from the mid-ocean ridge.

Conclusion: A paper and cardboard box model was used to model movement of plates at mid-ocean ridges and subduction zones.

Experiment 2

Analysis: See Figure A.7.

Volcanically active zones:

- Pacific Rim
- Indonesia
- Mid-ocean ridges (Iceland)
- Hot spots (Hawaii)
- Eastern Mediterranean
- Caribbean

Conclusion: The major volcanic zones are located near plate boundaries.

Experiment 3

Analysis:

Major earthquake plot: See Figure A.8.

1. Yes; Earthquakes are generally located at the margins of the tectonic plates.
2. As the plates move against each other forces are created which are ultimately released in the form of an earthquake.

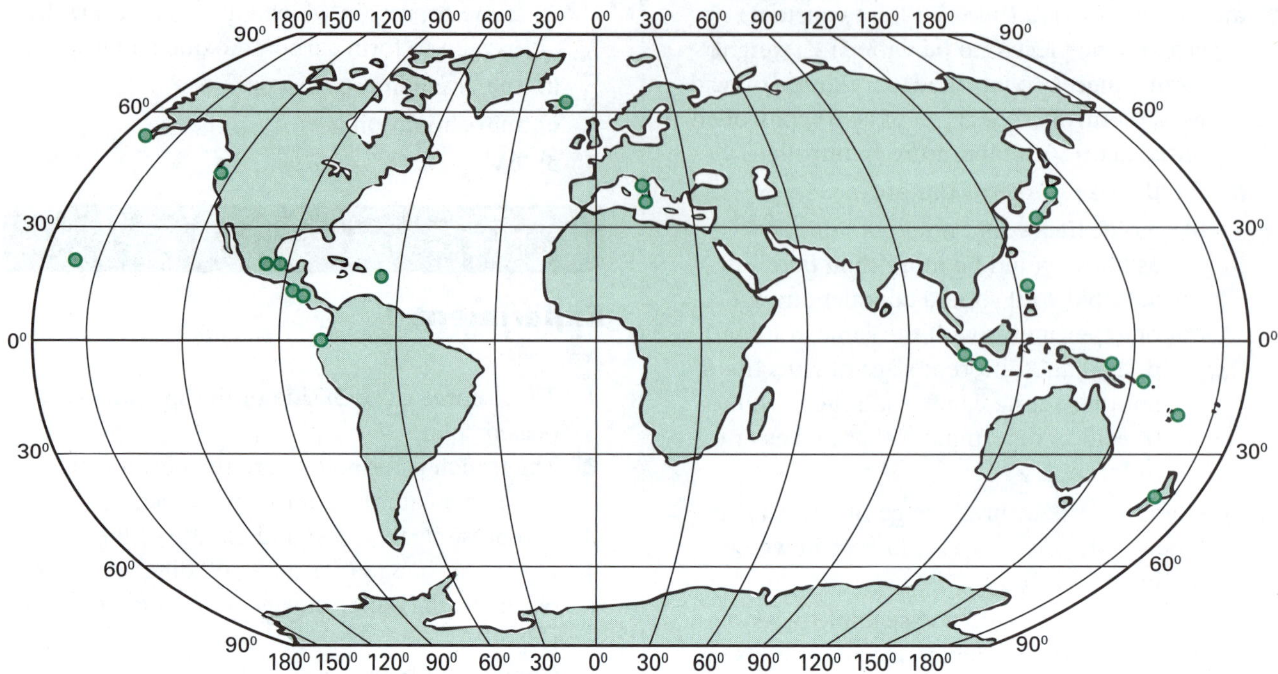

Figure A.7 Volcano plot

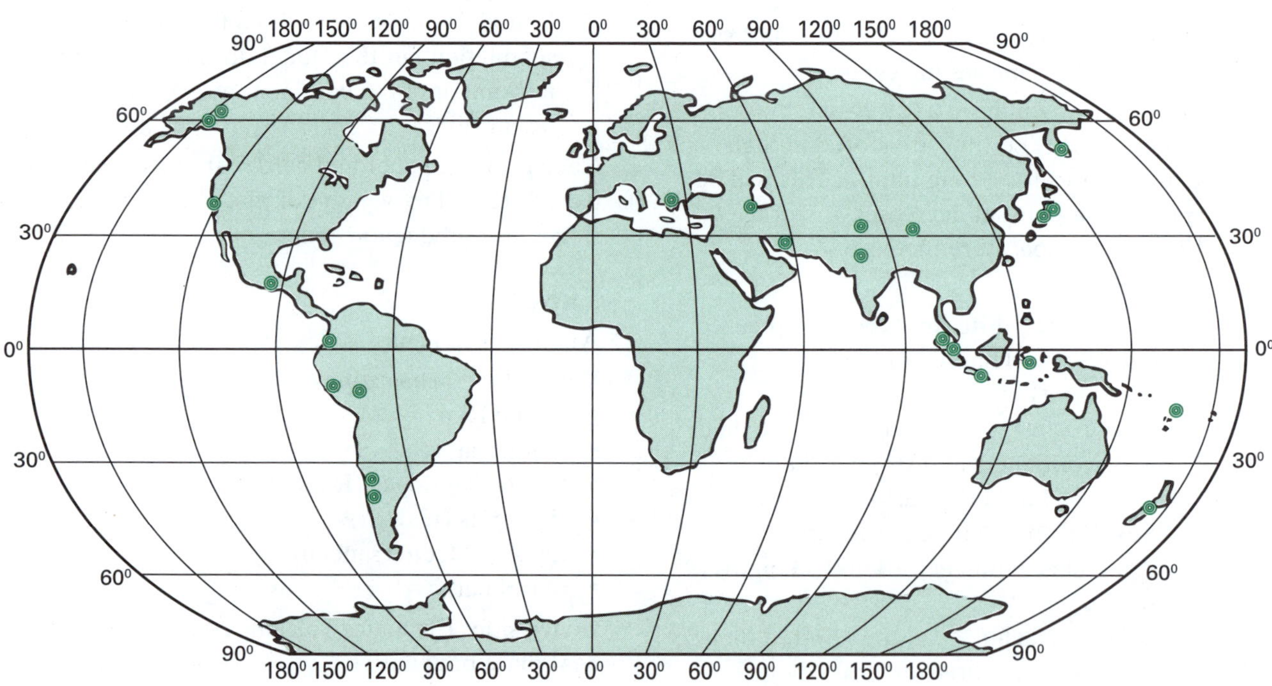

Figure A.8 Earthquake plot

Conclusion: The sites of major earthquake activity are generally at the margins of the tectonic plates.

Test yourself 1

Part A: Knowledge

1. **D ✓** It is a hypothesis as his collected data is consistent with the proposal. It cannot be a fact as many of these changes happened in the distant past and no-one was there to record them, so A is incorrect. It is not an observation as no-one was there to observe the changes, so B is incorrect. It is not a prediction as he is not suggesting what will happen in the future, so C is incorrect.

2. **B** ✓ As there was a spreading zone between India and Australia, then they moved apart but now we are not moving apart and this indicates we are on the same plate now. Thus A, C and D are wrong.
3. **A** ✓ B, C and D are therefore wrong as they are different types of zones.
4. **D** ✓ Collision zones occur between continental plates, so A, B and C are wrong.
5. **B** ✓ New crust is created here and so it is a convergent boundary. A is wrong as this is a destructive boundary and C is wrong as this is a conservative boundary. No subduction occurs here and so D is incorrect.
6. **a)** crust ✓; **b)** asthenosphere ✓; **c)** iron ✓; **d)** Pangaea ✓; **e)** tectonic ✓
7. A/J ✓; B/F ✓; C/I ✓; D/H ✓; E/G ✓

Part B: Skills

8. **a)** spreading zone (divergent boundary) ✓
   b) $67 \times 1\,000\,000 = 67$ million mm ✓ = 67 km ✓
   c) There is a collision zone between the two plates. This leads to mountain building. ✓
9. **a)** The rim of the Pacific contains many plate boundaries (e.g. subduction zones, transform fault zones). Earthquakes and volcanoes are produced due to these plate interactions. ✓
   b) These three continents were once joined as part of Gondwana. The ancestors of these birds were isolated when the continents split apart. They then evolved in isolation to their current forms. ✓
   c) Australia was once joined to Antarctica and the combined land mass occupied cold Antarctic latitudes. The continents then split apart and Australia moved north towards the equator. The climate became warmer and temperate. ✓
   d) The plate boundary between the African Plate, Arabian Plate and the Eurasian Plate runs through this area. Consequently, earthquakes and volcanic activity occur here. ✓
10. **a)** spreading zone (divergent plate boundary) ✓
    b) A = 2 million years old ✓; B = 4 million years old ✓
    c) New oceanic crust is forming at the mid-ocean ridge. The closer the rock is to the mid-ocean ridge, the younger is the rock. ✓
    d) asthenosphere ✓
    e) The old oceanic crust is removed and recycled at the subduction zones. This prevents Earth's crust from expanding. ✓
11. India, Australia and Antarctica had already split away from the rest of Gondwana before the reptile colonised these areas. ✓ This split occurred 135 million years ago. Africa and South America remained as one land mass until 105 million years ago. ✓
12. **a)** Scientists are human and some have closed minds about new ideas, particularly if the ideas are radically different from popular scientific view. ✓
    b) He did this to help like-minded scientists present their evidence so that continental drift could be better understood and become more widely accepted. These scientists could then also avoid the scorn of others. ✓
13. The Indo-Australian Plate has a boundary which runs through New Zealand to the east and New Guinea and Indonesia to the north. ✓ These regions should experience frequent and violent earthquakes and volcanic eruptions because they lie across plate boundaries where movement is constantly happening. ✓
14. The fossils provide evidence that the land masses where the animals once lived were much closer together. ✓

Test yourself 2

Part A: Knowledge

1. **A** ✓ The Pacific Plate dives down beneath the South American Plate along the coast line. This is an example of subduction. B, C and D are incorrect.
2. **D** ✓ Richter values between 6 and 7 are indicative of strong earthquakes. Therefore, A, B and C are incorrect.
3. **B** ✓ A volcano releases molten rock, and other material, that can harden on the Earth's surface. Earthquakes and tectonic plates don't release molten rock although they can be associated with volcanoes, so A and C are incorrect. Tsunamis may result from earthquakes, but don't cause lava to flow, so D is incorrect.
4. **C** ✓ A seismograph, or seismometer, is the instrument. The Richter scale is not an

instrument, so A is incorrect. B and D have nothing to do with earthquakes.

5. **A** ✓ P waves are push–pull waves. They are longitudinal waves, not transverse waves, and so D is incorrect. S waves are shear waves, so B is incorrect. L waves cause a circular shearing of the ground, so C is incorrect.
6. **a)** S ✓; **b)** decrease ✓; **c)** older ✓; **d)** regulations ✓; **e)** fault ✓
7. A/I ✓; B/J ✓; C/G ✓; D/F ✓; E/H ✓

Part B: Skills

8. **a)** There has been erosion between S (and R) and T. ✓ Erosion is a constant feature. It can be seen at the surface where the river is cutting through layer V, but the question asked for 'between two layers'.
   b) **i)** The fossils in Q are older than those in U. ✓
   ii) Both rocks are sedimentary ✓ as fossils may be found in these rocks.
9. A = magma chamber ✓; B = laccolith ✓; C = dyke ✓; D = sill ✓; E = lopolith ✓; F = vent ✓; G = lava flow ✓
10. Gases are dissolved within magma because they are under extreme pressures. ✓ This pressure is reduced at the surface, allowing dissolved gases to expand and escape. ✓ Water vapour makes up between 70 to 95% of these gases. The remainder is mainly carbon dioxide, sulfur dioxide and traces of nitrogen, hydrogen, carbon monoxide, sulfur, argon, chlorine and fluorine. ✓ Fine ash particles are also ejected from an eruption but fall out of the atmosphere too quickly to significantly cool the atmosphere over a long time period (although they do in the short term). ✓ CO_2 is abundant, but not enough to contribute significantly to the greenhouse effect. ✓ The greatest volcanic impact is caused by sulfur dioxide gas. In the cold lower atmosphere, the sun's rays reacting with stratospheric water vapour convert SO_2 to sulfuric acid forming suspended sulfuric acid aerosol layers. There is a relationship between the amounts of sulfur dioxide in these layers and the decrease in average temperature over several years. ✓ Chlorine is released from volcanoes as hydrochloric acid. This breaks down producing reactive chlorine atoms, which then proceed to destroy ozone. ✓
11. X = Adelaide; Y = Perth; Z = Melbourne ✓✓ P waves travel faster than S waves, so the further they have to travel, the greater is the time difference between these two waves.
12. shale; coal; sandstone; limestone; conglomerate; shale; sandstone *(3 marks, 1 mark off for each out of sequence)*
13. **a)** A seismogram is a record of the seismic waves from an earthquake. ✓ A seismograph or seismometer is the measuring instrument that creates the seismogram. ✓
    b) Inertia is the tendency of an object to resist any change in its motion. ✓ A heavy, suspended mass tends to remain still when the ground moves suddenly. ✓ The relative motion between the suspended mass and the ground is a measure of the ground's motion. ✓ This motion can be recorded by a moving pen.
14. **a)** dyke ✓
    b) shale; limestone; sandstone; volcanic tuff; dolerite ✓✓
    c) Heat from the cooling magma will bake the surrounding rocks, leading to contact metamorphism. ✓ The sandstone will turn to quartzite, limestone to marble, and shale to hornfels at the contact zone.
    d) The crystal size is much larger in the dyke than in the extrusion onto the surface. The slower the cooling of the magma, the larger are the crystals that form. ✓

Chapter test

Part A: Multiple-choice questions

1. **A** ✓ B is not correct as the lithosphere includes the crust and upper mantle. C is wrong as this statement refers to the core. D is wrong as they are more dense.
2. **C** ✓ Wegner was the first to use this name. Thus, A, B and D are incorrect.
3. **D** ✓ The Nazca Plate is an oceanic plate that is subducting with the South American Plate. A is wrong as the Juan de Fuca Plate is in the northern hemisphere. B is wrong as the Antarctic Plate only interacts with South America at the southern end. C is wrong as the Pacific Plate is to the west of the Nazca Plate.
4. **A** ✓ Asia was part of Laurasia. B, C and D were all part of Gondwana.

5. **A** ✓ The Himalayas are formed as the Indo-Australian Plate pushes against the Eurasian Plate. This is a collision zone and thus B, C and D are incorrect.

6. **A** ✓ A cinder cone is a steep conical hill of volcanic fragments that accumulate around and downwind from a volcanic vent. The types of volcanoes given in B, C and D do not support the information given.

7. **A** ✓ Earthquakes produced by stress and pressure changes in solid rock are due to the injection or withdrawal and unsteady transport of magma into surrounding rock. When magma injection is sustained a lot of earthquakes are produced, indicating that a volcano is about to erupt. So there is both an increase in frequency and increase in intensity.

8. **C** ✓ While earthquakes can occur elsewhere, as given in A, B and D, most earthquakes occur around plate boundaries.

9. **D** ✓ While earthquakes can occur elsewhere, as given in A, B and C, most earthquakes occur in the Pacific Ring of Fire. About 90% of all the earthquakes and 80% of the world's largest earthquakes occur along the Ring of Fire.

10. **B** ✓ Reverse faults occur in areas undergoing compression and are easily recognised as a layer is pushed up over itself. Normal faults occur in areas undergoing extension (stretching), while strike-slip faults involve motion in a horizontal plane. There are no hanging wall faults.

Part B: Short-answer questions

11. As more and more physical evidence about an idea is collected, the idea can be modified until it is accepted by most scientists. The work of science is ongoing and it builds on the work of past scientists. ✓

12. No, they used the best information and knowledge available to them at the time. ✓ Their views were later proved incorrect but fitted into the information of the day. ✓

13. **a)** Large convection currents in the asthenosphere (soft, partly molten layer) move the solid continents on top of them. ✓

    b) Earthquakes and volcanoes occur:
    - at parts where the plates move apart
    - when one plate pushes under a second
    - when two plates collide
    - when two plates slide past each other in response to the convection current. ✓

14. **a)** The Pacific and Philippines plates are moving under the Eurasian Plate at a subduction zone (the trench); this causes earthquakes and volcanic eruptions. ✓

    b) The Nazca Plate is subducting at the west side of the South American Plate and a chain of mountains are formed as well as volcanoes and earthquakes. ✓

    c) There are three plate boundaries in this zone where movement can occur. ✓

    d) Iceland lies along the Atlantic mid-ocean ridge; molten rock emerges from the earth at this spreading zone. ✓

    e) The Eurasian and Indo-Australian plates are forming a collision zone; this is a site of mountain building. ✓

    f) The Pacific, Juan de Fuca and North American plates form a transform fault zone. They are moving in the same direction but at different speeds. Sometimes the trailing plate 'slips' to catch up and a large earthquake is the result. ✓

15. **a)** discovered ✓; **b)** speed ✓;
    c) measurements ✓; **d)** movement ✓;
    e) found ✓; **f)** occurring ✓; **g)** faster ✓;
    h) region ✓; **i)** south ✓

16. It shows that these regions were once close together in Gondwana ✓ and are now further apart because of the process of continental drift that has occurred due to tectonic plate interaction ✓.

17. **b)** 135 mya ✓; **c)** 225 mya ✓;
    d) 65 mya ✓; **e)** 200 mya ✓

18. **a)** I ✓; **b)** II ✓; **c)** IV ✓; **d)** III ✓; **e)** V ✓

19. **a)** **i)** about 650 000 years ✓

    ii) 650 000 to 800 000 years ago ✓

    iii) six reversals ✓

    b) **i)** about 20 km ✓

    ii) $\frac{20}{650\,000}$ km/year = 3 cm/year ✓

    iii) X is 70 km to the left of the ridge in the normal magnetism zone. (This corresponds to an approximate age of 2.2 to 2.3 million years.) ✓

    iv) The pattern of magnetic reversals is symmetric about the mid-ocean ridge. As magma escapes from the mid-ocean ridge, the sea floor spreads and pushes continents apart. ✓

20. The size of the magnetic bands is increasing, indicating that the rate of spreading is changing. ✓

21. **a)** 43 283 582 years or approximately 43 million years ✓

b) **i)** The answer in a) is much less than 96 million years ✓

ii) The assumption that it moves a constant 6.7 cm a year is not valid. ✓

c) 3 cm per year ✓

22. **a)** Hypothesis B ✓: the evidence of ice would occur over a wide range of modern areas which were once close together at the South Pole in a previous geological age. ✓

b) evidence of glacial tillite in all parts of the southern hemisphere ✓

c) Hypothesis B ✓: the ice age was obviously not so severe ✓.

d) No ✓ This evidence supports but does not prove the idea of continental drift. ✓

23. The Atlantic Ocean will increase in size ✓ due to the creation of new sea floor at the mid-ocean ridge. ✓

24. east and under the western edge of the South American Plate ✓

25. These parts of the world were once close together and formed part of Gondwana. ✓ While the land masses were close together, the Proteaceae developed and spread via land, not over the sea. This is evidence for continental drift. ✓

26. Total CO_2 released is 500 + 17 600 = 18 100 million tonnes.

a) from volcanoes: $\frac{500}{18100} \times 100 = 2.76\%$ = 3% (to nearest whole number) ✓

b) from anthropogenic (human) impact on the environment: $\frac{17600}{18100} \times 100 = 97.24\% = 97\%$ (to nearest whole number) ✓; alternatively, 100% – 2.76% = 97.24%

27. D, F, C, E, A, G, B *(3 marks; take 1 mark off for each sentence out of sequence)*

28. **a)** There were more eruptions in the 19th century than in the 20th century. ✓ There was a period in the 19th century when eruptions were clustered closely together. There have only been two eruptions in the 20th century, about 40 years apart. ✓

b) Since 1872 to 1944 eruptions of Vesuvius occurred every 30 to 40 years. It has now been some 70 years since the last eruption, so one is due. Volcanic eruptions cannot be timed. Between 1850 and 1870 there could have been a period of uncharacteristically high volcanic activity, and now we are going through an unusual lull period. ✓✓ (For residents in and around Naples, there was a calm of about a century prior to the activity that prompted the eruption of 79 AD.)

c) There were six eruptions in the 19th century (100-year period). From 1631 to 1800 is 169 years, so we can estimate 8 to 11. ✓ (The principal eruptions occurred in 1660, 1694, 1698, 1707, 1737, 1760, 1767, 1779 and 1794, a total of nine.)

29. **a)** about 4 million years ✓

b) Using the straight line, a distance of 5000 km (5 000 000 m) takes around 58 million years. ✓

So $\frac{5000000}{58000000} = 0.86$ m/year
= 8.6 cm/year✓.

(This is an average over the 5 million years. Scientists have calculated that around Hawaii, the plate is currently moving at about 7 cm/year.)

30. **a)** **i)** 5.00 minutes ✓; **ii)** 4.50 minutes ✓

b) P waves travel the fastest ✓, arriving first; then S waves arrive while L waves are the slowest, arriving last ✓.

c) The time difference would be less ✓ and they haven't travelled all that far for the two waves to spread out in time ✓. (Think of two runners in a race. They start off together but the further they go, the greater is the distance between them.)

31. **a)** 2.2% ✓

b) See Figure A.9 (on the next page).

32. **a)** At a subduction zone, an oceanic plate collides with a continental plate. Due to differing densities, the oceanic plate moves down beneath the continental plate to form the subduction zone and ocean trench. Frictional heating occurs as the oceanic plate moves down towards the asthenosphere.

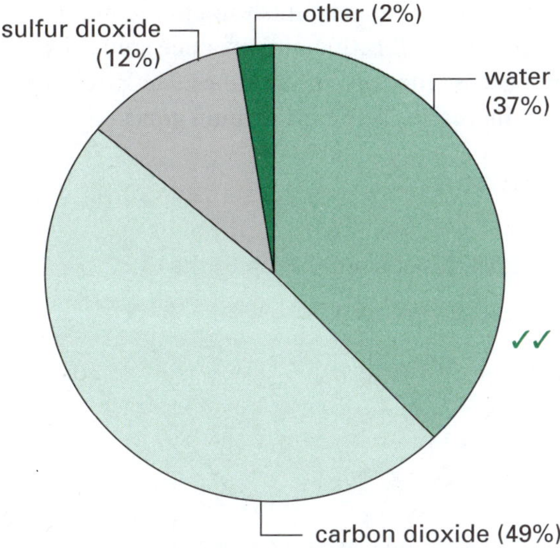

Figure A.9 Gases from volcano

Earthquakes result from this frictional contact. Magma is also formed from the heating and it moves upwards under the pressure through faults. The release of hot melted rock produces volcanism. ✓✓

b) the west coast of South America where the Nazca Plate collides with the South American Plate; Japan where the Pacific and Philippine plates and Eurasian Plate collide ✓✓

33. a) Sudden earth movements can lead to the formation of rift valleys and block mountains. Avalanches may occur in mountainous areas. New volcanic landforms such as cinder cones, lava shields and volcanoes may be formed. ✓✓

b) Earthquakes can lead to tsunamis; lava emerging from ocean vents can lead to mineralisation of the ocean. Volcanic gases can produce acid rain. ✓✓

c) Volcanic gas and ashes are released into the atmosphere. This ash can block sunlight and cause atmospheric cooling. SO_2 forms acids and helps deplete ozone. ✓✓

34. Core Z has the most recent fossil. ✓ The shell fossil at the top of Y is older than the bone fossil in the middle of X as it is underneath the bone. ✓

35. a) $\frac{4.5}{5} = 0.9$ billion years ✓

b) the percentage of lead-206 that forms from the breakdown of uranium-238 ✓

c) The intersection of the two curves indicates the half-life of U-238 ✓, which is 4.5 billion years. This is when 50% of the original U remains.

d) about 6.5 billion years ✓

e) around 60% ✓

36. a) Tsunamis begin by a sudden displacement of the ocean. This often occurs when huge blocks on the ocean floor are moved vertically, such as during earthquakes. This uplifts a huge volume of water. ✓✓

b) Gravity sets this body of water in motion as it moves to return to a stable position. It moves in all directions from where it was initiated and can cover huge distances across the ocean. ✓✓

c) In deep water, tsunamis have wave lengths in the hundreds of kilometres with an amplitude of around only a metre. So at sea they are not very large and pose no danger. Most ships can mistake them for ordinary waves. ✓✓

d) When tsunamis approach shallows, such as near the shore of an island or country, the wave slows down, bunches up and grows very high, perhaps some tens of metres as it surges on land. The tremendous energy in a tsunami can push water many kilometres inland and devastate many coastal communities. ✓✓

(In deep water, tsunamis travel at some 700 km/h. Compared to seismic waves, this is slower so alerts can be put out before the tsunami arrives. However, if the tsunami begins not very far from shore, as in many places along the Pacific Ring of Fire, there is not very much warning given. So the death toll will be higher, especially in populated areas, since people won't have time to evacuate.)

37. 4500 km ✓ (Use the vertical scale to measure 6 minutes. Then transfer this vertical distance to determine where these curves have a 6-minute gap between them.)

38. See Figure A.10 (on the next page). (Use a compass to draw circles, having appropriate radii to determine where they intersect.)

39. Australia is an island that has been cut off by oceans from the rest of the world for millions of years, and being geologically stable has

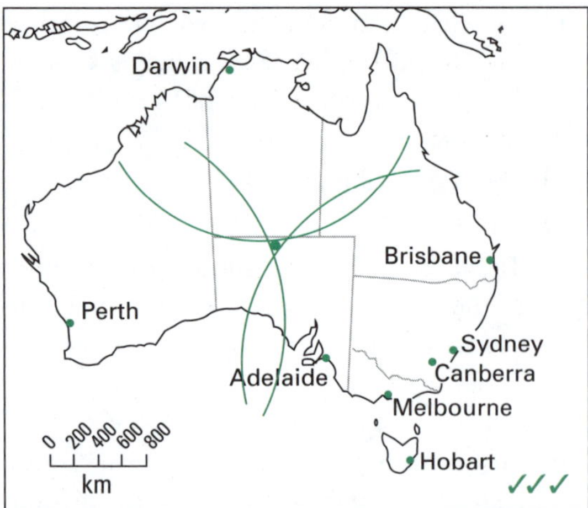

Figure A.10 Epicentre of quake

permitted organisms to evolve without major upheavals. ✓ This has allowed the animals and plants to be able to develop in their own special way without mixing with other species or types of animals from other parts of the world. Marsupials became the dominant animal species, and evergreen plants such as eucalypts dominated the plant world. ✓ Over many thousands of years, Australia as a land has completely changed; so has its climate and the weather patterns, as it has become a much drier continent. The plants and animals have changed with it and because of it. ✓

40.
- Deposition of sandstone ✓
- Deposition of shale ✓
- Deposition of conglomerate ✓
- Deposition of sandstone ✓
- Folding of sedimentary layers ✓
- Intrusion of a dolerite dyke ✓
- Faulting ✓
- Uplift and extensive erosion (unconformity) ✓
- Extrusion of flood basalt ✓
- Deposition of sandstone ✓
- Erosion to present-day landscape ✓

Chapter 4 answers

Experiment 1

Analysis:

1. The pitch of the note increases as the length of the air column decreases (i.e. as more water is added).
2. The wavelength of the sound wave decreases as the air column gets shorter.

Conclusion: This experiment shows that columns of air can be set vibrating and produce sound by blowing across the tops of the tubes. The pitch of the note increases as the air column gets shorter.

Experiment 2

Analysis:

1. **a)** Rays travel from the tip to the eye.
   b) See Figure A.11.

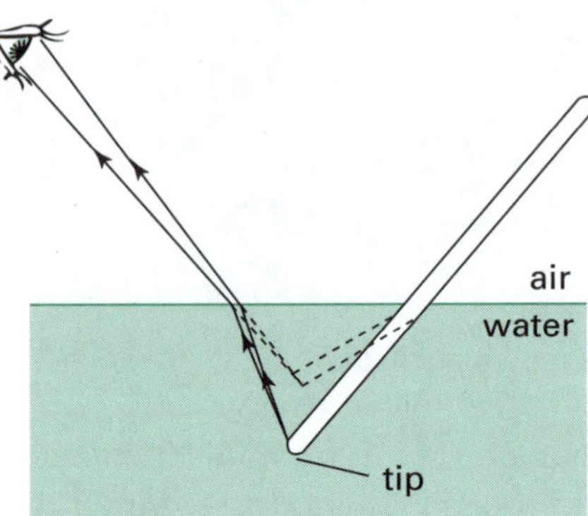

Figure A.11 Rod in water

2. **a)** Light rays travel from the coin to the eye.
   b) See Figure A.12.

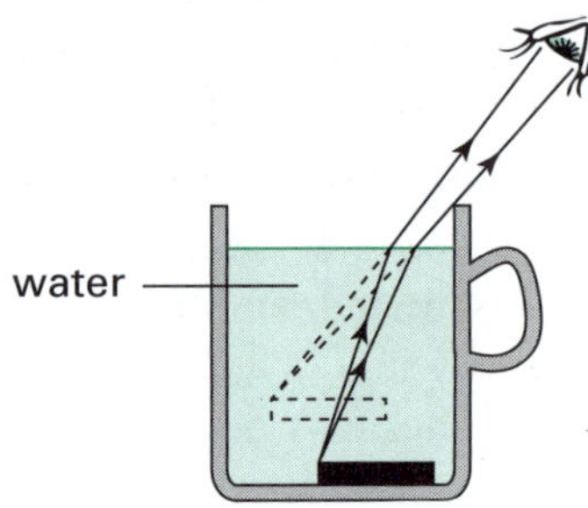

Figure A.12 Coin in a cup

Conclusion: Light refraction in water can cause objects to appear bent and to appear to be at a more shallow depth.

Experiment 3

Analysis:

1. Increasing the voltage on the power pack increased the brightness of the lamp. This is because a higher voltage (potential difference) is pushed through a greater current and so the lamp is brighter.
2. No, heat was also produced. Lamps can get quite hot.
3. With two lamps in the circuit, the lamps were dimmer than the single lamp. Lamps act as resistors, so more lamps means more resistance and less current flows through the circuit.

4. Removing the lamp effectively opens the circuit. That is, the path is no longer complete and so current cannot flow around.
5. Increasing the voltage increased the current that flowed through the circuit and so the reading on the ammeter increased.

Conclusion: For a current to flow through a circuit, there must be a continuous path. The greater the voltage applied, the greater is the current. When the resistance is increased (for a given voltage), such as adding more lamps, less current flows around the circuit.

Test yourself 1

Part A: Knowledge

1. **A** ✓ Black objects absorb heat energy faster than light-coloured or silvery-coloured objects but they also emit heat at a faster rate. Therefore B and D cannot be correct. C is incorrect as the heat is lost by radiation.
2. **B** ✓ Gamma rays are electromagnetic whereas A, C and D are all examples of mechanical waves.
3. **A** ✓ Infra-red rays have shorter wavelengths and therefore higher frequencies than radio waves. B is not correct as X-rays are used to treat cancer. C is wrong as all electromagnetic rays travel at the same speed. D is incorrect as visible light is used by green plants.
4. **C** ✓ A is incorrect as there is a small loss in energy. There is very little scattering and the glass is transparent and not translucent, so B is wrong. D is wrong as the window is not opaque.
5. **D** ✓ Refraction refers to waves bending when they travel from one medium to another. Thus A, B and C cannot be correct.
6. **a)** transverse ✓; **b)** amplitude ✓;
   c) electromagnetic ✓; **d)** towards ✓; **e)** faster ✓
7. A/H ✓; B/F ✓; C/J ✓; D/G ✓; E/I ✓

Part B: Skills

8. **a)** The lowest value printed in the remaining floating spheres is the temperature of the day. ✓
   b) No ✓ The whole container has to adjust to the temperature. It is a slow process. ✓
9. Liquid water is a poor conductor of heat and so the water at the top of the tube can be boiled rapidly without the ice at the bottom of the tube melting. ✓ Eventually the ice will melt as a convection current will be set up.
10. **a)** Station A (702 kHz) ✓
    b) i) much shorter wavelength than station A (and also B) ✓
    ii) same velocity (speed of light) for all electromagnetic waves ✓
11. A = 2.5 cm ✓; λ = ~6.7 cm ✓
12. **a)** The vibrations cause the air particles to vibrate about a mean position, producing a compression wave. The sound compression waves move out until they reach the student's ear. ✓
    b) i) Loudness may stay the same but the pitch would increase. In order to achieve this, the student must shorten the ruler if it is to be pulled it down the same distance before letting go. The loudness is related to the amplitude of vibration, which is assumed to be constant. ✓
    ii) Loudness may stay the same but the pitch would be lower. ✓
13. Diagram B is correct. ✓ Violet light refracts more than red light does as it has a greater frequency (shorter wavelength). ✓ When light enters a glass prism from the air, it refracts towards the normal. ✓ When light leaves a glass prism and enters the air, it will refract away from the normal. ✓ Thus, A is wrong as the rays are bending up and not down and C is wrong as violet light is showing less refraction.
14. time for echo to travel across the valley

    $$\frac{0.860}{2} = 0.430 \text{ s}$$ ✓

    distance = speed × time

    $d = 350 \times 0.430 = 150.5$ m ✓

Test yourself 2

Part A: Knowledge

1. **D** ✓ Copper is the best conductor. Iron (C) does conduct but not as well as copper. A and B do not conduct.
2. **A** ✓ As the car is moving towards the radar, the returning waves are scrunched up, hence the wave's frequency increases (the wavelength gets shorter). Therefore, B is incorrect. C and D are incorrect as amplitude is not used to measure speed.

3. **A** ✓ Ammeters measure current (in amps). The other alternatives are incorrect as the voltmeter (B) measures potential difference (voltage), an ohmmeter (C) measures resistance and a battery (D) is the source of power and not a measuring device.

4. **C** ✓ They are all in parallel, so there is the same voltage drop across them all. A would be correct if the resistors were connected in series. B is incorrect as the current flow depends on the resistance. D is incorrect as the current splits at the junction where two or more wires meet.

5. **D** ✓ It is the wave's amplitude that is altered (modulated). The other options are incorrect. A refers to am/pm time.

6. **a)** electrons ✓; **b)** voltmeter ✓; **c)** series ✓; **d)** higher ✓; **e)** ampere (amp) ✓

7. A/I ✓; B/H ✓; C/J ✓; D/G ✓; E/F ✓

Part B: Skills

8. See Figure A.13.

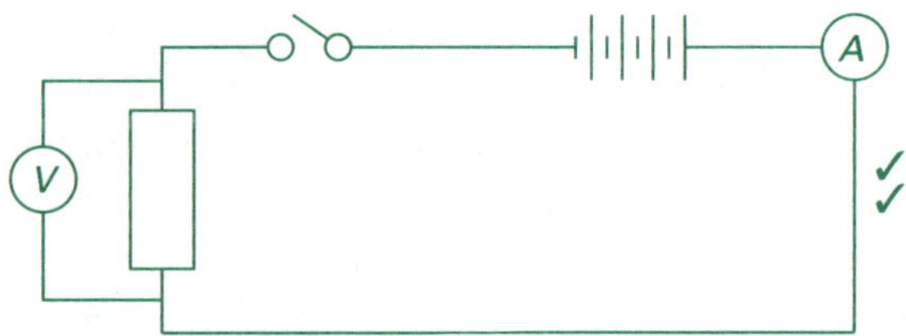

Figure A.13 Circuit

9. Model Y shows the movement of negative electrons only. The positive metal ions are fixed and do not move, as shown in the incorrect model X. ✓✓

10. **a)** The cell phone network can be considered to be like a blanket of radio waves spread over a large area. This blanket is really more like a quilt made of small, hexagonal pieces called cells. ✓ Each cell has its own tower and base station.

b) In spite of all the bells and whistles, a cell phone really is just a radio that can send and receive information. All the applications a phone can do are transmitted by radio waves to the tower and information is received from the tower by radio waves. ✓

c) This is full-duplex communication. ✓ People at either end of the line can talk simultaneously. In half-duplex communication (such as CB radio), people can only talk one at a time.

11. See Figure A.14.

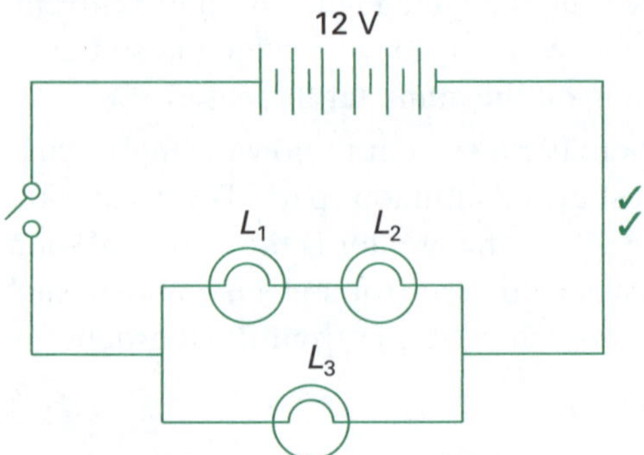

Figure A.14 Electrical circuit

12. **a)** Smoke signals and fires were used with certain codes to convey messages. In some cultures people developed a system of signalling using mirrors to reflect sunlight and send 'flashes'. ✓ Lighthouses produce flashes of different timing and length so that sailors can tell which lighthouse they are seeing many kilometres away and so guide ships at sea. ✓

b) Even very powerful lights can only be seen from a relatively short distance. This is partly because the Earth is curved and so light sources dip below the horizon ✓, and partly because as light radiates out from a source it gradually becomes weaker with distance, and especially due to the effects of dust, smoke and fog ✓.

c) When light passes down an optical fibre, it continues to travel in straight lines bouncing off its mirror-like side. ✓ Lasers generate a suitable light source ✓ that is very tightly focused and can produce pulses ✓, using a digital code, many times a second. These pulses are then picked up at the other end of the fibre by a light-sensitive cell which converts them into pulses of electricity. ✓ A computer is then used to decode and reveal the message. ✓ The pulses are generated so fast that it is possible to send the entire contents of about 40 large book volumes through an optical connection in about 1 second!

13. **a)** E, B, ✓ D, C, ✓ F, A, G ✓

b) The range of a distant object is determined by comparing the time when the signal was sent to when it was received. This is the time taken for the radio wave to reach the object and return back to the antenna. Knowing how fast they travel (the speed of light) allows a computer to almost instantly calculate the object's distance. ✓✓

14. a) See Figure A.15.

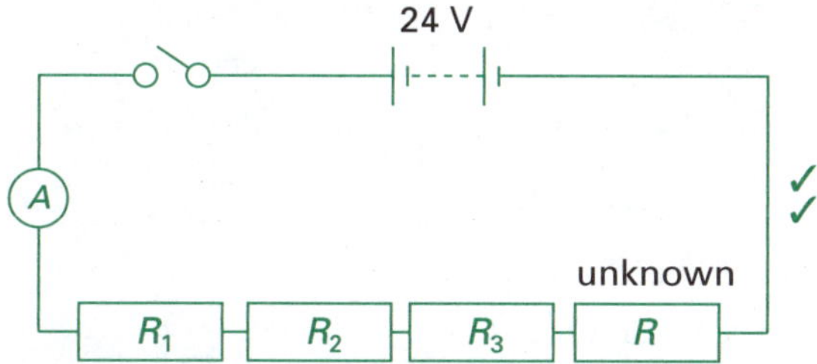

Figure A.15 Four resistors

b) Total resistance = 3 + 4 + 1 + R

= 8 + R ohms.

V = 24 V; I = 1.5 A.

From Ohm's Law:

$V = IR$. Solve for R.

$24 = 1.5(8 + R)$

$R = 12/1.5 = 8$ ohms ✓✓

c) The lowest voltage drop will be across R_3 as it has the lowest resistance. ✓

Chapter test

Part A: Multiple-choice questions

1. **A** ✓ B is wrong as water waves are transverse at the surface. C is wrong as the oscillation is at right angles to the axis of propagation. D is incorrect as infra-red rays have very small wavelengths (of the order of micrometres).
2. **D** ✓ There are 0.4 waves in 1 second so in 10 seconds there will be 4 waves. A is not correct as we do not know the velocity of the waves. B cannot be calculated without a knowledge of wave velocity. C is wrong as microwaves have much higher frequencies.
3. **A** ✓ B is wrong as the cardboard is shiny and so some light is reflected. C is wrong as no light is refracted as there is no change in media. D is wrong as more light is reflected and most is absorbed.
4. **A** ✓ B and C are incorrect as the mirror would be concave. D is wrong as this reflector is parabolic.
5. **B** ✓ Candles are made of wax, which is a poor conductor of heat and so A is incorrect. B is wrong as not all the heat is lost upwards. D is wrong as the melting point of icing is quite low.
6. **B** ✓ Electric currents leave the positive battery terminal and travel through the external circuit back to the negative battery terminal. A current flows only when the circuit is closed, so A is incorrect. C and D do not relate to an electric current.
7. **A** ✓ The greatest current (and therefore greatest heat) occurs when the resistance is least. This occurs when the resistors are in parallel.
8. **B** ✓ Geostationary satellites appear to be fixed over one spot above the equator. They are Earth-orbiting satellites, so A is wrong. C is wrong because these satellites are placed at an altitude of approximately 35 800 km directly over the equator and revolve in the same direction the Earth rotates (west to east). At this altitude, one orbit takes 24 hours. If the satellite was placed at a smaller altitude, its orbit would take less than 24 hours. Not all geostationary satellites are for communication, making D incorrect.
9. **D** ✓ Radar has been used for all of these and other tasks such as astronomy and geological mapping. (Students sometimes think 'all of the above' or 'none of the above' is some code that the examiner couldn't think of a fourth alternative. This just shows that isn't so.)
10. **C** ✓ These signals are forms of electromagnetic radiation, so D is wrong. Signals sent using this type of radiation travel at the speed of light and are assigned different wave bands (frequencies), so B is incorrect. These signals are in the radio/microwave range and can't be detected with our senses, making A incorrect.

Part B: Short-answer questions

11. See Figure A.16.

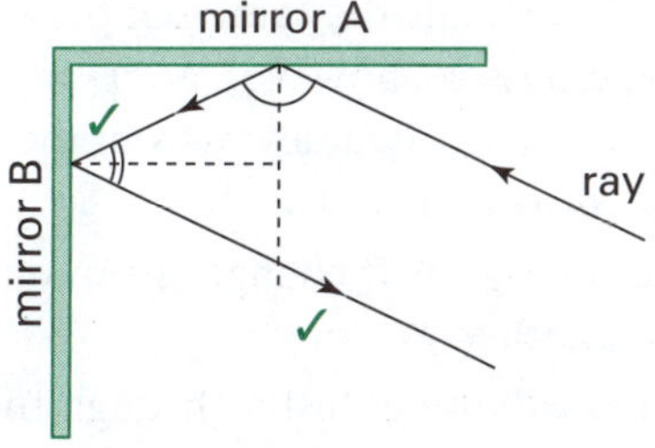

Figure A.16 Double reflection

12. a) 60° ✓

b) See Figure A.17.

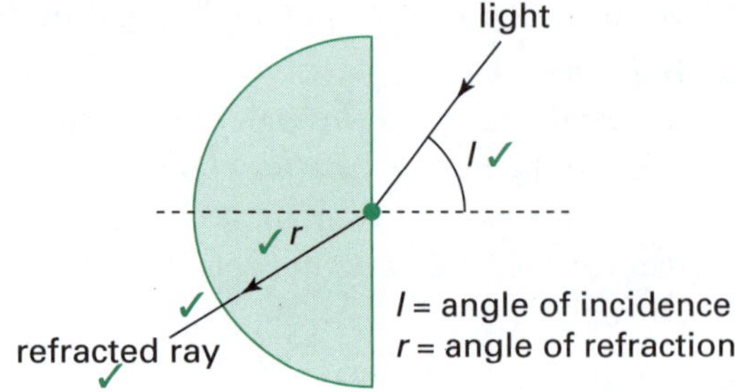

Figure A.17 Refracted ray

c) Some of the light energy has been absorbed by the plastic ✓ and some light may have been reflected at each surface ✓.

13. 6000 m/s = 6 km/s ✓; time = $\frac{60}{6}$ = 10 seconds ✓

14. $v = \frac{d}{t} = \frac{1000}{2.86}$ = 350 m/s ✓

15.

$$\text{speed} = \frac{\text{distance}}{\text{time}}$$

$$= 2 \times \frac{\text{depth}}{\text{time}} \checkmark$$

$$1500 = 2 \times \frac{\text{depth}}{0.54}$$

$$\text{depth} = 1500 \times \frac{0.54}{2}$$

$$= 405 \text{ m} \checkmark$$

16. The flanges are flexible and can move (compress or expand) to ensure the steel girders do not make contact when they expand or contract. Bridges change in length by up to a metre between winter and summer. Such changes in length must be allowed for in designing bridges. If the structure could not alter its length with changes in temperature, huge forces would develop and severe damage might result. ✓

17. The paper is a poor conductor of heat so the heat is not carried away from the hand. Therefore, the book feels the same temperature as the air. ✓ The scissors, however, are good conductors and conduct heat away from the hand, so the scissors feel cold. ✓

18. a) Sound loses energy as it changes from one medium to another. ✓

b) Yes ✓ Sound will travel faster through the water and there is no change of medium other than the air in the ear canals. ✓

19. a) T ✓; **b)** F ✓; **c)** T ✓; **d)** F ✓; **e)** T ✓; **f)** F ✓

20. a) See Table A.4.

Table A.4 Speed of sound through different types of gas at 0 °C

Gas	Speed (m/s)
helium	965
nitrogen	334
air	331
oxygen	319
carbon dioxide	259

✓

b) helium ✓

c) carbon dioxide ✓

d) See Figure A.18.

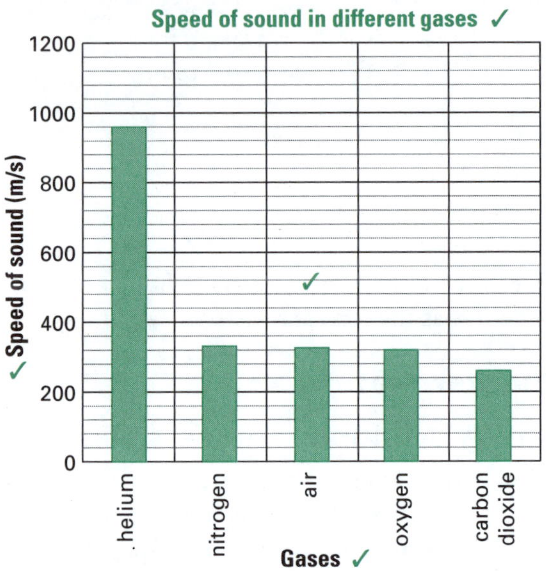

Figure A.18 Speed of sound in different gases

e) The speed of sound is almost the same in air as in nitrogen. As air is about 80% nitrogen and 20% oxygen, the speed of sound in air is $\frac{80}{100} \times 334 + \frac{20}{100} \times 319 = 331$ m/s. ✓

f) If there were more CO_2 in the air, then the speed of sound would be much lower. ✓

21. The outside of the glass expands too quickly but the cold inside does not because glass is a poor conductor. Stresses build up, and these are relieved by cracking. ✓

22. No ✓ Heat moves from areas of hot to areas of cold. The bar took heat away from Jim's hand. ✓

23. See Figure A.19 (on the next page).

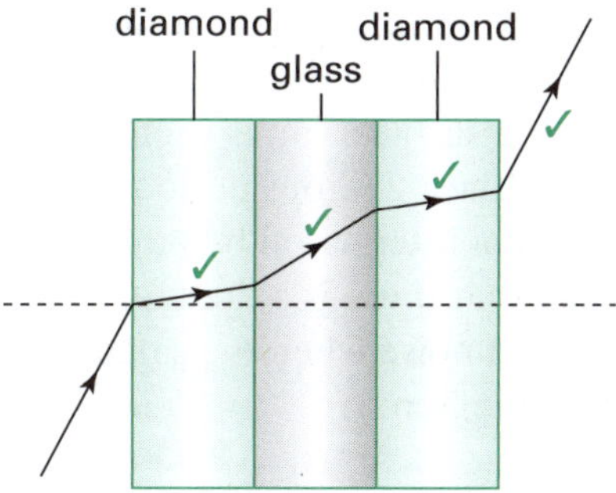

Figure A.19 Glass and diamond

24. a) See Figure A.20.

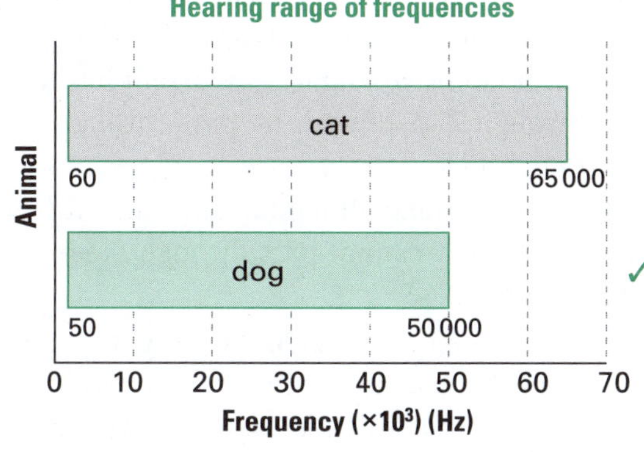

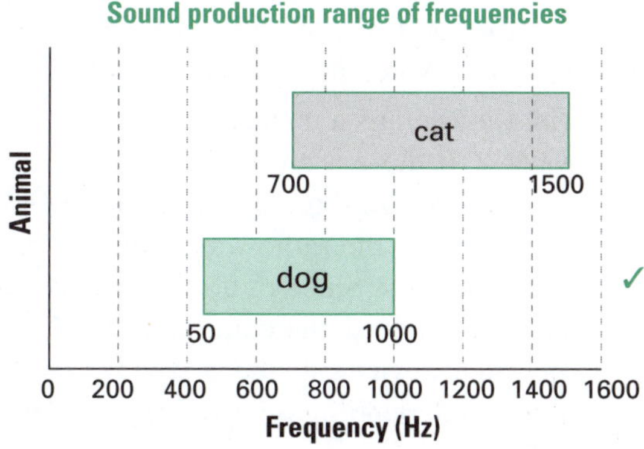

Figure A.20 Sounds production and hearing

b) i) cats ✓; ii) cats ✓

c) No, this sound is above their hearing ranges. ✓

25. a) See Figure A.21.

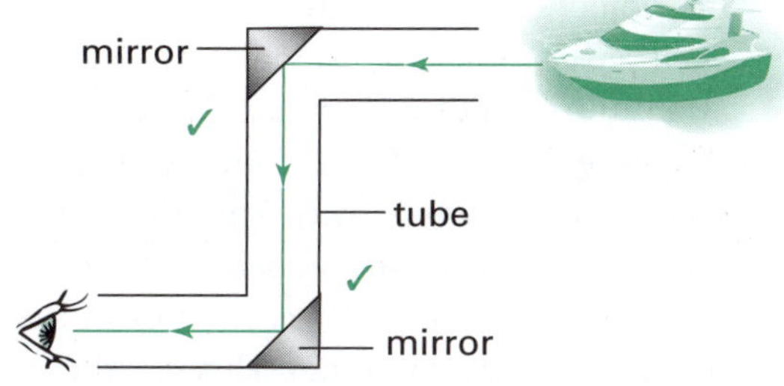

Figure A.21 Periscope

b) The thicker glass will refract and absorb some light and so the image will be blurred and fainter. ✓

26. See Table A.5.

Table A.5 Conduction, convection and radiation

	Conduction	Convection	Radiation
How heat travels	particles vibrate and bump into their neighbours ✓	particles move from one place to another ✓	travels as an invisible ray ✓
Can heat travel through empty space?	no ✓	no ✓	yes ✓

27. a) direct current (DC) ✓

b) The four batteries provide a total voltage of 6 V. The total current flow through the lamp will therefore be four times greater and thus the lamp will glow more brightly. ✓✓

28. a) See Figure A.22.

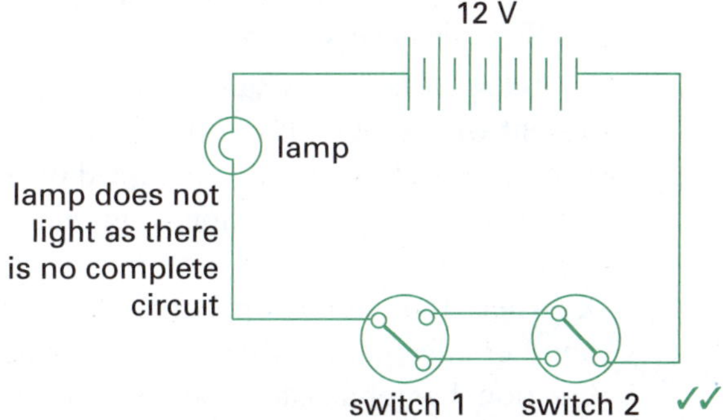

Figure A.22 Switches

b) These two-way switches are used in corridors or long rooms so that you can switch lights on/off as you enter/leave at either end. ✓

29. a) $6.0 - 2.0 = 4.0$ V ✓

b) $R_{AB} = \frac{V_{AB}}{I} = \frac{2.0}{1.0} = 2.0\ \Omega$ ✓

c) $R_{CD} = \frac{V_{CD}}{I} = \frac{4.0}{1.0} = 4.0\ \Omega$ ✓

d) $R_T = R_{AB} + R_{CD} = 2.0 + 4.0 = 6.0\ \Omega$ ✓

30. a) The signals from each tower don't travel far beyond their range. Advantage: cell providers can use the same frequencies over

and over in different (but not adjoining) cells, so more people can talk at the same time. ✓ Disadvantage: cell phone users at the end, or outside, the coverage area can experience dropout or no signal. ✓

b) This is done automatically by the Mobile Telephone Switching Office which coordinates when your phone changes frequencies from one tower to another. ✓ This happens as you move from one cell to the next since the signal received by the tower of the cell you are in is getting weaker, but that being received by the tower in the cell you are moving to gets stronger. ✓

c) The signals don't need to travel very far, only as far as the nearest tower. ✓ Sending a signal doesn't drain the battery. ✓ A high-powered signal would mean short battery life and possible interference with distant phone calls.

31. a) **i)** tap ✓; **ii)** pump ✓

b) The rate of water flow will decrease as the diameter of the pipe decreases. ✓

c) The resistance to flow is greatest in the 1-cm diameter pipe because narrow pipes have greater resistance. ✓

32. a) A single geostationary satellite is on a line of sight with about 40% of the Earth's surface. This allows it to be accessed over a large area using a directional antenna, usually a small dish, aimed at the spot in the sky where the satellite appears to hover. ✓ Another advantage is that an earthbound directional antenna can be aimed and then left in position without further adjustment. ✓ Since highly directional antennas can be used, interference from surface-based sources, and from other satellites, is minimised.

b) The zone of orbit is an extremely narrow ring in the plane of the equator so there is a limited number of satellites that can be maintained in geostationary orbits without causing problems or colliding. ✓ The distance that an electromagnetic signal needs to travel to and from a geostationary satellite is at least 72 000 km. Thus, a time delay (240 milliseconds or about ¼ second) is introduced when making a round trip from the surface to the satellite and back. ✓ Other, but less serious, problems include that the exact position of the satellite, relative to the surface, varies slightly over the course of each 24-hour period because of gravitational interaction by other celestial objects; and that there is a dramatic increase in background noise when the satellite comes near the Sun.

33. a) **i)** greater ✓

ii) lower ✓

b) $R = \frac{V}{I}$ ✓ (check by substitution)

34. a) Only Y will light up; when closed, switch B allows the current to bypass Z. ✓✓

b) Both lamps will light up. Current has to flow through Z as there is no path through open switch B. ✓

c) Neither lamp will light as the current from the battery cannot pass through the open switch A. ✓✓

35. There will be no change in the brightness of the lamp, as the potential difference across it stays constant as the switch is moved. It will continue to draw the same current through it even though the total current in the whole circuit drops as the switch moves from Y to X. ✓✓

36. a) The radar emits a concentrated radio wave. ✓ If there is an object in the path of the radio wave, some of this energy will be reflected and the radio wave bounces back to the radar. ✓ Since radio waves travel at a constant speed, the computer attached to the radar device calculates how far away the object is depending on how long it takes the radio signal to return. ✓

b) Radio waves are emitted at a certain frequency ✓; this is the number of oscillations for each unit of time. When the radar gun and the object are both standing still, the returning signal will have the same frequency as the original signal. ✓ When the object (relative to the radar) is moving, each part of the radio signal gets reflected at a different point in space. This changes the wave pattern. ✓ For example, when an object is moving away from the radar, the second part of the signal has to travel a greater distance to reach the object than the first part of the signal. This has the effect of 'stretching' the wave (lowering its frequency). ✓ If the object is moving towards the radar device, the second part of the wave travels a shorter distance than the first part before being reflected. This effectively 'squeezes' the wave (increasing its frequency). Depending on how much the frequency changes, a computer attached to the radar device can calculate how quickly an object is moving towards it or away from it. ✓

37. a) A geostationary orbit is circular, with satellites revolving from west to east being exactly the same as that of the Earth, so that it appears stationary from any point on the Earth's surface. ✓

b) Further away and the orbit speed will be slower ✓, making the satellite appear it is falling behind. Closer and the orbit speed will be faster ✓, making the satellite appear it is running ahead. (Space stations and shuttles in low Earth orbit, typically a thousand kilometres above the Earth's surface, make around 15 revolutions of the Earth each day. The Moon, at an altitude of about 384 400 km, takes about 27 days and 7 hours to make one complete revolution. This is a bit like in the children's story of Goldilocks, where her problem is whether the porridge is too hot, too cold or just right; the satellite must be at the correct height to remain geostationary.)

c) Like a stone being whirled around on the end of a string, the tug is always towards the centre. For a satellite the tug of gravity is towards the centre of the Earth. In order to remain at exactly above the same point on the Earth's surface as the Earth rotates, it needs to be above the equator. ✓ If a satellite was located elsewhere, it would move north and south of the equator during its orbit, while a geostationary satellite does not. ✓

d) i) $23 \times 60 \times 60 + 56 \times 60 + 4$
$= 82\,800 + 3360 + 4$
$= 86\,164$ seconds ✓✓

ii) $3.07 \times 86\,164 = 264\,523$ km ✓ (correct to nearest km)

38. radios ✓; data ✓; communicates ✓; transceivers ✓; geographic ✓; network ✓; transmission ✓; phone ✓; stations ✓; moving ✓; connected ✓; efficient ✓

39. a) i) A_1 has the highest reading as the total circuit current flows through this ammeter. ✓

$V_1 = V_2 = V_3 = 12$ V;
$V_2 = I_2R_2$; $12 = (I_2)(3)$;
$I_2 = 4$ amps $= A_2$;
$V_3 = (I_3)(R_3)$; $12 = I_3(4)$;
$I_3 = 3$ amps $= A_3$;
$I_T = 4 + 3 = 7$ amps $= A_1$

ii) A_3 has the lowest reading as less current flows through the larger resistor. ✓

b) The readings are the same in all voltmeters as this is a parallel circuit. ✓✓

c) The resistance increases and consequently the total current will decrease. A_1 and A_3 will now read the same value (3 A) and A_2 will read zero as no current flows through this arm when the new switch is off.

$(R_T = 4\ \Omega; I_T = \frac{12}{4} = 3$ A;
$A_1 = 3$ A; $A_3 = 3$ A; $A_2 = 0$ A) ✓

40. a) See Figure A.23.

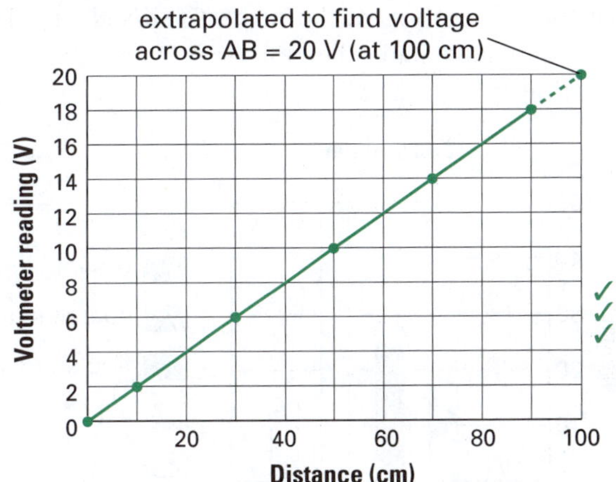

Figure A.23 Voltmeter reading

b) 20 V (see graph) ✓

c) $R = \frac{V}{I} = \frac{20}{4} = 5\ \Omega$ ✓✓

Chapter 5 answers

Experiment 1

Analysis: See Figure A.24. If the line is extrapolated to 8 weights, then the volume of gas is estimated to be 36 mL.

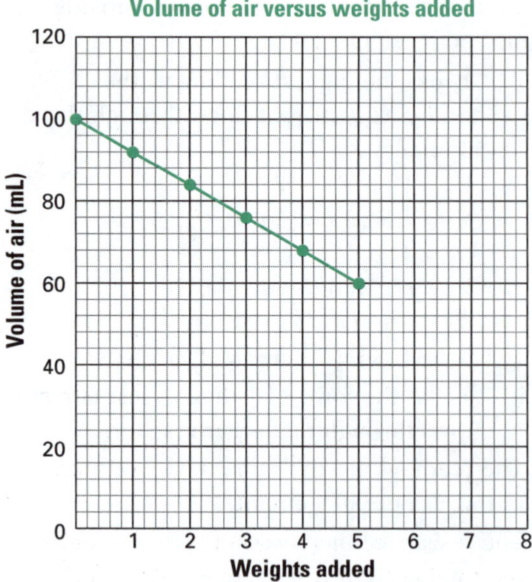

Figure A.24 Graph of volume versus weight

Conclusion: The volume of gas in the syringe decreases linearly as the number of added weights increases.

Experiment 2

Analysis: See Figure A.25. Zone B is the best match for the soil analysis of the suspect's shoe. Note: The shoe may have had other soil on it from a different location.

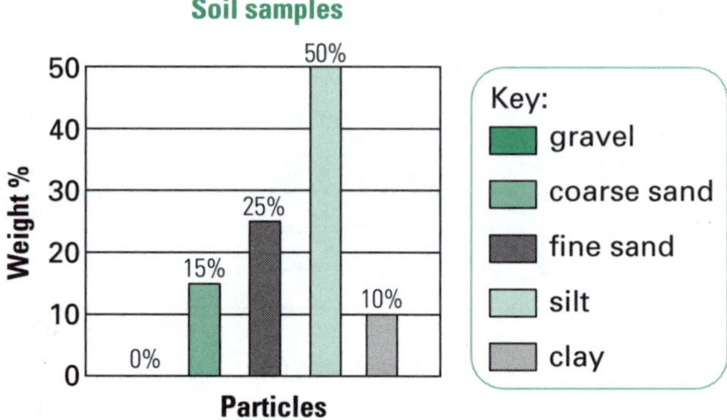

Figure A.25 Soil samples

The sample is small and unless the soil is well mixed, it will be variable in its composition. The soil may be common to many regions and so this analysis by itself will not prove that the suspect was at the victim's home.

Conclusion: The soil sample from the suspect's shoe closely matches the soil in zone B of the victim's garden.

Experiment 3

Analysis:

1. So that they are as alike as possible, and from the same production batch.
2. Yes. Not only are the batches different, but rubber bands deteriorate as they age. Therefore, the tensile strength would not be similar for all samples. Remember, in experiments you are trying to keep all other factors constant.
3. The factor that is being investigated is the tensile strength of a type of rubber. The outcome is the maximum length of the rubber before it breaks.
4. By measuring this length for several samples and then calculating the mean and range, you get a better estimate of the tensile strength of the rubber. You can then also determine how reliable the data is. If several samples are used, incorrect values (such as outliers) can be ignored. The others are used to calculate the mean, which is then used as the best estimate.
5. In this investigation there are several factors, also called variables, which may affect the outcome. These include the type of rubber, width of rubber, thickness of rubber, length of rubber, age of the rubber band and temperature.
6. If the other factors are not controlled, they may have an effect on the outcome. This means that any measured effect could be the result of one of these other factors, rather than the factor that is being examined. This could lead to unreliable results. If these other factors were not controlled, it would be a design flaw in the experiment.
7. It is difficult to tell exactly where the rubber snapped as this happens so fast. Another factor is that rubber bands are not produced to exacting standards. They are to be used for specific purposes, so the thickness may not be all that constant.
8. Shopping bags are designed to carry groceries. By attaching the shopping bag handles to some support, you can start to add weights to the bag until it fails. The weight needed just before the bag tears can be used as a measure of the bag's tensile strength.

Conclusion: The rubber band extended to around six to seven times its original length before snapping. Different thicknesses and widths (i.e. cross-sectional areas) extended a different amount before snapping.

Test yourself 1

Part A: Knowledge

1. **A** ✓ B is wrong as predictions are used to test inferences. C is wrong as this refers to an inference. D is wrong as inferences and hypotheses are developed to explain an observation.
2. **B** ✓ A is wrong as the burn must be cooled rapidly to minimise damage. C and D are wrong as this would take too long. Quick action is essential.
3. **C** ✓ A is incorrect as this will be too low to heat the tube. B is wrong as metal tongs will slip on the glass. D is wrong as the tube must be only partly filled otherwise acid will boil out; also, a blue flame is used for heating.
4. **C** ✓ A is wrong as the angle is the independent variable. B is incorrect as a plant of the same age should be used. D is wrong as the direction of growth is the dependent variable.
5. **D** ✓ A is wrong as the 55 °C experiment showed that plants do not grown well under these conditions. B is wrong as there is no evidence for this conclusion in the results. C is wrong as the results have not shown this.
6. **a)** title ✓; **b)** personal ✓; **c)** reliable ✓; **d)** floor ✓; **e)** inferred ✓
7. A/G ✓; B/I ✓; C/J ✓; D/F ✓; E/H ✓

Part B: Skills

8. **a)** 3.5 kW/km ✓
   b) 1.6 km/h ✓
   c) Do not light fires in the bush. If using a barbecue, clear a sight so that it is free of fuel and sparks cannot jump into the bush. ✓ Follow the bushfire warnings. Never light fires on hot, windy days with low humidity. ✓
9. **a)** very alkaline ✓
   b) crimson ✓
   c) Both are alkaline ✓; the drain cleaner is more alkaline than the baking soda. ✓
   d) These three substances are bases. ✓
10. It is not scientific to rely on the word of others simply because they are well known and have a good reputation. It is not scientific to make conclusions about what God can or did do. Science advances by facts and evidence, not on the beliefs of experts, no matter how well-respected they are. ✓
11. It is often difficult for scientists to give up a belief they have had for many years. The ability to discard a belief that has been shown to be incorrect is vital for scientific knowledge to advance. ✓
12. **a)** Ask an unexpected question and you stimulate discussion and discovery. New ideas are vital for real breakthroughs in knowledge to be made. ✓
    b) Facts must be accumulated but without some unifying theory they are rather useless. ✓
    c) Science is an enterprise that uses logic and skill. Those who are highly skilled and can think and reason in a logical manner will be good scientists. ✓
13. **a)** We don't know whether the wax will effectively seal the hole, especially during incubation. ✓ The piercing of the shell may damage the embryo. ✓ Alcohol may cause damage to the embryo. ✓
    b) There are insufficient numbers of eggs used. No control is used. No effect of alcohol alone is studied. ✓
    c) There are a number of controls to eliminate other variables. There is more than one egg tested. ✓
    d) It is to determine whether alcohol alone will affect the growth of bird embryos. ✓
    e) It is to determine whether piercing the shell has an effect on the growth of bird embryos. ✓
    f) It is to allow a comparison with other eggs that have been altered in some way. ✓
    g) It is to ensure that the results are not just due to chance. ✓
    h) Researcher B has conducted a fair test by controlling the variables and used a number of runs to eliminate chance. ✓
14. **a)** Original hypothesis: something in sunlight was attracting the bees to the yellow rather than the white daisies. ✓

b) no ✓

c) The experiment indicated that as many bees were attracted to real yellow as real white daisies. The bees respond to the white and yellow daisies because they 'see' the daisies in the ultraviolet region where they have a different pattern. ✓

d) Repeat the experiment comparing:

- sunlight filtered to remove the ultraviolet light with electric light ✓
- sunlight with UV light. ✓

Test yourself 2

Part A: Knowledge

1. **B** ✓ Anyone can publish material on the internet; it is not always checked by reliable people. Answer A is not true as there is considerable relevant information available. C is not correct as the author's name is often present. D is wrong as many articles are not untrue.
2. **A** ✓ The information must come from outside Australia. The other answers are wrong because information from South America (B), New Guinea (C) and around the world (D) is all relevant.
3. **C** ✓ The procedural recount allows a student to use the past tense to describe how the experiment was performed. A is wrong as a narrative is used in describing an event or telling a story. B is wrong as a discussion is not used in an experimental method. D is wrong as an exposition is a presentation.
4. **D** ✓ Selecting only data that fits a hypothesis is poor science. A, B and C are wrong as they are all critical thinking strategies.
5. **C** ✓ No date of publication; place (New York) is not at the end. A, B and D are all wrong as they have all the required information.
6. **a)** newtons ✓; **b)** valid ✓; **c)** data (or information) ✓; **d)** collated ✓; **e)** half ✓
7. A/J ✓; B/H ✓; C/F ✓; D/G ✓; E/I ✓

Part B: Skills

8. A = 6.5 ✓; B = 4.8 ✓; C = 0.325 ✓
9. See Figure A.26.

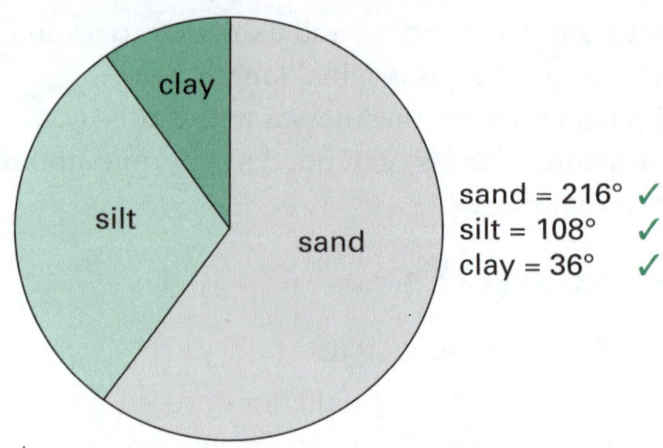

Figure A.26 Components of sandy loam soil sample

10. **a)** See Figure A.27.

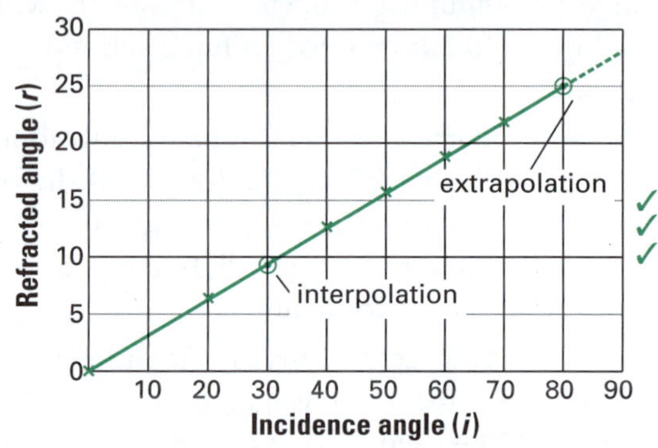

Figure A.27 Refracted and incidence angles

b) 9° ✓

c) 80° ✓

11. $L = 3.5 \times 0.5 = 1.75$ mm ✓
12. **a)** This is a straight line graph $v = u + at$. Substitute values from the graph to show that the equation is consistent. For example:
    At $t = 4$, $v = 5$ and $u = 3$
    $5 = 3 + 4a$; solve for a
    $a = \frac{5-3}{4} = 0.5$
    At $t = 8$, $v = 7$ and $u = 3$
    $7 = 3 + 8a$
    $a = \frac{7-3}{8} = 0.5$ ✓✓

    b) $u = 3$ m/s ✓ from intercept on vertical axis

    c) $a = 0.5$ m/s^2 ✓ as calculated in part a)
13. **a)** 8 mL ✓

    b) 18 mL ✓

    c) true value = 1½ × measured value ✓ or measured value = 2/3 × true value

    d) systematic error ✓

14. A meandering river will change its course over a long time period. Twain is using values measured over this brief time, less than two centuries, to extrapolate ✓ lengths of the Mississippi River between the two cities, Cairo and New Orleans. Extrapolating its length backwards and forwards in time to these extremes is not valid ✓. He is using the notion that the same conditions prevail ✓ as in the 176 years of measurements he uses. For example, stating that the river would have been over 1 300 000 miles (2 080 000 km) is ridiculous. (And that is for just the section between the two cities mentioned!) Compare this number with the circumference of the Earth, around 40 000 km long! And, given that the river continues to flow between Cairo and New Orleans, the shortest distance between these two cities it could ever possibly get is the straight-line distance, not 1¾ miles. (Incidentally, the distance from Cairo to New Orleans is around 790 km, or 490 miles.) He is also using language to coerce readers to his view: 'any calm person, who is not blind or idiotic can see' and 'any person can see'. ✓

Chapter test

Part A: Multiple-choice questions

1. **D** ✓ A placebo does not contain the new drug and is used to compare results. Thus A is wrong. B is wrong as a drug is not a person. C is wrong as the term 'blind' here refers to the fact that the volunteers do not know if they are receiving the drug or the placebo.
2. **B** ✓ This procedure eliminates any biases of the scientist. A is not correct as the control group receives the placebo. C is wrong as no masks are used. D is wrong as the number of scientists involved is irrelevant.
3. **C** ✓ A is wrong as accuracy depends on the equipment chosen and the skills of the experimentalist. B is wrong as no amount of repeating can improve an invalid experimental method. D is wrong as safety is not related to repetition.
4. **C** ✓ There will be more divisions on a 250-mL measuring cylinder. Thus A, B and D are wrong as they have fewer scale divisions.
5. **B** ✓ Articles in science journals are peer reviewed and cannot be published without scrutiny from practicing scientists. A is wrong as newspapers often contain personal opinions. C is wrong as some bloggers may not fully understand the scientific principles. D is wrong as creation 'science' is not real science but only a matter of faith without scientific proof.
6. **B** ✓ The cause comes before the effect. Statements A, C and D are true.
7. **A** ✓ As seen from the graph, this error occurs randomly; it is not systematic, so B is wrong. Validity (option C) and reliability (option D) are important in science, but do not answer this question.
8. **C** ✓ Just like a divided bar graph, a sector graph or pie diagram is used to show how parts of a whole are divided. A pie chart is divided into sectors, illustrating proportion. In a pie chart, the area of each sector (and consequently its central angle and arc length), is proportional to the quantity it represents. A and B are best shown using line graphs. D is incorrect as a circular graph, like this one, is not related to cycles.
9. **D** ✓ Options A, B and C are incorrect.
10. **A** ✓ Both statements P and Q are correct, so answers B, C and D are wrong.

Part B: Short-answer questions

11. a) local hospital, football clubs, sports physiotherapists ✓
    b) No ✓ Thirty-four students is a very small sample compared with the whole school population. ✓
12. a) Method:
    1. Select a piece of cotton fabric, 15 cm × 15 cm. ✓
    2. Drop 20 of the test seeds (from a fixed height) onto the fabric held horizontally. ✓
    3. Then hold the fabric vertically. ✓
    4. Count the number of seeds that drop off immediately. ✓
    5. Shake the fabric five times in 5 seconds with a constant amplitude of shaking. Count the number of seeds that drop off. ✓
    6. Repeat step 5 another five times. ✓
    7. Repeat steps 1 to 6 with the other types of fabric. ✓

b) Variables held constant:
- size of the pieces of fabric ✓
- height from which the seeds are dropped ✓
- amplitude of the shake ✓
- number of seeds originally dropped onto the fabric. ✓

13. a) Aim: To see if (1) bath crystals and (2) washing soda added to hard water will make soap lather. ✓

b) Variables controlled: *(any 6)* ✓✓✓✓✓✓
- constant concentration of hard water (made by adding the same amount of calcium chloride or magnesium sulfate to a given amount of water)
- type of bath crystals used
- type of washing soda used
- water temperature
- type of soap used to create bubbles
- demineralised water used as the control
- number of drops of soap used
- amount of water used in the test
- number of shakes used to create bubbles.

c) Method:
1. Place 5 mL of demineralised water in a large test tube, add three drops of detergent, stopper the test tube and shake 50 times. ✓
2. Record the height of the foam formed. ✓
3. Place 5 mL of the hard water in a test tube with three drops of detergent, stopper test tube and shake 50 times. ✓
4. Record the height of the foam formed. ✓
5. Repeat steps 1 to 4 with the addition of a rice-grain-size quantity of bath crystals. ✓
6. Repeat step 5 using washing soda. ✓
7. Repeat each experiment five times. ✓

d) Data: Use a table to enter results; the hypothesis is supported if the height of foam in the demineralised water is the greatest and in the hard water it is the least, ✓ and that the presence of bath crystals or washing soda causes the foam height to be similar to the demineralised water. ✓

14. a) See Table A.6.

b) Make further and more careful observations about the possible causes of the noise. ✓

They could set up a monitoring video camera with sound. ✓

Table A.6 Inferences and observations

Inferences	Observations
noise seemed to come from the mangrove tree ✓	ripples ✓
could have been a mangrove seed ✓	lots of seeds floating in the water ✓
cormorant dived into the water ✓	wind blowing ✓
wind could have caught the door ✓	two cormorants on the handrail ✓
water rat dived into the water ✓	family of water rats in the roof ✓

15. a) yes, the total number of asthma attacks drops ✓

b) what the subjects did during the day which could induce asthma attacks ✓; not all subjects are experiencing the same controlled environment ✓

c) tests on animals ✓; toxicity tests (is it poisonous?) ✓ long-term contraindications ✓

16. a) Variables controlled:
- the way the fabric dries (rolled into identical cylinders) ✓
- the size and shape of the pieces of fabric ✓
- the force applied to the fabric (the mass of the iron). ✓

b) It is suitable for a home experiment where one experimenter is judging. ✓

c) The sole plate seems to allow the silk to be ironed on the hottest settings without scorching, but seems to be unable to remove the wrinkles from the other two fabrics. ✓

d) The iron has only its weight acting down on the material. The addition of a mass on the handle of the iron may improve the results of the experiment as the usual way to use an iron is to apply downwards pressure. ✓

17. You cannot 'prove' a hypothesis, you can only disprove it. If all the experiments continue to support the hypothesis then it is likely to be elevated to a theory. In the future, if an experiment shows it to be false then the theory must be modified. ✓

18. A = observations ✓; B = hypothesis developed ✓; C = experiment performed ✓; D = experiment redesigned ✓; E = report ✓

19. A = hypothesis ✓; B = law ✓; C = theory ✓

20. a) All one needs to do is capture one of the little yellow men. If one never succeeds in this quest then the statement is not falsifiable. ✓
    b) If the microbes are undetectable then there is no experiment that can show their existence and thus the statement is not falsifiable. ✓
    c) There is no experiment that can be done to reveal the existence of God. Thus the statement is not falsifiable. ✓

21. a) We can test the theory by going to the Moon and searching for these little men. ✓
    b) We can test the theory by doing experiments in Loch Ness to determine whether the monster exists. ✓
    c) We can go to the Himalayas and search for the abominable snowman in order to test the theory. ✓
    d) We can use spacecraft to move out from the Earth's surface to test this theory. ✓
    e) We can send probes to each planet to test this theory. ✓

22. Modern scientists do not accept the views of ancient scholars as the absolute truth. ✓ They look for alternative explanations for observations in an attempt to find some more universal truth or law. ✓

23. High root growth and little shoot growth occur when the night temperature and day temperatures are low. ✓ When the night and day temperatures are high, the shoot growth is high but the root growth is not as great. ✓

    This data suggests that in the winter (low night and day temperatures) the pine seedlings will use their energy to grow roots but not shoots. ✓ In the summer, however, when there is abundant sunshine as the days are longer, the plants will grow shoots and leaves to maximise photosynthesis and food production. ✓

24. There are various types of models used in chemistry. One of the most common types is a structural model of atoms and molecules.
    - These models are useful as they help to visualise the arrangement of subatomic particles in the atom or the ways in which atoms combine in three dimensions to form a molecule. ✓
    - Molecular models are also helpful in visualising how particles interact in a chemical reaction. The way that ions or atoms are arranged in a crystal lattice is best shown by making a 3-D model based on X-ray analysis. ✓
    - Without models, the observations made by chemists in their experiments would be difficult to interpret. Computer modelling and visualisations have improved the ways chemists can interpret and understand chemical processes. ✓
    - These models have their limitations, however, as they are constructed from wood, plastic and metal and consequently cannot represent the true nature of atoms, particularly the low-density electron cloud. ✓
    - The simple ball and stick models of compounds do not show how electrons interact between atoms to produce bonds. ✓
    - Space-filling models are better indicators of the shape of the molecule but do not show the bond angles easily. ✓
    - Some models in chemistry are mathematical. The kinetic theory of matter contains a mathematical description of the way matter behaves. Mathematical models and theories help us to calculate chemical properties or behaviour. The experimental measurements are then compared with theoretical models to see if they are consistent. These models are limited in that they often only approximate the true behaviour of matter. ✓

25. a) The northern and western sides of the house are sunnier than the other two sides. ✓
    b) One student has infected the other nine boys. ✓
    c) The refrigerator is too cold to allow bacteria to reproduce. ✓
    d) Water is less dense than mercury. ✓
    e) The children's parents have olive skin and brown eyes. ✓

26. a) An outlier is an observation that lies outside the overall pattern of a distribution. ✓
    b) Outliers can have many causes. A physical apparatus for taking measurements may

have suffered a momentary malfunction. ✓ There may have been an error transmitting or transcribing (writing down) the data. ✓ Outliers can also arise due to human error, ✓ instrument error ✓ or a slight change in the variables being kept constant ✓.

27. a) These experimental errors affect each measurement differently and are often due to variations in the control variables that are difficult to keep exactly the same. ✓

b) human error (carelessness) ✓; faulty technique in taking measurements ✓; faulty equipment ✓

c) Yes ✓, by taking a large number of readings .✓

28. a) effect/cause ✓; **b)** effect/cause ✓; **c)** cause/effect ✓; **d)** cause/effect ✓

29. a) For many years there was no hard and fast rule of what a planet should be. The definition was fairly much arbitrary. There are many objects circling the Sun, some of them larger than Pluto, but are not considered planets. As more and more information is collected, a point comes when someone asks 'What exactly is a planet?' These discoveries challenged long-perceived notions of what a planet could be. Scientists were not wrong, since for a long time there was no clear definition; a planet was whatever an astronomer said it was. Now that there have been guidelines set, Pluto does not meet the criteria to be a planet and so has been demoted. ✓✓✓

b) The International Astronomical Union's decision (the worldwide organisation that decides these things) has not resolved all controversies. While many scientists have accepted the definition, there will be astronomers who have grudgingly accepted it while others have rejected it outright. ✓ After all, astronomers are people too and they have their points of view and biases. But at least there is now a working definition in place. ✓

30. a) There is a linear (straight line) relationship between variables. ✓ All the readings are shifted in one way from the true value. ✓ In this example, all readings are higher than they ought to be. ✓

b) by using a wrongly calibrated instrument ✓

c) by subtracting 10 milliampere from each measurement ✓

31. a) Since 1900 there has been an increase in the atmospheric CO_2 concentration over time. ✓ It is not linear but curves upwards. ✓

b) The outlier occurs in 1980 ✓, with a value of 300 ppm ✓.

c) The points do not exactly follow a curve. ✓ Samples of air are taken at regular intervals and there are many random factors ✓ that can influence the exact CO_2 concentration at any point in time. Of course, scientists try to minimise these factors by taking the samples at locations well away from polluting industries and other sources that can locally affect CO_2 concentration; samples are taken in the same location, and at the same time of day.

d) around 380 to 390 ppm ✓

e) extrapolation ✓

f) that the same trend will continue into the future ✓

32. a) measuring ✓; **b)** modelling ✓;
c) classifying ✓; **d)** predicting ✓;
e) investigating ✓; **f)** communicating ✓;
g) observing ✓; **h)** inferring ✓

33. a) 6% ✓ 100% – (20 + 40 + 8 + 4 + 15 + 4 + 3)% = 6%

b) $\frac{20}{100} \times 360° = 72°$ ✓

c) No, it does not show how much greenhouse gas is emitted overall ✓, nor how much each category emits. It shows the proportion of the total emitted by each category. ✓

d) $\frac{15}{100} \times 517 = 77.55$ million tonnes CO_2 ✓

e) increase is 580 – 517 = 63 million tonnes ✓,

so percentage increase $= \frac{63}{517} \times 100$
$= 12.2\%$ ✓

f) Not necessarily ✓, although if the time period is not all that great they should be similar. Different government regulations, changes in technology and community requirements over time will affect the size of each of these sectors. ✓✓

34. a) divided bar graph ✓

b) 0.9% ✓ 100% – (78.1 + 20.9 + 0.1)% = 0.9%

c) CO_2 forms part of the trace gases present in the atmosphere. ✓ (In 2011, the CO_2 global average concentration in Earth's atmosphere was about 0.0391% by volume. So it doesn't take much carbon dioxide to affect temperature.)

d) $\frac{78.1}{100} \times 15 = 11.7$ cm ✓

35. Claims A and B can be tested scientifically. The concentration of free radicals in the blood can be monitored in the presence of Oxideze and without it. The ingredients can be determined to be natural or synthetic. Claim C is not easy to test as it would be impossible to test all other alternative products on the market. ✓✓✓

36. a) condition X: average length = 5.1 cm ✓✓; condition Y: average length = 6.9 cm ✓✓

b) *(any three ✓✓✓)* same age of seedlings; same amount of water; same amount of exposure to light; same temperature; same concentration of each gas tested

c) He needed to do a control experiment in which the seedlings were not exposed to either gas. ✓

d) No. It could be due to natural variability. He tested very few seedlings and he did not do a control. ✓

37. a) 12 mL ✓

b) The pH was initially low and rose slowly as the base was added. At the point of neutralisation, the pH changed rapidly upwards. With excess base there was a rise in pH but there was a slower change in pH. ✓✓

c) 7 mL ✓ because the pH = 7

d) 3.0 ✓

e) 5.35 mL ✓

38. a) 9% ✓

b) See Figure A.28.

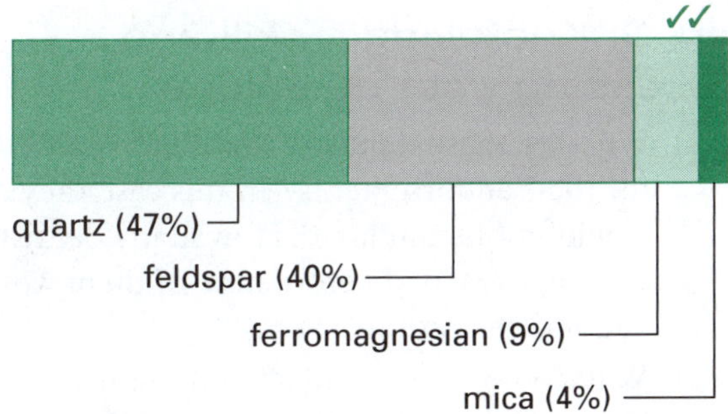

Figure A.28 Minerals

39. a) See Figure A.29.

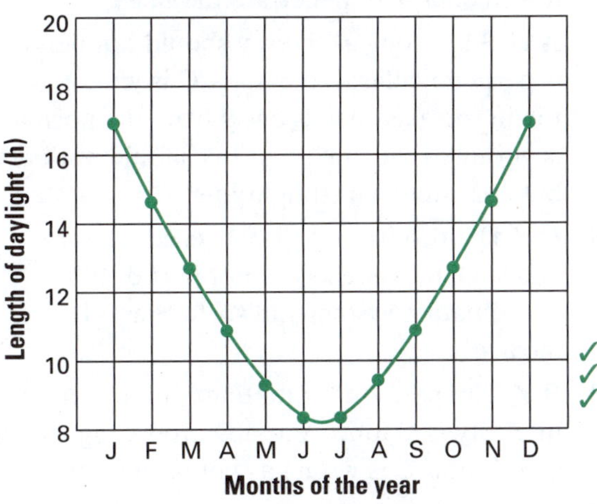

Figure A.29 Length of daylight

b) i) winter or early spring ✓

ii) October, November, December, January, February, March ✓

40. a) concentration of A $= \frac{3.6 \text{ g}}{150 \text{ mL}}$

$= \frac{3600 \text{ mg}}{0.150 \text{ L}}$ ✓

$= 24000$ mg/L ✓

b) D, A, C, B ✓

Test 1 answers

Part A: Multiple-choice questions

1. **A** ✓ Each scale division is 0.2 degrees and so the edge of the mercury is at 42.8 degrees. In a mercury thermometer the reading is taken from the top of the convex meniscus. Thus, B, C and D are wrong.
2. **C** ✓ The percentage of zinc = 85 – 60 = 25%. Thus, A, B and D are wrong.
3. **A** ✓ B is wrong as there are 12 protons. C is wrong as the number of protons equals the number of neutrons. D is wrong as the nuclear charge is +12.
4. **B** ✓ Universal indicator has different colours at different pHs. A is wrong as chlorine is a disinfectant. C is wrong as sodium chloride is neutral. D is wrong as this is as strong base and never added to a pool.
5. **C** ✓ The wave is transverse and 4 scale units in wavelength which is equivalent to 400 mm. Thus, A, B and D are wrong.

6. **C** ✓ A and B are wrong as these waves are not electromagnetic. D is wrong as infrared is too low in energy to penetrate the body.
7. **A** ✓ B is wrong as the ray should not bend as it is perpendicular to the glass. C is wrong as the ray in the glass bends away from the normal. D is wrong as the emergent ray should not bend but continue in a straight line.
8. **C** ✓ The respiratory, digestive, circulatory and skeletal systems are not part of the coordination system and so A, B and D are incorrect.
9. **B** ✓ Competition for mating is biotic and light intensity is abiotic. A and D are wrong as both are abiotic. C is wrong as both are biotic.
10. **A** ✓ The Pacific Ring is where plate boundaries are located. B, C and D are wrong as there no plate boundaries at these locations
11. **D** ✓ A is wrong as compression waves require a medium. B is wrong as these waves are mainly absorbed by the atmosphere. C is wrong as some visible light is absorbed by the atmosphere.
12. **D** ✓ Mumps is a viral disease; therefore, A, B and C are wrong.
13. **C** ✓ The plague was bacterial; therefore, A, B and D are wrong.
14. **A** ✓ Lister used carbolic acid so B, C and D are wrong. You wouldn't want to use hydrochloric acid as it could cause chemical burns and other irritations.
15. **A** ✓ As a lot of heat and light are produced, the reaction is exothermic. Thus B is wrong. C is wrong as the white powder is magnesium oxide; magnesium + oxygen $\rightarrow$ magnesium oxide. D is wrong as a chemical separation rather than a physical separation method is required.
16. **A** ✓ B is wrong as it is a constructive boundary. C is wrong as subduction does not occur at a ridge. D is wrong as the mid-ocean ridges are also found in the Southern ocean and the Indian ocean.
17. **C** ✓ Ammonia has four atoms (one N and three H); tetra = four. A is wrong as carbon dioxide is triatomic and B is wrong as methane (CH_4) is penta-atomic. D is wrong as oxygen is diatomic.
18. **B** ✓ A is hydrochloric acid, C is sulfuric acid and D is carbonic acid.
19. **B** ✓ A is wrong as barium chloride is a salt. C is wrong as calcium chloride will form. D is wrong as barium nitrate will form.
20. **B** ✓ A is wrong as CO is not formed. C is wrong as oxygen is a reactant. D is wrong as CO is formed rather than CO_2.
21. **D** ✓ Thus A, B and C are wrong. The activity order is K, Zn, Fe then Cu.
22. **B** ✓ A and D are collision zones. C is a mid-ocean ridge.
23. **C** ✓ The decreasing order of frequency is UV, visible (D), microwave (B) and radio (A).
24. **D** ✓ A is wrong as light rays do not enter a new medium but bounce off the boundary. B is wrong as light rays scatter in all directions. C is wrong as a translucent object causes scattering.
25. **D** ✓ Sound waves propagate as a series of compressions and rarefactions through a medium.

Part B: Restricted-response questions

26. age ✓
27. forces ✓
28. mantle ✓
29. Eurasian ✓
30. glands ✓
31. bloodstream ✓
32. cancerous ✓
33. positron ✓
34. immunity ✓
35. atomic ✓
36. ions ✓
37. evaporating ✓
38. oxygen ✓
39. concentration ✓
40. medium ✓

Part C: Knowledge, skill and processing data questions

41. a) Who are these dentists? Were they paid for their endorsements? In this case they would not be unbiased. How many dentists were approached before nine of them would endorse the toothpaste? ✓
    b) Which university conducted these tests? Using the word 'university' sounds impressive but adds little weight if payment

depended on finding a favourable outcome. Sometimes advertisers make it sound even more impressive by stating a 'leading university'. Does this mean that other universities are just run-of-the-mill degree factories? And 22% whiter … whiter than what? Mud? How was this measured? ✓

c) 'Leading' is misleading. Is the definition of 'leading' if they sell their product? It is there to give the impression that high-quality stores wouldn't sell anything cheap and nasty, so it follows the product must be good. ✓

42. a) Inference. Coughing can be caused by many conditions. A virus is just one possibility. ✓

b) Generalisation. Examination of a wide variety of copper compounds shows that most of them are blue and so this is a useful generalisation. ✓

c) Observation. The colour of the different minerals can be observed. ✓

d) Prediction. The statement suggests that the student has already tried the dilute acid and is now predicting that a more concentrated solution will react even faster. ✓

43. a) Don't use your hands; some chemicals may be harmful. Besides, you are contaminating the contents of the jar. Use a spatula or plastic spoon. ✓

b) Tests tubes are clear so you can look through the side. Sometimes a reaction can be violent and shoot out, causing eye damage. ✓

c) All substances being heated need monitoring. Leaving the beaker unattended may lead to a fire or the contents of the beaker bubbling over. ✓

d) Some chemicals have very strong or poisonous odours. Wafting some of the chemical towards her nose with her hand is a better way to smell the contents. ✓

e) It's easy to knock the glassware over and cause it to smash onto the floor, especially if someone walks by and brushes by the bench. Keep glassware well away from the edge. ✓

f) Contents may be harmful to the skin. A better way is to hold the neck of the test tube tightly with thumb and finger and, with the other hand, drum your fingers across the side near the bottom. ✓

g) Ouch! The contents will be hot. Leave the beaker to cool down before doing this. Alternatively, use wooden tongs or wrap a cloth around the beaker before picking it up. ✓

h) Solids should go into the bin. Liquids, depending on what they are, can be flushed down the sink using plenty of water or placed in a special disposal container. ✓

44. a) suspect 2 = flowerbed print ✓

b) Any evidence of wear on the shoe would be important evidence to compare with the flowerbed print. ✓

45. a) i) 8.1 ppm at 25 °C ✓ (see Figure A.30)

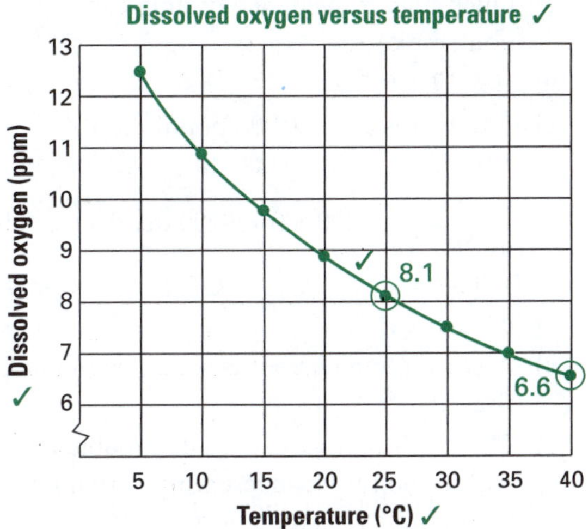

Figure A.30 Dissolved oxygen versus temperature

ii) extrapolate, about 6.6 ppm at 40 °C ✓

b) See Figure A.31.

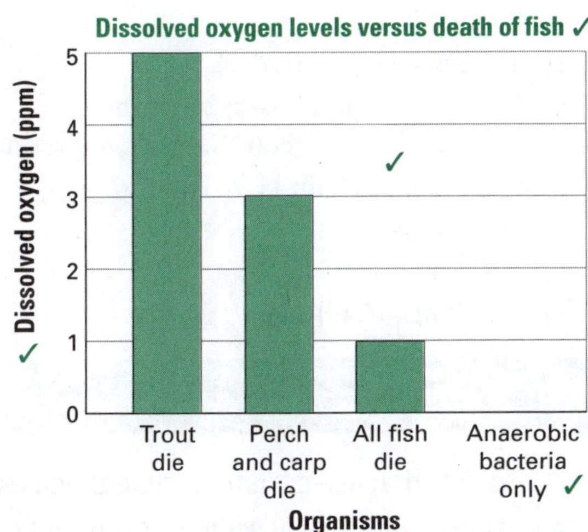

Figure A.31 Dissolved oxygen and death of fish

c) about 7.5 °C ✓

46. P/E, Q/B, R/D ✓; S/C, T/A ✓

47. a) > 10 units ✓

b) $\frac{540\,000}{10\,000} = 54 \text{ mJ/cm}^2$ ✓

UV reading = $\frac{54}{9} = 6$ UV units ✓

c) 6 UV units ✓; 10 UV units ✓ The reading is increasing as the intensity of the Sun's energy increases because the Sun is closer to being overhead and the UV rays do not have to pass through as great a thickness of atmosphere. ✓

d) ~22.5 UV units ✓

48. a) Yes ✓ The smokers of both sexes have a much higher incidence of lung cancer than non-smokers. ✓

b) 103:17 = 6:1 ✓

c) 120 deaths per 100 000 population ✓ Therefore, of 8 000 000, there will be $\frac{120}{100\,000} \times 8\,000\,000 = 9600$ deaths from lung cancer. ✓

49. a) infra-red ✓

b) A convection current is set up as the hot air rises. ✓

c) These oils and waxes are flammable and add to the fuel load, creating very high temperatures as they burn. ✓

d) The burning fragments can be carried by winds to new locations and start spot-fires at these places. ✓

50. a) in order to see the passage of light through the medium ✓

b) The image is inverted. ✓

c) 1 = contact lens/glasses; 2 = cornea; 3 = lens ✓; 4 = vitreous humour; 5 = retina ✓

51. a) i) 12 ✓; ii) 22 ✓; iii) 11 ✓; iv) 45 ✓

b) diagram B ✓

c) i) 6 ✓; ii) 6 ✓

d) i) 1 ✓; ii) 4 ✓; iii) 2 ✓

Test 2 answers

Part A: Multiple-choice questions

1. B ✓ Both atriums receive blood (so A and C are incorrect) while the right ventricle (D) pumps blood to the pulmonary artery that then goes to the lungs to be oxygenated.
2. A ✓ B, C and D are incorrect as these can all vary but it does not change what the element is.
3. D ✓ Emphysema is a disease of the lung tissue caused by the destruction of structures feeding the alveoli. Smoking is one major cause of this, where the small airways in the lungs collapse during forced exhalation. The others are all hereditary diseases.
4. C ✓ This is the retina. The other labelled parts are: A, vitreous humour (the clear gel that fills the space between the lens and the retina); B, fovea (the pit in the retina allowing for maximum acuity of vision); and D, iris (the thin, circular, coloured structure in the eye responsible for controlling the diameter and size of the pupils and hence the amount of light reaching the retina).
5. B ✓ Thomson's atomic theory stated that atoms as a whole were neutrally charged. He also suggested that the atom was made up of electrons embedded into a cloud of positive particles. This was known as the Chocolate Chip Cookie Model or the Plum Pudding Model, among other names. The other scientists mentioned developed different atomic models.
6. D ✓ All statements are true. (Don't think that 'all of the above' or 'none of the above' are inserted as fillers and can never be answers.)
7. A ✓ Intestinal villi are tiny, finger-like projections that protrude from the epithelial lining of the intestinal wall. Each villus is approximately 0.5 to 1.6 μm long, increasing the surface area of the intestinal wall. B is in the lungs, C is located outside the body and D does not absorb nutrients.
8. B ✓ Hence, A, C and D are incorrect statements.
9. C ✓ The years of discovery were: electron, 1897; proton, 1919; neutron, 1932; quark, 1964.
10. A ✓ A major threat is from humans taking over their habitats for other uses. B, C and D should all show an increase in populations.
11. C ✓ The outer core of the Earth is a liquid layer about 2266 km thick composed of iron and nickel. Layers A, B and D are solid.
12. D ✓ L-waves, the slowest of the earthquake waves, travel along the surface of the Earth rather than down into it so C is wrong. L-waves usually cause more damage than either P or S waves, so A and B are wrong.

13. **C ✓** Hawaii lies over a hot spot, a point where hot magma currents rise from deep within the Earth. While most volcanic activity occurs along tectonic plate boundaries, powered by the plates' movements (B and D), hotspots can occur far from such geological boundaries. Underwater volcanoes are formed when two plates move apart (A).

14. **A ✓** The cerebrum is the largest and most highly developed part of the human brain. It has about 2/3 of the brain mass and lies over and around most of the structures of the brain. It controls the functions listed. One of the most important functions of the hypothalamus is to link the nervous system to the endocrine system via the pituitary gland, so B is wrong. C is wrong, as the cerebellum is a region of the brain playing an important role in motor control. It is also involved in some cognitive functions such as attention and language. Functions of the brain stem include those necessary for survival (breathing, digestion, heart rate, blood pressure) and for arousal (being awake and alert), so D is wrong.

15. **C ✓** The more predators there are, the less prey there will be. A, B and D will allow prey numbers to increase.

16. **D ✓** Since most α-particles went straight through the gold foil, Rutherford concluded that atoms were mostly empty space. A refers to Thomson's model. In B, Rutherford's experiment showed that α-particles were deflected by gold nuclei, not that they were broken up. As for C, Rutherford came to the opinion that an atom consisted of a nucleus of size 10^{-15} to 10^{-14} m diameter.

17. **A ✓** Mid-ocean ridges are places where the Earth's tectonic plates are gradually moving apart and, as they do, magma rises up to fill the gap sometimes leading to submarine volcanic eruptions. Hence, B, C and D are incorrect.

18. **C ✓** CT scans use X-rays, but no radiation is left in your body after the scan is finished. Statements A, B and D are true.

19. **A ✓** Chemical energy in the battery is changed to electrical energy, which flows along the wires to the bulb. Here the electrical energy is transformed to light and heat. Thus, B, C and D are wrong. While heat is not what you want your battery energy changed into, it is an unfortunate side product.

20. **A ✓** The continent is old and volcanically inactive, even though there has been some volcanic activity in the recent past in South Australia. Antarctica has a very active volcano, Mount Erebus, so A is wrong. Europe and Asia have many active volcanoes, so C and D are wrong.

21. **B ✓** Convection transfers heat by the mass motion of a fluid such as air or water when the heated fluid moves away from the source of heat, carrying energy with it. Conduction (A) transfers heat by agitating the molecules within a material without any motion of the material as a whole. Radiation (C) occurs when energetic particles or energy or waves travel through a medium or space. Vaporisation (D) is the transition of matter from a solid or liquid phase into a gaseous (or vapour) phase.

22. **D ✓** All these different forms of evidence support continental drift.

23. **B ✓** X-rays are a form of high-energy electromagnetic radiation. They are shorter in wavelength than UV rays but longer than gamma rays. Radio waves (A) and microwaves (D) are on the other side of the visible light spectrum, while sound waves (C) do not feature on this spectrum as they are not electromagnetic in nature.

24. **D ✓** The oesophagus is the food pipe, and is part of the digestive system. A, B and C are shown in the diagram.

25. **C ✓** A communications satellite needs to have its own power source for proper functioning. A combination of solar cells and storage batteries is used. This is a practical choice, even though the result is far from ideal since only about 10% of the sun's energy that strikes the solar cells is converted to electrical power. Modern communications satellites have about 32 000 solar cells mounted on their surfaces, supplying about 520 watts. A nickel cadmium battery is used for backup power during periods of low light such as eclipses. B and D are incorrect as eventually any supplied fuel would run out (as well as adding to the weight of the satellite, making it more costly to launch), whereas the Sun shines indefinitely. A is wrong as power is not transmitted from Earth.

Part B: Restricted-response questions

26. circulatory ✓
27. diaphragm ✓
28. positive ✓
29. Eustachian ✓
30. modulation ✓
31. voltage ✓ (or potential difference)
32. respiratory ✓
33. increase ✓
34. dyke ✓
35. asthenosphere ✓
36. lava ✓
37. amplitude ✓
38. resistance ✓
39. Pangaea ✓
40. translucent ✓

Part C: Knowledge, skill and processing data questions

41. **a)** The aorta carries blood to the lower part of the body. From the aorta, the renal artery enters the kidney carrying blood to be filtered. ✓ Filtered blood leaves through the renal vein and joins up with the vena cava. ✓ Kidneys intervene in many processes and balances in the body to control many vital body functions. The major role of the kidneys is to remove waste from the blood and eliminate it in the urine. ✓ The ureter carries urine from the kidney ✓ to be stored in the urinary bladder ✓, which is periodically emptied. The urethra is a tube connecting the urinary bladder to the genitals for removing fluids, such as urine, out of the body. ✓

b) Wastes in the blood come from the normal breakdown of active tissues, such as muscles, and from food. ✓✓ The body uses food for energy and self-repairs. After the body has taken what it needs from food, wastes are sent to the blood. If the kidneys did not remove them, these wastes would build up in the blood and damage (poison) the body.

42. **a)** milligrams per decilitre = a thousandth of a gram for each one-tenth of a litre ✓ (decilitre = 100 mL)

b) It could indicate the possible presence of kidney disease, a sign that the kidneys are not working at full strength. ✓ As the kidneys become damaged for whatever reason, the creatinine levels in the blood will rise due to poor removal by the kidneys. Abnormally high levels of creatinine thus warn of possible problems or failure of the kidneys.

c) **i)** $w = 4; x = 7$ ✓; $y = 3; z = 1$ ✓ *(½ mark each)*

ii) 4 pairs (a total of 8 electrons) ✓

iii) C and N share 2 pairs (4) electrons ✓

43. **a)** $1.938 \times 10^{-18} - 1.635 \times 10^{-18} = 3.03 \times 10^{-19}$ J ✓

b) 2.043×10^{-18} J ✓

c) False ✓ It takes less energy to move an electron from orbit 4 to orbit 5.

d) true ✓

44. **a)** 2010 – 1880 = 130 years ✓

b) number of rabbits in the population ✓

c) Permission was granted somewhere around the early to mid-1960s. ✓ There had been a slight increase in rabbit numbers before this time, but this could be due to a number of factors such as improvements in weather or some illegal shooting of foxes in anticipation of the rules being changed.

d) Without any (or fewer) predators, the rabbit population increased dramatically. ✓ This stayed high for a while but then started to decrease (possibly due to rabbits destroying their environment by overeating plant cover). ✓

45. **a)** It is a nerve pathway controlling an action reflex. In higher animals like humans, most sensory neurons do not connect directly to the brain, but to the synapse in the spinal cord. This feature allows reflex actions to occur relatively quickly by activating spinal motor neurons without the delay of routing signals through the brain. ✓✓ However, the brain will receive sensory input while the reflex action occurs.

b) Sensory fibres in the skin detect the painful stimulus (the pin). ✓ This information is relayed by a single sensory neurone to nerve cells (interneurones) in the spinal cord. ✓ Detection, processing and response occur entirely within the spinal cord. ✓ The interneurones link with motor neurones controlling skeletal muscles. ✓ In this case, flexor muscles in the front of the arm contract while extensor muscles at the rear of the arm relax.

c) Reflex arcs assist organisms in avoiding excessive injury. They allow for the immediate withdrawal of that body part from dangerous stimuli. ✓ All sensory information does reach the brain for analysis; however, the reflex arc can process and give a rapid response without the time delay required in obtaining instructions from the brain. In this manner a response is in progress even before pain is perceived at a conscious level.

46. a) i) fungi ✓; **ii)** bacteria ✓; **iii)** virus ✓; **iv)** protozoa ✓

b) *(one disease from each only)*

virus: flu (influenza), common cold, rabies, measles, German measles (rubella), cowpox, chickenpox, smallpox, HIV, mumps, ebola, dengue fever, yellow fever, hepatitis ✓

bacteria: anthrax, cholera, tuberculosis (TB), septicaemia (blood poisoning), bacterial meningitis, bacterial pneumonia, leprosy, diphtheria, legionella, syphilis, methicillin-resistant Staphylococcus aureus (MRSA) ✓

fungi: athlete's foot, ringworm, thrush, tinea ✓

protozoa: malaria, dysentery, sleeping sickness (trypanosomiasis), toxoplasmosis, amoebic meningitis, leishmaniasis, diarrhoea (caused by the protozoan giardia) ✓

47. a) A tsunami is a sea wave (or may even be a series of waves) produced by a large disturbance of the sea floor over a relatively short period. ✓ Such disturbances cause the water column to move vertically. The resulting wave energy spreads outwards across the ocean at high speed ✓ (up to 950 km/h).

b) A steep-sided volcano bordering an ocean can have those sides weakened by volcanic activity. Rising magma super-heating trapped water creates considerable pressures. These waters, and other trapped gases, move the flank of the volcano and any resulting earthquakes can collapse the flank into the ocean. ✓ This sudden submarine slide will displace many thousands of litres of water providing much energy to the overlying sea water. The moving wave begins travelling out from where the disturbance occurred. ✓

c) Most tsunamis are caused by earthquakes ✓ generated in a subduction zone. This is a region where an oceanic plate is being forced down into the mantle by plate tectonic forces. The friction between the plates is enormous, with energy accumulating in the overriding plate until it exceeds the frictional forces. This may take decades or even centuries. When it happens, the overriding plate snaps back into an unrestrained position, with an enormous release of energy. The sea floor abruptly deforms and vertically displaces the overlying water. This sudden motion is the cause of a tsunami.

48. a) B ✓

b) In a series circuit the current flows through the first globe and then through the second. ✓ Each globe can only make use of part of the energy carried by the current. If one of the globes blows (burns out) in a series circuit, the current can no longer pass through it and so no current flows through the circuit at all. ✓ In a parallel circuit the current can pass simultaneously through each globe ✓ (the current splits at the junction of the wires) and so energy is available to each globe. If one of the globes blows in a parallel circuit, no current passes through it; however, current can continue to flow through the rest of the circuit. ✓

c) House lights are connected in parallel. If one globe blows, the others remain lit. ✓ If the lights were connected in series, they would all go out if one globe blew. Also, you can turn one light on without the rest of them going on. Similarly, you can turn one light off without the rest of them going off. As the current passing through them isn't split, each globe will be brighter in a parallel circuit compared to those in a series circuit (with the same voltage). ✓ Generally, parallel circuits make each globe independent of the others, whereas in a series circuit all globes are dependent on the actions of the others.

49. a) Q and R only ✓ Closing or opening switch P makes no difference to whether the globes light up or not.

b) The remaining three globes will now glow equally as bright but not as bright as before globe L burnt out. ✓ This is because before L burnt out, the total resistance of the circuit was less. Now that L does not provide another pathway, the resistance is greater and hence the remaining globes will glow less brightly.

50. a) Radar works by transmitting a microwave or radio wave beam on a given frequency. Any targets that are hit by the beam reflect this electromagnetic energy to the antenna. A computer then analyses any changes in frequency and displays this information. ✓ (Radio waves are used as they travel far, are invisible to humans and are easy to detect even when they are faint. Air traffic control uses radar to track planes both in the air and on the ground; geographers and astronomers use radar to map the Earth and other planets, and to track satellites and space debris; military uses it to detect the enemy and for weapons guidance; meteorologists use radar to track storms, hurricanes and tornadoes.)

b) A transmitter on the police vehicle sends out a radio wave of a given wavelength (or frequency) that reflects off a car back to a detector on the radar gun. If the radar gun and car are stationary with respect to each other, the returning signal will be at the same wavelength (or frequency). The car's speed is determined from the difference in the wavelength (or frequency) of the transmitted beam and reflected beam. ✓ The faster the car is moving towards the radar gun, the more scrunched up the reflected waves will be. The faster the car is moving away from the radar gun, the more stretched out the reflected waves will be.

51. a) It converts harmful compounds in the exhaust gases of cars into harmless compounds. ✓

b) No. As catalysts, they facilitate the conversion of other substances. ✓

c) i) A hydrocarbon is a chemical compound consisting only of carbon and hydrogen. ✓

ii) oxygen ✓; CO, CO_2 and H_2O all contain oxygen which they get from air

d) i) 45 ✓ (2 + 8 + 18 + 16 + 1, the same as the number of electrons)

ii) 103 – 45 = 58 ✓ There are several isotopes, but ^{103}Rh is the most common, and stable, with 58 neutrons.

Index

Page numbers in **bold** are definitions of terms.

D

E

F

G

H

I

J

K

L

M

N

O